14th International Conference on Future Environment and Energy (ICFEE 2024)

AA004179

E3S Web of Conferences Volume 530 (2024)

Matsue City, Japan
15 – 17 March 2024

Editor:

R. San Jose

ISBN: 978-1-7138-9805-4

Printed from e-media with permission by:

Curran Associates, Inc.
57 Morehouse Lane
Red Hook, NY 12571

Some format issues inherent in the e-media version may also appear in this print version.

For additional information, please contact EDP Sciences – Web of Conferences
at the address below.

EDP Sciences – Web of Conferences
17, Avenue du Hoggar
Parc d'Activité de Courtabœuf
BP 112
F-91944 Les Ulis Cedex A
France

Phone: +33 (0) 1 69 18 75 75
Fax: +33 (0) 1 69 28 84 91

contact-edps@webofconferences.org

Additional copies of this publication are available from:

Curran Associates, Inc.
57 Morehouse Lane
Red Hook, NY 12571 USA
Phone: 845-758-0400
Fax: 845-758-2633
Email: curran@proceedings.com
Web: www.proceedings.com

TABLE OF CONTENTS

AIR POLLUTION MONITORING AND CONTROL

ANALYSIS AND RESTORATION OF AQUATIC ENVIRONMENT

ENVIRONMENTAL BIOTECHNOLOGY AND WASTE-TO-ENERGY

ECOLOGICAL ENVIRONMENT POLLUTION CONTROL AND SUSTAINABLE DEVELOPMENT

CLEAN ENERGY TECHNOLOGY AND BUILDING ENERGY MANAGEMENT

Preface

It is a great pleasure to welcome you all to the 14th International Conference on Future Environment and Energy (ICFEE 2024), hosted in the vibrant city of Matsue City, Japan from 15-17 March, 2024. As we gather once again for this prestigious event, we are excited to explore and discuss the latest advancements, innovations, and challenges at the Future Environment and Energy. We believe that the participants have gained rich knowledge and experience through this conference, and this knowledge and experience will have a positive impact on their work and learning in their respective fields.

We are delighted to present this compilation of research papers that showcase the latest advancements and discoveries in the field of Future Environment and Energy.
ICFEE serves as a premier platform for researchers, scientists, engineers, and industry professionals from around the globe to come together and exchange ideas, insights, and experiences.

The papers included in these proceedings represent the highest standards of academic excellence and originality. Each submission underwent a rigorous peer-review process, ensuring that only the most exceptional and innovative research finds its place in this esteemed collection. The contributions featured in this volume span a wide spectrum of topics, including but not limited to Water Resources Management and Water Pollution Control; Atmospheric science and air pollution control; Ecosystem Management and Sustainable Development ; Clean and Renewable Energy.

Finally, we would also like to express our gratitude to the organizing committee, session chairs, reviewers, and everyone involved in making this conference a resounding success. Your dedication and hard work have been instrumental in creating an inclusive and enriching environment for the exchange of ideas and the cultivation of new connections.

We hope that you find these conference proceedings insightful and inspiring. We believe it will contribute significantly to the further development of these fields. We extend our warmest thanks to all the authors for their valuable contributions.

Conference Chair
Prof. Akira Kondo
Osaka University, Japan

Conference Committee

Conference Co-Chairs

Guangwei Huang, Sophia University, Japan
Akira Kondo, Osaka University, Japan

Conference Program Chairs

Eric Wolanski, James Cook University, Australia
James T. Anderson, Clemson University, USA

Conference Program Co-Chairs

Roberto San Jose, Technical University of Madrid, Spain
Rodney Stewart, Griffith University, Queensland, Australia
Abbas Taheri, Queen's University, Canada

Conference Advisory Chair

Hui Tong Chua, University of Western Australia, Australia

Conference Publicity Chairs

Tanakorn Wongwuttanasatian, Khon Kaen University, Thailand
Jiayuan Ji, The University of Tokyo, Japan

Technical Committee

Lee Hwang Sheng, Universiti Tunku Abdul Rahman, Malaysia
Low Sheau-Ting, Universiti Teknologi Malaysia, Johor, Malaysia.
Md Maruf Mortula, American University of Sharjah, UAE.
Qing Zhu, Lawrence Berkeley National Lab, USA.
Ramani Bai V., UCSI University Kuala Lumpur Campus, West Malaysia.
Wannawit Taemthong, King Mongkut's University of Technology North Bangkok, Thailand.
Willyanto Anggono, Petra Christian University, Indonesia.
Yao Yu, North Dakota State University, USA.
Joseph Dob Retumban, National University, Manila, Philippines.
Carmelo Muñoz Ruiperez, Universidad de Burgos, Spain.
Masoud Taghavi, Chung-Ang University (CAU), Seoul, South Korea
Baba Imoro Musah, Xishuangbanna Tropical Botanical Garden,Chinese Academy of Sciences, Yunnan, China.
Enrique Romero, Cadaval University of Extremadura, Spain.
Satya Harpalani, Southern Illinois University, United States.
Chao-Heng Tseng, National Taipei University of Technology, Taiwan.
Louise Grace Avena, Mapúa University, Manila, Philippines.
Amnart - Suksri, Khon Kaen University, Thailand.
Worapong Lohpaisankrit, Khon Kaen University, Khon Kaen, Thailand
Mohammad Arif Kamal, Aligarh Muslim University, India
S. Joseph Antony, University of Leeds, UK
Anizahyati Alisibramulisi, Universiti Teknologi MARA, Malaysia
Yanan Liu, Chongqing Jiaotong University, China
Shamsul Rahman Mohamed Kutty, Universiti Teknologi Petronas, Malaysia
Xiaohua Li, Chongqing University, China
Bashir M Suleiman, University of Sharjah, UAE

Evaluation of the effectiveness of various pollution removal mechanisms of indoor air purifiers in removing suspended particles

Yi-Cheng Lo [1], *Ya-Si* Liao [1*] and *Chao-Heng* Tseng [1]

[1]National Taipei University of Technology, Taipei, Taiwan

Abstract. This study comprehensively investigated the efficacy of five distinct air purifiers in the removal of both particulate matter and gaseous pollutants. The analysis was conducted within a consistent test environment, employing a standardized benchmark to ensure accurate and comparable results. A rigorous methodology was devised to evaluate the efficiency of each air purification mechanism. The experimental setup involved placing the air purifiers inside a controlled test chamber, carefully designed to replicate typical indoor conditions. The focus of the analysis was on the removal efficiency of PM2.5, and the evaluation utilized the Clean Air Delivery Rate (CADR) values, measured in cubic feet per minute (ft^3/min or CFM). Importantly, the study maintained uniform specifications, including unit filter area and unit electrostatic volume, to facilitate a fair and meaningful comparison of the removal efficiency for each purification mechanism. The CADR values served as a quantitative metric to assess the air purifiers' effectiveness in removing PM2.5 particles. This standardized approach allowed for a direct comparison of their performance under identical conditions. By considering unit filter area and electrostatic volume, the study aimed to provide insights into the relative merits of each mechanism, enabling consumers to make informed decisions based on the specific removal capabilities of the air purifiers. In summary, the research established a robust methodology for evaluating air purifier performance, facilitating a thorough analysis of their effectiveness in removing particulate matter and gaseous pollutants within a controlled environment. The use of CADR values and standardized specifications ensures a reliable basis for comparing the efficiency of different air purification mechanisms.

Keywords: Particulate Matter 2.5, Air Purifier, Clean Air Delivery Rate

1.Introduction

In recent years, heightened health consciousness among the Taiwanese population, coupled with the demand for pandemic prevention, has driven a substantial 20% to 30% increase in

* Corresponding author: ariel10050003@gmail.com

the sales of air purifiers from leading brands. However, these air purifiers employ various pollution removal mechanisms, and manufacturers employ different methods to verify their efficacy.

Research indicates that while most of these pollution removal mechanisms are generally effective, they often achieve only a modest 10% to 20% actual removal rate. Average consumers typically lack a clear understanding of how these air purifiers operate and their specific pollution removal mechanisms. Moreover, consumers have limited knowledge about the indoor air quality in their homes, which makes it challenging to establish precise criteria for selecting the right air purifier.

Commercially available air purifiers typically employ composite pollution removal mechanisms. This research aims to compare the performance of air purifiers with market representation and suitability for smaller indoor spaces, while ensuring they can operate a single pollution removal mechanism. The air purifiers are placed inside an air quality testing chamber (referred to as the testing chamber) to simulate a typical indoor environment, allowing for the analysis of their PM2.5 removal efficiency using CADR (Clean Air Delivery Rate, ft^3/min, CFM) values. [1] This quantification facilitates a comparative assessment of the efficiency of different mechanisms, enabling a deeper exploration of the factors contributing to the varying air purifier removal efficiencies.

2.Methods and materials

The removal techniques one main types in this experiment: inhalation-type. Inhalation-type air purifiers primarily function by actively drawing external air into the purifier using an internal motor. Subsequently, the air is purified using built-in filters or electrostatic dust collection techniques, including HEPA filters, electrostatic filters and electrostatic precipitators (ESP).

Table 1 Experiment related parameters.

Experimental parameters	
Environment Control System Volume	$1.5m^3$ ($53ft^3$)
Temperature	25±1 °C
Relative Humidity	65±5 %
Time	1 hr
PM2.5 Concentration	70±5 μg/m^3

Before conducting the experiment, it's important to maintain the testing parameters in Table 1. Chamber's temperature at 25±1 ℃ and humidity within the 65±5% range. To introduce PM2.5 into the 1.5 m^3 testing chamber (Fig 1) at a concentration of 70 μg/m^3, we use an aerosol generator with controlled high-pressure air flow through an inlet valve. When introducing other air pollutants, the exhaust valve and internal chamber fan are activated to ensure even mixing. After achieving the desired concentration, the exhaust valve is closed, and the fan is turned off for the PM2.5 experiment. [2] Then, the external air purifier's power switch is turned on to begin the experiment.

https://doi.org/10.1051/e3sconf/202453001001

(a)

(b)

Fig 1. Testing chamber (a:open; b:close).

This study is to explore the comparison of PM 2.5 removal efficiency of air purifiers operating with different removal mechanisms, and must exclude the influence of other factors, such as cabin adsorption, cabin sealing degree, and other related chemical interference reactions, so air purification is carried out Before turning on the air purifier to remove PM 2.5, a natural decay numerical measurement experiment will be conducted. The results of the above two experiments are the natural decay curve and the removal rate of the air purifier when it is turned on. The calculation methods are as follows: formulas (1) and (2).[3]

$$\text{Natural decline rate (\%)} = \left(\frac{\text{PM 2.5 } \textit{initial concentration } C_0 - \text{PM 2.5} \textit{final concentration} C_1}{\text{PM 2.5} \textit{initial concentration } C_0}\right) \times 100\% \qquad (1)$$

$$\text{Start}-\text{up removal rate}(\%) = \left(\frac{\textbf{PM 2.5} \textbf{\textit{self}} - \textbf{\textit{decay final concentration}} C_1 - \textbf{PM 2.5} \textbf{\textit{final concentration}}\ C_2}{\textbf{PM 2.5} \textbf{\textit{final concentration}}\ C_1}\right) \times \mathbf{100\%} \quad \mathbf{(2)}$$

CADR is based on the AHAM AC-1 method, [4] which is an indicator for evaluating the removal energy of an air purifier. The calculation method is to use the mass conservation of pollutant concentration and solve the differential equation to obtain the characteristic equation, such as Equation (3).

$$C_t = C_0 e^{-kt} \quad (3)$$

C_t : The pollutant concentration at time t , ppm

C_0 : initial pollutant concentration , ppm

K : Decay constant , min^{-1}

t : Time , min

Calculate decay constant Kn and Ke :

$$K = Kn + \frac{\eta E_d Q}{V} = Ke \quad (4)$$

When natural decay experiment , K = Kn

Since it is difficult to calculate the air purifier efficiency (η) and the short cycle coefficient (Ed), we conducted unstarted experiments and start-up experiments of air purifiers with each air pollution removal mechanism, and the experimental data obtained were compared to ln Ct Do linear regression with t to obtain Ke and Kn values and then substitute them into equation (3) to know the CADR value of the air purifier. [5]

$$CADR = (Ke - Kn) \times V \quad (3)$$

CADR : Clean Air Delivery Rate , ft^3/min

V : Air quality test cabin volume , ft^3

Ke : Power-on attenuation constant , min^{-1}

Kn : natural decay constant , min^{-1}

3. Results and Discussion

3.1 Comprehensive comparison of the removal efficiency of different removal mechanisms for air pollution removal

This study uses five air purifiers with eight air pollution removal mechanisms each to compare their PM 2.5 removal efficiencies. In the comprehensive comparison of the PM 2.5 removal efficiencies of each mechanism, it can be observed in Figure 2 that HEPA filter, HEPA composite The removal efficiency of powdered activated carbon filter, electrostatic filter, electrostatic dust collector A, electrostatic dust collector B and negative ions can reach 100% within 60 minutes after starting up. Among them, the operating air volume is the fastest electrostatic filter. The PM2.5 concentration of the filter returns to zero as quickly as possible after starting up, and has already returned to zero after 9 minutes after starting up. The granular activated carbon filter, powdered activated carbon filter, and photocatalyst filter that are primary filters remove PM2 .5 has a low efficiency and cannot be reset to zero within 60 minutes after startup. The lowest removal efficiency is the operating air volume of photocatalyst filter A. The removal efficiency at the 60th minute after start up is only 30.2%. Use Figure 2 to show the relationship between them.

https://doi.org/10.1051/e3sconf/202453001001

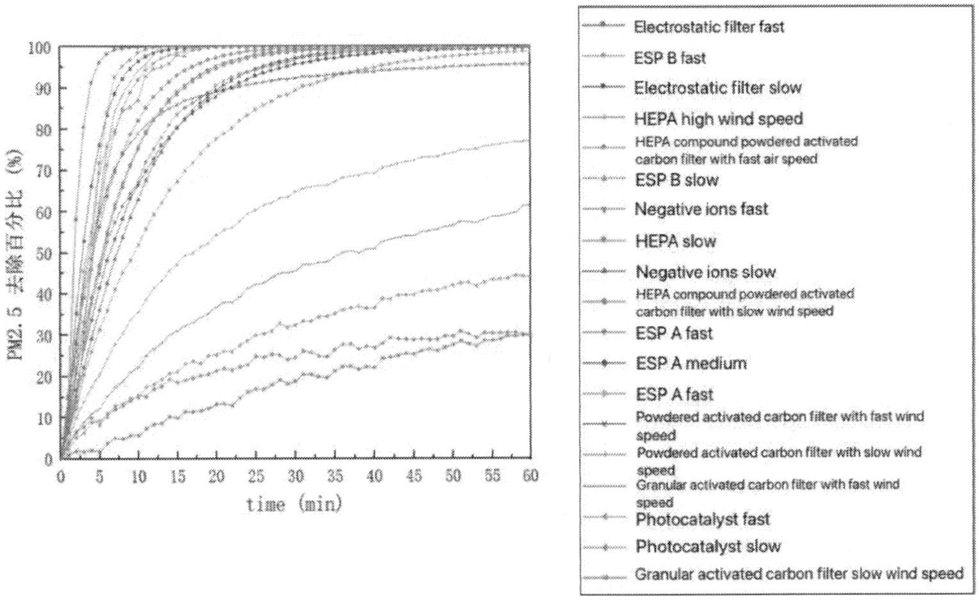

Fig 2. Comprehensive overlay comparison of PM2.5 removal efficiency by each mechanism

3.2 Comprehensive comparison of CADR values for air pollution removal by different removal mechanisms

After quantitative and unified comparison of PM2.5 removal efficiency, it can be clearly observed that the CADR value of fast operating air volume of each removal mechanism is greater than the CADR value of slow operating air volume. The largest CADR value is electrostatic filter (fast).

It means that the electrostatic filter has the best PM2.5 removal efficiency when the operating air volume is high, and the results are compared with the removal efficiency; the CADR values of HEPA filter, HEPA composite powdered activated carbon filter, and electrostatic filter B are all greater than that at fast operating air volume. 10, which means it has a very good removal efficiency of granular gas pollutants; the CADR value of each operating air volume of electrostatic dust collector B is greater than that of electrostatic dust collector A; the granular activated carbon filter, powdered activated carbon filter, which belongs to the primary filter, The CADR value of the photocatalyst filter for removing PM2.5 is low, even lower than 1; if you want to reduce the concentration of indoor particulate air pollutants, this study recommends setting the operating air volume of the air purifier to fast, and it is not recommended to use a primary filter , as shown in Table 3.

Table 2 Comparison of PM 2.5 removal efficiency and CADR at the 60th minute of the PM 2.5 removal experiment of air purifiers with various air pollution removal mechanisms.

Removal mechanism / Conditions and Specificatio	Granular activated carbon filter	HEPA filter	HEPA powdered activated carbou filter	Powdered activated carbon filter	Electrostatic filter	Photocatalyst	ESP A	ESP B
Operating air volume (m^3/min)	3.00(fast) 0.97(slow)	1.43(fast) 0.37(slow)	1.5(fast) 0.42(slow)	2.7(fast) 0.7(slow)	2.24(fast) 1.3(slow)	0.78(fast) 0.39(slow)	1.23(fast) 1.06(medium) 0.88(slow)	2.8(fast) 1.70(slow)

Operating air volume (ft^3/min)	106.1(fast) 34.2(slow)	50.5(fast) 13.1(slow)	53(fast) 14.75(slow)	95.4(fast) 24.7(slow)	79.1(fast) 45.9(slow)	27.6(fast) 13.8(slow)	43.5(fast) 37.4(medium) 31.1(slow)	98.9(fast) 60.0(slow)
Decay constant Kn	0.00367							
Natural decline rate(%)	20.5							
Initiation Rate constant Ke	0.0185(fast) 0.0099(slow)	0.3321(fast) 0.1567(slow)	0.3150(fast) 0.1262(slow)	0.0462(fast) 0.0272(slow)	0.7660(fast) 0.3784(slow)	0.0131(fast) 0.0088(slow)	0.1017(fast) 0.0985(medium) 0.0778(slow)	0.5204(fast) 0.2978(slow)
Power - On Removal rate t=60	61.2(fast) 30(slow)	100(fast) 100(slow)	100(fast) 100(slow)	95.7(fast) 77.2(slow)	100(fast) 100(slow)	44.1(fast) 30.2(slow)	100(fast) 100(medium) 98.8(slow)	100(fast) 100(slow)
CADR (ft^3/min)	0.78(fast) 0.33(slow)	17.4(fast) 8.1(slow)	16.5(fast) 6.5(slow)	2.16(fast) 1.23(slow)	40.4(fast) 19.8(slow)	0.50(fast) 0.27(slow)	5.19(fast) 5.02(medium) 3.9(slow)	27.4(fast) 15.6(slow)

3.3 The CADR value of each removal mechanism of the suction-type purifier is converted to the same specifications for comparison.

In order to make the calculated CADR values more uniform and standard for comparison, this study tried to convert the CADR of each removal mechanism into the same air volume using the internal and external difference method and then conducted a further in-depth comparison and evaluation under the conversion to the same specifications. Table 2 shows that after the CADR value of the removal mechanism is converted to the same base air volume, it is divided into two groups: filter type and electrostatic type. The CADR value is converted into specifications per unit filter area or corona volume.

The CADR value benchmark selected in this study is 1 m^3/min = 35.3 ft^3/min for low air volume and 2 m^3/min = 70.6 ft^3/min for high air volume. The CADR value of each removal mechanism is interpolated using the two points of operating air volume speed. Method to convert the CADR value to the same air volume. The PM2.5 CADR value of the filter-type removal mechanism is converted into data equivalent to low air volume 35.3 (ft^3/min) and high air volume 70.6 (ft^3/min) using the internal difference method, and divided by the filter area, it can be obviously It was observed that under the same filter area ($ft^3/min/cm^2$), the CADR value of the electrostatic filter dropped significantly. It is speculated that the PM2.5 removal efficiency is positively related to the filter contact area. HEPA composite powdered activated carbon filter and HEPA filter both have the highest CADR. The CADR value of the electrostatic dust collector needs to be divided by the volume of the corona dust collector after interpolating the air volume. It can be observed that the CADR value of the electrostatic dust collector B with denser corona dust collection is higher than that of the electrostatic dust collector A. It is inferred that for electrostatic dust collector Generally speaking, the removal rate of suspended particles increases with the density of the dust collection plate and electrode lines. The detailed CADR value conversion information is shown in Table 3.

When the air volume is doubled, the unit area of each filter and the CADR of each unit of corona dust collection volume are almost twice the converted low air volume of 1 m^3/min, which proves that the experimental data is reasonable and the unit efficiency of each mechanism is proportional to the air volume. It is recommended that when turning on the purifier to remove indoor PM, it is recommended to operate with a high air volume, which is expected to have a better removal effect.

Table 3 Converted data with the same specifications excluding PM2.5 CADR value (low air volume)

Removal mechanism	CADR in Same Flow rate (=35.3ft3/min)	filter area (cm2)	CADR per filter area (ft3/min/cm2)	CADR in Same Flow rate (=70.6ft^3/min)	CADR per filter area (ft^3/min/cm^2)
HEPA Composite Powdered Activated Carbon cleaner	11.87	64.34	0.1803	21.1	0.328
HEPA	13.62	64.34	0.2117	22.4	0.348
Electrostatic cleaner	13.22	276.00	0.0479	35.13	0.127
Powdered activated carbon cleaner	1.37	64.34	0.0213	1.83	0.0284
Photocatalyst	0.63	49.00	0.0129	1.22	0.0249
Granular activated carbon cleaner	0.337	64.34	0.00528	0.558	0.00867
Removal mechanism	CADR in Same Flow rate (=35.3ft3/min)	filter area (cm3)	CADR per filter area (ft3/min/cm3)	CADR in Same Flow rate (=70.6ft3/min)	CADR per filter area (ft3/min/cm3)
ESP B	8.11	2400	0.003379	18.82	0.007842
ESP A	4.65	1600	0.002906	-8.01	0.005006

4. Conclusion

This study was simulated in a test chamber and is more suitable for indoor air purifiers with smaller square footages. Several single air pollution removal mechanisms are independently operated to remove PM2.5. The following conclusions can be drawn:

In the PM2.5 removal experiment, the air volume of HEPA filter, HEPA composite powder activated carbon filter, electrostatic filter, electrostatic dust collection A, and electrostatic dust collection B at each level can be achieved within 60 minutes of the experiment. 100%, among which the high operating air volume electrostatic filter has the highest removal rate. It is speculated that the reason is that the electrostatic filter has the largest filter area; the powdered activated carbon filter, photocatalyst, and granular activated carbon filter have lower PM2.5 removal efficiency, equivalent to a primary filter.

The unit filter area analysis is to convert the PM2.5 CADR value of the filter type into the same reference air volume using the internal difference method and divide it by the filter area (ft^3/min/cm^2). It is known that in the PM2.5 removal experiment, the HEPA filter The CADR value per unit filter area of the net and HEPA composite powdered activated carbon filter is the highest, followed by the electrostatic filter, and the PM2.5 CADR value is approximately proportional to the filter contact area; the CADR value of the electrostatic dust collection type has been After interpolation and conversion to the standard air volume, divide it by the corona volume ($ft^3/min/cm^3$). From the unit corona volume CADR value, we can know that the unit corona volume CADR value of electrostatic dust collector B with dense dust collecting plates and electrode lines is as high as electrostatic Collect dust twice as much as A.

References

1. Angus Shiuea, Shih-Cheng Hua, Chao-Heng Tseng, En-Huai Kuob, Chao-Yun Liub, Chia-Te Houc, Tuan Yuc," **Verification of air cleaner on-site modeling for PM2.5 and TVOC purification in a full-scale indoor air quality laboratory** ", Atmospheric Pollution Research vol:10, pp.209-218
2. Congrong He, Lidia Morawska, Dale Gilbert," **Particle deposition rates in residential houses**", Atmospheric Environment vol:39, pp. 3891-3899
3. Elizabeth U Finlayson, Ashok J Gadgil, Tracy L Thatcher, Richard G Sextro,"**Pollutant dispersion in a large indoor space: Part 2-Computational Fluid Dynamics (CFD) predictions and comparison with a scale model experiment for isothermal flow**", Submitted to Indoor Air vol:2, pp 102-130
4. Frequently Asked Questions about Testing of Portable Air Cleaners, ANSI/AHAM AC-1-2020.
5. M Jamriska, L Morawska, D S Ensor," **Control strategies for sub-micrometer particles indoors: model study of air filtration and ventilation**",Indoor Air vol:13(2),pp.96-1

Modeling study of $PM_{2.5}$ pollution episode of early spring 2019 in Hokkaido, Japan caused by biomass burning in Northeast China

Katsushige Uranishi[1*,3], *Hikari* Shimadera[2], *Fumikazu* Ikemori[3], *Kyohei* Takami[4], *Atsushi* Nogami[1], and *Seiji* Sugata[5]

[1]Department of Life and Environment Engineering, The University of Kitakyushu, Kitakyushu, Japan
[2]Graduate School of Engineering, Division of Sustainable Energy and Environmental Engineering, Osaka University, Suita, Japan
[3]Nagoya City Institute for Environmental Sciences, Nagoya, Japan
[4]KANSO TECHNOS CO. LTD., Osaka, Japan
[5]National Institute for Environmental Studies, Tsukuba, Japan

Abstract. Biomass burning (BB), in particular agricultural waste burning (Agri-BB), occurs at random locations, scales, and times. These factors make it challenging to detect Agri-BB accurately through satellite observations. Thus, the BB emission inventories using satellite observation data have uncertainties for their emission estimation approach and cause poor model performance for air pollutants including $PM_{2.5}$. We utilized the two BB emission inventories, GFEDv4.1s and FINNv2.5 with the CMAQ model to simulate the $PM_{2.5}$ heavy pollution episode in Hokkaido 2019. To estimate Agri-BB contributions, we conducted three simulation cases for each BB emission inventory: with and without Agri-BB emission, and the boosted Agri-BB emission cases. The baseline simulation failed to capture the temporal and spatial variation patterns of $PM_{2.5}$. Meanwhile, the boosted Agri-BB case could show favorable performance for $PM_{2.5}$ concentrations. These results indicated that the two BB emission inventories underestimated Agri-BB emissions. In the two boosted Agri-BB cases, the $PM_{2.5}$ contributions from Agri-BB accounted for more than 50% during the episode. Moreover, high $PM_{2.5}$ emissions were found in Northeast China and its surrounding regions similar to the two boosted Agri-BB cases. Consequently, the results revealed that Agri-BB emissions during the episode were significantly derived from the agricultural areas in Northeast China.

1 Introduction

Biomass burning (BB) refers to the combustion of organic resources derived from living and dead vegetation. BB, such as forest fires and agricultural waste burning (hereinafter referred to as Agri-BB) has caused significant atmospheric pollution and health concerns worldwide [1-2]. Agri-BB is one of the anthropogenic BB and is widely practiced in Asian countries as

* Corresponding author: uranishi@kitakyu-u.ac.jp

part of agricultural activities due to its cost-effectiveness and convenience in waste disposal [3-5]. However, the timing of Agri-BB is difficult to determine accurately due to post-harvest season variations influenced by annual climate and weather patterns. Moreover, when extensive Agri-BB occurs within a short timeframe before planting, substantial air pollutants are released into the surrounding areas, which lead to haze formation and trigger respiratory and cardiovascular health issues [6-8]. This phenomenon has emerged as a critical issue in Asian countries such as China and India [6-8].

Northeast China is one of the major granaries in China and extensive BB events have been observed in the region. It is reported that the intensity of Agri-BB in Northeast China surpassed what has been observed in other regions [6] When an extensive Agri-BB event occurs during the post-harvest season and releases lots of air pollutants under stagnant weather conditions, it leads to severe haze events and exerts a myriad of impacts on regional air quality [9-10]. In addition, BB including Agri-BB in Northeast China partly contributed to the transboundary air pollution episode of autumn 2014 in the northern part (Rishiri Island) [11] and the central part (Noto peninsula) [12] of Japan. Hence, BB exerts damaging effects across various spatial scales from local to regional. Thus, it's important to understand how BB affects air quality in East Asia and suggest mitigation strategies to reduce its level. However, there are few studies simulating $PM_{2.5}$ transboundary pollution in Japan that originated from BB in Northeast China.

In Sapporo City, Hokkaido, Japan, hourly $PM_{2.5}$ concentration increased drastically from the 27th of February to the 3rd of March 2019, recording up to 200 $\mu g/m^3$ (hereinafter referred to as the $PM_{2.5}$ heavy pollution episode). In late February 2019, significant hotspots were observed in Northeast China and its surrounding regions via the Fire Information for Resource Management System (FIRMS, https://earthdata.nasa.gov/firms) (Figure 1). We reported that the boosted BB emission on the crop land was able to well reproduce daily $PM_{2.5}$ concentrations in Northeast China and Hokkaido, Japan during the PM2.5 heavy pollution episode using the Community Multiscale Air Quality (CMAQ) model and the Global Fire Assimilation System (GFAS). GFAS was the sole available global BB emission inventory in 2020 when we worked on our study [13]. The GFAS is a BB emission inventory converting fire radiative power (FRP) observations from MODIS satellites into smoke constituents [14]. However, the BB emission inventory does not provide BB emissions for different land use/cover categories (savanna, boreal forest, temperate forest, and agricultural land etc.). Therefore, it was not possible to accurately evaluate the $PM_{2.5}$ contribution from Agri-BB in Northeast China.

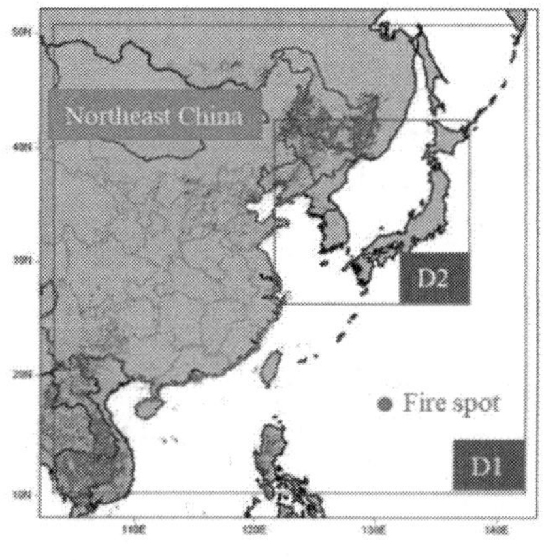

Fig. 1. Modeling domains for CMAQ and fire spots (MCD14DL) obtained from FIRMS during the $PM_{2.5}$ heavy pollution episode from the 27th of February to the 3rd of March in 2019. Green lines indicate Northeast China region in China.

On the other hand, both the Global Fire Emissions Database (GFED) [15] and the Fire INventory from NCAR (FINN) [16] are some of the available global BB emission inventories for air quality simulations and provide several types of BB emission for corresponding land use/cover categories. FINN calculates the dry matter burned using fire hotspot data to estimate the burned area, while GFED uses the burned area product for its calculations [17]. Moreover, GFED v4.1s includes newly burned area estimates with contributions from small fires, with boosted BB emissions in agricultural areas in Asia, including Northeast China [15]. Meanwhile, the latest version of FINN v2.5 utilizes active fire detections from the Visible Infrared Imaging Radiometer Suite (VIIRS) at 375 m spatial resolution to allow smaller fires such as Agri-BB to be included in the emission processing. Both emission inventories show similar temporal and spatial variations in $PM_{2.5}$. However, the quantities of total $PM_{2.5}$ emissions differ substantially because each BB inventory was developed with a different approach for emission estimation. Therefore, assessing the model performance with other BB emission inventories should be taken into consideration.

In this work, we utilized the two BB emission inventories, GFED v4.1s and FINN v2.5 with the CMAQ model to simulate the $PM_{2.5}$ heavy pollution episode and to evaluate the model performance as well as the Agri-BB contributions on $PM_{2.5}$ concentrations in Hokkaido, Japan.

2 Materials and Methods

CMAQ v5.2.1 was employed to simulate the $PM_{2.5}$ heavy pollution episode. Meteorological fields were prepared using version 3.9.1 of the Weather Research and Forecasting (WRF) model. Figure 1 shows the modeling domains from East Asia (D1: 45 × 45 km) to Japan (D2: 15 × 15 km). The vertical layers consisted of 34 sigma-pressure coordinate layers ranging from the surface to 100 hPa with the vertical middle height of the first layer. Table 1 summarizes configurations of WRF and CMAQ models. The details of chemistry and physics options are available in the documentation for the models.

The emission data for CMAQ simulations were derived from various emission inventories as shown in Table 1. In particular, BB emissions were from GFED v4.1s and FINN v2.5. It is assumed that BB emissions were uniformly distributed from the surface to the Planetary Boundary Layer (PBL) height in accordance with our previous study [13]. Note that there were differences between the simulation period and the reference years in the anthropogenic emission data available for the simulations. Therefore, the uncertainties associated with these differences may affect the CMAQ model performance.

In order to investigate the BB emission divergence between the two BB inventories and the influence of Agri-BB pollutants during the $PM_{2.5}$ heavy pollution episode, the following simulations with each BB emission inventory were executed for 2 months from the 1st February 2019 to the 31st March 2019 (hereinafter called to as the target period) with spin-up period of 10 days: (1) the baseline cases with regular BB emission (base_GFED and base_FINN), (2) GFED_Agri-BB10, a simulation case with 10 times boosted Agri-BB emission in GFED v4.1s, (3) FINN_Agri-BB05, a simulation case with 5 times boosted Agri-BB emission in FINN v2.5 so that the emission level is comparable with that GFED_Agri-BB10. An additional simulation without Agri-BB sources (noAgri-BB) was conducted to estimate the $PM_{2.5}$ contribution from Agri-BB sources.

For evaluation of the CMAQ model performance, the simulation results were compared to the ground-level $PM_{2.5}$ concentration data from 17 sites in Northeast China obtained from Air quality data of China (https://www.aqistudy.cn/historydata/index.php), and to the data from 14 sites in Hokkaido, Japan provided by Air pollution monitoring data of Japan (https://tenbou.nies.go.jp/download/). We used time-series comparison and the index of agreement (IA). IA is a statistical measure to evaluate the model performance and varies between 0 and 1, where 1 indicates a perfect model performance [18].

Table 1. Model configurations for WRF and CMAQ

WRF configuration	
Meteorology Model	WRF v3.9.1
Geography data	USGS 30sec topography and land use (East Asia) MLIT-GIAJ land use (100m, Japan)
Analysis data	NCEP FNL, RTG_SST_HR
Explicit precipitation scheme	WRF Single–moment 6–class scheme
Radiation	RRTM scheme (long wave), Dudhia scheme (short wave)
Surface-layer option	Revised MM5 scheme
Land-surface	Noah Land-Surface Model
Planetary boundary layer scheme	Yonsei University scheme
Cumulus option	Kain-Fritsch sheme
FDDA	$G_{t,\,q,\,uv} = 3.0 \times 10^{-4}\ s^{-1}$ (D1, D2)
CMAQ configuration	
Chemistry Model	CMAQ v5.2.1
Emission	Asia: REAS v3.2.1 (2017), Japan: MOE PM2.5 EI (2015) & OPRF2010 (Ship), Biogenic: MEGANv2.04, Volcano: Aerocom, Biomass burning: GFED v4.1s or FINN v2.5,
Initial/Boundary concentrations	CAM-chem (D1)
Advection	Yamartino/WRF-based scheme, Multiscale/ACM2
Vertical diffusion	Asymmetrical Convective Model version 2 (ACM2)
CMAQ chemistry	SAPRC07 and AERO6 with Aqueous chemistry

3 Results and Discussions

3.1 Model simulation performance

Figure 2 shows the comparison of the results of baseline and boosted Agri-BB cases versus the observation data in terms of the mean concentrations and IA values of $PM_{2.5}$ during the target period. Compared to the two boosted Agri-BB cases (GFED_Agri-BB10 and FINN_Agri-BB05), both GFED_base and FINN_base cases showed relatively large differences between simulated and observed $PM_{2.5}$ concentrations in Northeast China and Hokkaido regions. Meanwhile, the favorable performance in the boosted Agri-BB cases was also supported by the higher IA values for the two regions (0.74 in GFED_Agri-BB10 case and 0.76 in FINN_Agri-BB05 versus 0.65 in GFED_base and 0.72 in FINN_base). Figure 3 shows the comparison of the observed and simulated daily mean $PM_{2.5}$ concentrations during the target period in Hokkaido, Japan. The CMAQ model in the two boosted Agri-BB cases effectively simulated the day-to-day variation patterns for the site-averaged products in Hokkaido, including the occurrence of high concentration peaks around the 1st of March 2019. These results indicate that the two BB emission inventories (GFED v4.1s and FINN v2.5) underestimate Agri-BB emissions.

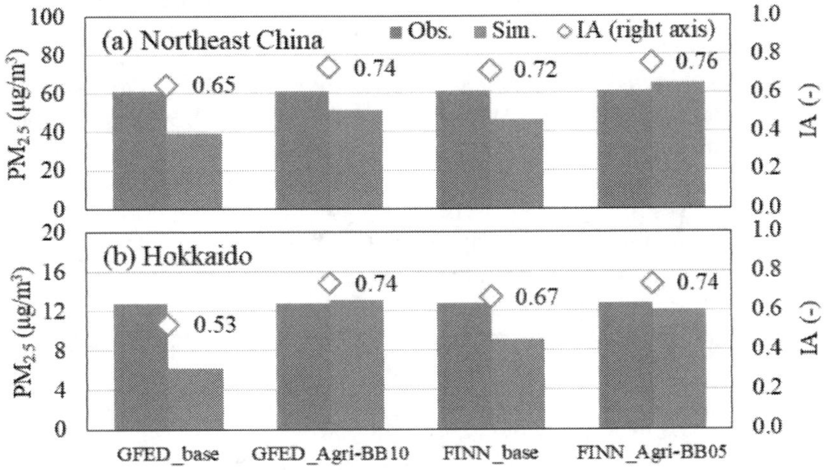

Fig. 2. Comparison of the results of baseline and boosted Agri-BB cases versus the observation data in terms of the mean concentrations and IA values of PM2.5 during the target period (from 1st Feb to 31st Mar 2019).

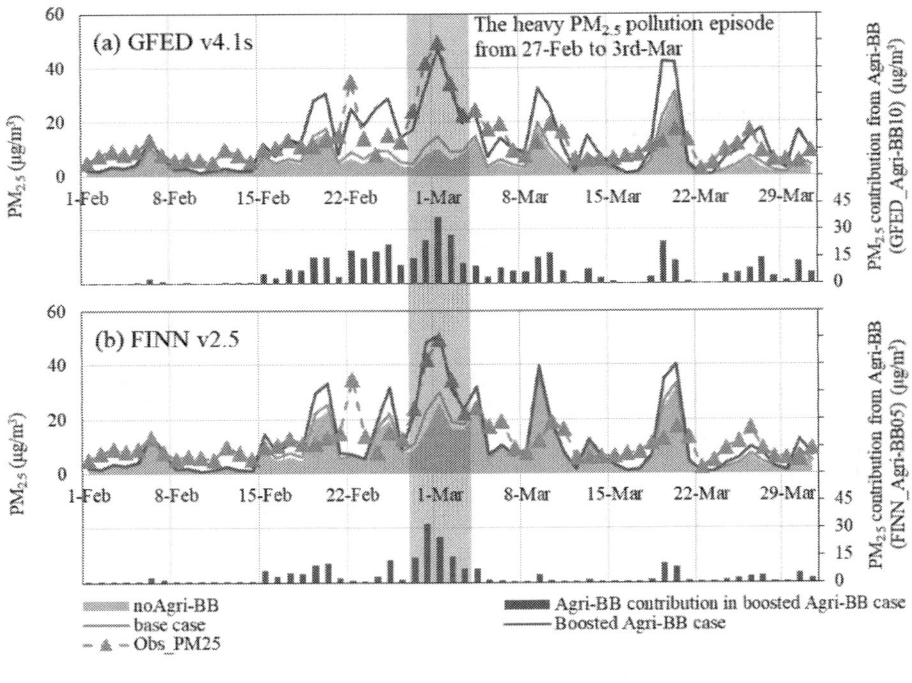

Fig. 3. Comparison of the observed and simulated daily mean $PM_{2.5}$ concentrations in the base, boosted Agri-BB, and noAgri-BB cases as well as $PM_{2.5}$ contributions in the boosted Agri-BB cases in Hokkaido, Japan

3.2 Agri-BB source analysis during the $PM_{2.5}$ heavy pollution episode

The Agri-BB contributions on $PM_{2.5}$ concentrations in Hokkaido in the two boosted Agri-BB cases were estimated during the target period as shown in Figure 3. The results in the two simulation cases were consistent in which the $PM_{2.5}$ contribution from Agri-BB was highest in Hokkaido during the PM2.5 heavy pollution episode (36.8 $\mu g/m^3$ on the 1st of March in GFED_Agri-BB10 and 32.1 $\mu g/m^3$ on the 28th of February in FINN_Agri-BB05) when the daily mean $PM_{2.5}$ concentration was maximum (46.9 $\mu g/m^3$ in GFED_Agri-BB10 and 51.3 $\mu g/m^3$ in FINN_Agri-BB05 versus 49.4 $\mu g/m^3$ as observed value), although trends of Agri-BB contributions in the two boosted Agri-BB cases differ in their day-to-day variation patterns due to differences of BB emission estimation approach between GFED v4.1s and FINN v2.5. In both cases, the $PM_{2.5}$ contribution from Agri-BB accounted for more than 50% during the $PM_{2.5}$ heavy pollution episode (66% in GFED_Agri-BB10 and 54% in FINN_Agri-BB05 cases), indicating that Agri-BB source was strongly involved in the episode.

Figure 4 shows the spatial distributions of mean $PM_{2.5}$ emissions from Agri-BB during the $PM_{2.5}$ heavy pollution episode in the two boosted Agri-BB cases. Both $PM_{2.5}$ emissions from Agri-BB have a similar spatial pattern to one another and show high spatial variation. In particular, high $PM_{2.5}$ emissions were found in Northeast China and its surrounding regions that are common with the two cases. These results illustrated that Agri-BB emissions during the $PM_{2.5}$ heavy pollution episode was substantially derived from agricultural area in Northeast China and also indicated that the $PM_{2.5}$ heavy pollution in Hokkaido 2019 was strongly affected by Agri-BB occurred in Northeast China. Meanwhile, it was found that the existing BB emission inventories, such as GFED v4.1s and FINN v2.5 were found to underestimate emissions from Agri-BB sources in Northeast China, since the boosted Agri-BB emission cases enable us to effectively reproduce daily $PM_{2.5}$ concentrations during the $PM_{2.5}$ heavy pollution episode.

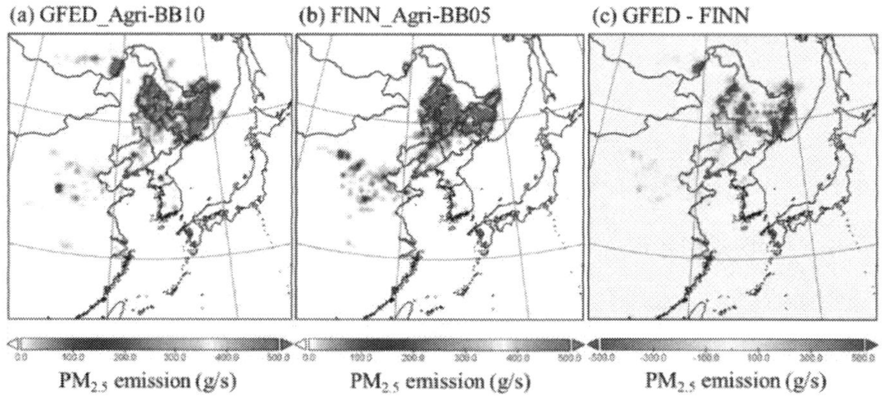

Fig. 4. The spatial distributions of mean $PM_{2.5}$ emissions from Agri-BB during the $PM_{2.5}$ heavy pollution episode: (a) GFED_Agri-BB10, (b) FINN_Agri-BB, and (c) GFED_Agri-BB10-FINN-FINN_Agri-BB05. Green lines indicate Northeast China region in China.

4 Conclusion

Agri-BB is a widespread practice to process crop residue in rural China and causes heavy air pollution from local to regional scales which significantly affects public health. This work investigated the $PM_{2.5}$ pollution episode in Hokkaido that occurred in early spring 2019 using the CMAQ model with the two BB emission inventories (GFED v4.1s and FINN v2.5). We also estimated Agri-BB impact during the heavy $PM_{2.5}$ pollution episode conducting three simulation cases for each BB emission inventory: with and without Agri-BB emission, and the boosted Agri-BB emission cases. The CMAQ performance for regional transport of $PM_{2.5}$ concentrations was evaluated using ground-level observation data in Northeast China and Hokkaido, Japan. The model in the baseline cases failed to capture the temporal and spatial variation patterns of $PM_{2.5}$ mass during the $PM_{2.5}$ heavy pollution episode. Meanwhile, the model in the boosted Agri-BB cases were able to show a favorable performance for $PM_{2.5}$ concentrations during the whole simulation period, including $PM_{2.5}$ concentration spikes during the $PM_{2.5}$ heavy pollution episode. These results indicated that the two BB emission inventories (GFED v4.1s and FINN v2.5) underestimated Agri-BB emissions. Moreover, the Agri-BB contributions on $PM_{2.5}$ concentrations in Hokkaido in the two boosted Agri-BB cases were estimated during the target period for Agri-BB source analysis. The results in the two simulation cases were consistent in that the $PM_{2.5}$ contribution from Agri-BB was highest in Hokkaido during the $PM_{2.5}$ heavy pollution episode when the daily mean $PM_{2.5}$ concentration was maximum. In both cases, the $PM_{2.5}$ contribution from Agri-BB accounted for more than 50% during the $PM_{2.5}$ heavy pollution episode, indicating that Agri-BB source was strongly affected in the episode. During the episode, high $PM_{2.5}$ emissions were found in Northeast China and its surrounding regions that were common with the two cases. The result illustrated that Agri-BB emissions during the $PM_{2.5}$ heavy pollution episode was substantially derived from agricultural area in Northeast China. In addition, GFED v4.1s and FINN v2.5 were found to underestimate emissions from Agri-BB sources in Northeast China. Thus, in conducting simulations using existing BB emission inventories, it is necessary to consider substantial underestimation of Agri-BB emissions, specifically during harvesting and post-harvest seasons.

This research was conducted as Type II joint research of the National Institute for Environmental Studies (NIES) and environmental research institutes of local government in Japan. The computational resources were provided by NIES. We acknowledged the use of data and imagery from LANCE FIRMS operated by the NASA/GSFC/Earth Science Data and Information System (ESDIS) with funding provided by NASA/HQ.

References

1. Andreae M O and Merlet P 2001 Glob. Biogeochem. Cycles **15** 955-66
2. Chen J, Li C, Ristovski Z et al 2016 Sci. Total Environ. **579** 1000-34
3. Lasko K, Vadrevu K P and Ngruyen T T N 2018 Plos One **13** e0196629
4. Lasko K, Vadrevu K P, Tran V T and Justice C 2018 IEEE J. Sel. Top. Appl. Earth Obs. Remote Sens **11** 498-12
5. Lasko K and Vadrevu K P 2018 Environ. Pollut. **236** 795–06
6. Yin S, Wang X, Xiao Y, H. Tani H, Zhong G and Sun Z 2017 Environ. Pollut. **220** 204–21
7. Wang Z, Zhao J, Xu J, Jia M, Li H and Wang S 2019 Int. J. Environ. Res. **16** 1379
8. Vadrevu K P, Csiszar I, Ellicott E, Giglio L, Badarinath K V S, Vermote E and Justice C 2013 IEEE J. Sel. Top. Appl. Earth Obs. Remote Sens **6** 224-38
9. Wang H, An J, Shen L, Zhu B, Pan C, Liu Z, Wang Y 2014 Sci. Total Environ. **490** 501-08
10. Long X, Tie X, Cao J, Huang R, Feng T, Li N and Zhang, Q 2016 Atmos. Chem. Phys. **16** 9675-91
11. Zhu C, Kanaya Y, Yoshikawa-Inoue H, Irino T, Seki O and Tohjima Y 2019 Environ. Pollut. **247** 55–63

12. Uranishi K, Ikemori F, Shimadera H, Kondo A and Sugata S 2019 Environ. Pollut. **244** 414–22
13. Uranishi K, Ikemori F, Shimadera H, Kondo A and Sugata S 2020 J. Jpn. Soc. Atmos. Environ. **55** 34–49
14. Kaiser J W, Heil A, Andreae M O, Benedetti A, Chubarova N, Jones L, Morcrette J-J, Razinger M, Schultz M G, Suttie M and van der Werf, G R 2012 Biogeosciences **9** 527–54
15. Van der Werf G R, Randerson J T, Giglio L, Van Leeuewn TT, Chen Y, Rogers B M, Mu M, Van Marle M J, Morton D C, Collatz G J and Yokelson R J 2017 Earth System Science Data 2017 **9** 697–720
16. Wiedinmyer C, Kimura Y, McDonald-Buller E C, Emmons L K, Buchholz R R, Tang W, Seto K, Joseph M B, Barsanti K C, Carlton A G and Yokelson R J 2023 Geosci. Model Dev. **16** 3873–91
17. Feng Z, Jun W, Charles I, Edward J H, Zhifeng Y, Cui Y G, Shenjian S, Xiaoyang, Z, Shobha K, Johannes W K, Christine W and Arlindo da S 2014 Environ. Res. Lett. **9** 075002
18. Willmott C J 1981 Phys. Geogr 2 184-94A. Mecke, I. Lee, J.R. Baker jr., M.M. Banaszak Holl, B.G. Orr, Eur. Phys. J. E **14**, 7 (2004)

E3S Web of Conferences **530**, 01003 (2024) https://doi.org/10.1051/e3sconf/202453001003
ICFEE 2024

Estimation of on-road mobile emissions based on the vehicle technology in a high-traffic avenue in Panama City, Panama.

Ana Valdes-Montenegro[1], *Franchesca* Gonzalez-Olivardia [1,2,*], *Sarawut* Thepanondh [3]
Cesar Pinzon-Acosta[1,2]

[1] Universidad Tecnologica de Panama, Panama City, Panama, Panamanian.
[2] Centro de Estudios Multidisciplinarios en Ciencias, Ingeniería y Tecnología AIP, Panama City, Panama.
[3]Faculty of Public Health, Mahidol University, Thailand.

Abstract. This study addresses the emissions from mobile sources in a busy avenue. Latest mobile emission inventories estimated pollutants based on fuel sales activities in the country. While this is an approved methodology, it does not consider the characteristics of vehicles and their emissions control systems. Applying the International Vehicle Emissions (IVE) model involves low costs, the vehicle fleet's technology features, and the study region's environmental parameters. The IVE model allows a better understanding of how vehicle technology impacts air quality. This study aims to generate information about vehicle emissions in Panama using a model that considers factors like vehicle technology, driving patterns, and local conditions, all of which influence air pollution in the region. The focuses are on passenger vehicles, the most common in the country, particularly in one of the densest districts. Carbon monoxide (CO), sulfates (SOx), nitrogen oxide (NO, and particulate matter (PMx) emissions were estimated with the IVE model, the first one stood out, represented 98% (6479.82 g/km) of the air pollutants emitted into the atmosphere by these four gases. Finally, it is established that the combination of emission control technologies and vehicle features is what will determine its emission reduction efficiency. Only in diesel vehicles, an increase in PMx emissions was identified.

1 Introduction

Outdoor air pollution is a major environmental health problem affecting everyone in low-, middle-, and high-income countries [1]. Air quality is affected by anthropogenic emissions from stationary and mobile sources, and these emission sources have varying impacts depending on the country or region and their economic activities. The economic activity in the Republic of Panama is primarily based on the service sector (banking, commerce, insurance, container ports, flagship registry, medical, and health tourism), which is closely

* Corresponding author: franchesca.gonzalez@utp.ac.pa

linked to the transportation sector, thus mobile emission sources. It is well known that emission inventories are used for air quality management, providing information on the quantity of air pollutants released into the atmosphere over a specific period in a specific area due to a particular activity.[2]

According to the WHO [1], mobile sources account for a significant portion of NOx, CO, and particulate matter PMx, that is being emitted into the atmosphere. The National Institute of Statistics and Census (INEC) of Panama indicated that mobile emissions are the primary source of atmospheric pollutants far exceeding emissions from industries and the energy sector. The latest updates of air pollutant emissions data in Panama date back to 2017, where mobile emissions account for a total of 732,457 metric tons of PMx, CO, SOx, NOx and hydrocarbons (HC) combined (Figure 1).

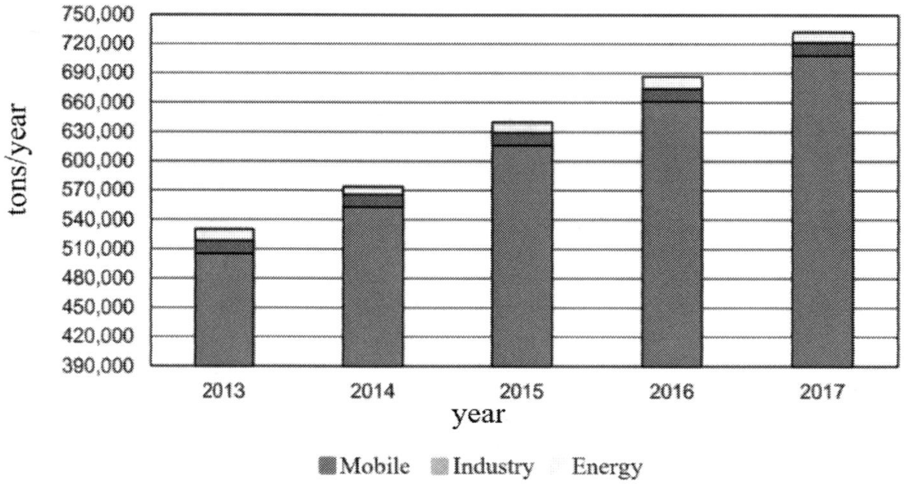

Fig. 1. Inventory of air pollutant emissions in the Republic of Panama, by emission source from 2013 to 2017 according to INEC. Red represents mobile sources, green is the industry sector, and yellow is the energy sector.

Of the Total air pollutant emissions in 2017, 96.7% came from mobile sources, mainly CO emissions (Figure 2).

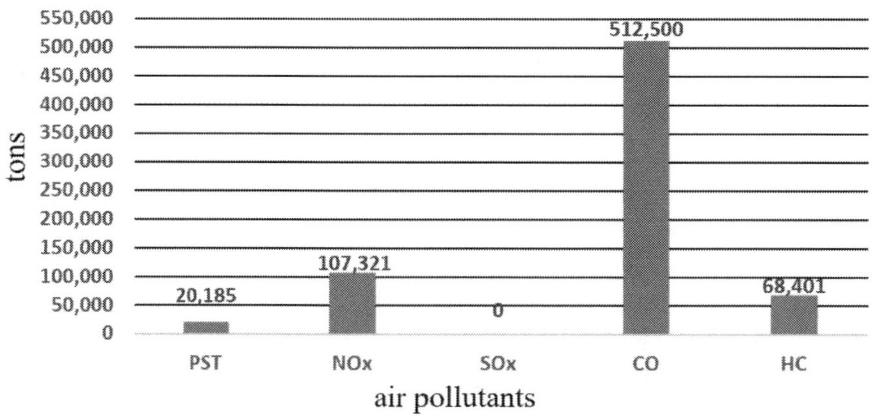

Fig. 2. Inventory of air pollutants emissions in the Republic of Panama from mobile sources by type of pollutants in 2017 according to INEC.

Despite the lack of information about the methodologies for the emission inventory estimation related to NO_x, CO, PMx, SO, and HC, an estimation of emissions from land transport in Panama's National Greenhouse Gas Inventory for 1994-2017 (INGEI) showed the use of the guidelines from the Intergovernmental Panel on Climate Change 2006, applying the Tier 1 method. This approach utilizes activity data from national statistics and default emission factors, considering the amount of fuel sold according to available records. However, this methodology is focused on the calculation of greenhouse gas emissions (CO2, CH4 and NO2) [3].

Nevertheless, a more comprehensive emission inventory will include other factors, such as driving cycles, the vehicle's emission control system, and urban traffic [4]. In a study conducted in Hanoi, Vietnam, that analyzed mobile emissions while considering driving patterns, it was identified that vehicle idling and frequent stops result in significant emissions, especially of toxic air pollutants such as fine particles, volatile organic compounds (VOC) and CO. These toxic exhaust gases are released at breathing levels in densely populated urban areas, posing a high risk of human exposure [4]. In Bangkok, it is concluded that vehicle kilometers traveled, number of vehicles, and emission factors influence the number of toxic emissions from vehicles, so improving the public transport system and decreasing travel time can mitigate emissions from the transport sector [5].

The information about the vehicles will allow better emission inventories; however, documenting each vehicle technology to reduce emissions is difficult because each brand has its features, and the national laws don't require companies to disclose that information. The emission control system, also named catalyzer, is in charge of reducing air pollutants at the tailpipe of internal combustion engines, such NO_x, CO, and HC through chemical reactions. [6].

2 Methodology

To quantify on-road mobile emissions, on-road measurement of exhaust emissions with real-world data using instantaneous real-time emission collection to correlate vehicle emissions with actual driving conditions is generally used. However, this method results in high costs for a fleet at the regional or national level. The most used option is based on dynamometer testing of many laboratory vehicles to estimate the emission rates in terms of the instantaneous or average data based on multiple simulated driving behaviors [7].

This work uses the IVE Model to estimate emissions from on-road mobile sources and create a guide for following emission inventories. The IVE is based on the Federal Testing Program of the USA [8] and allows quantifying the contribution of automobiles to air pollution, considering the vehicle technology present in the region and its environmental characteristics. The application of low-cost methods is required due to the lack of investment in this research area.

This work shows the transportation sector's contribution to air pollution from the highest population density district in the country, with the largest number of vehicles in circulation nationwide (Figure 3).

Only Panama district, located in Panama City, contains 68% of the fleet of vehicles in circulation compared to the nationwide fleet, with passenger cars representing 82% of the total in 2019 (Figure 4). According to INEC passenger cars are defined as those with a capacity of up to 13 persons, including vans, jeeps, sedans and coupes. Cabs, even though they have a different use and are given more travel during the day compared to private cars, are also classified as passenger cars [9].

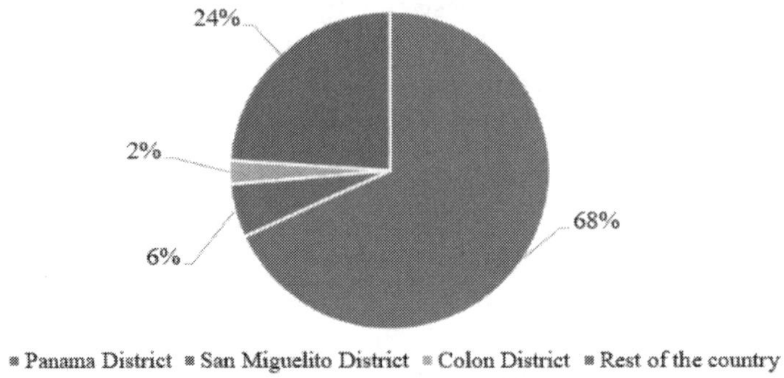

Fig. 3. The vehicles fleet in circulation per district according to INEC.

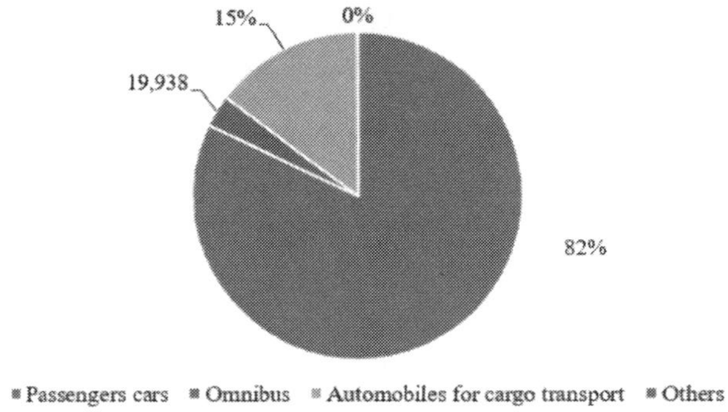

Fig. 4. The fleet of vehicles by category in the Panama District according to INEC.

2.1 Fleet Identification

One of the main traffic routes in the District of Panama is Domingo Díaz Avenue situated at latitude 9° 4' 5.197" N and longitude 79° 24' 33.336" W. To collect the data, a busy gas station located at Domingo Diaz Avenue (Figure 5) was selected.

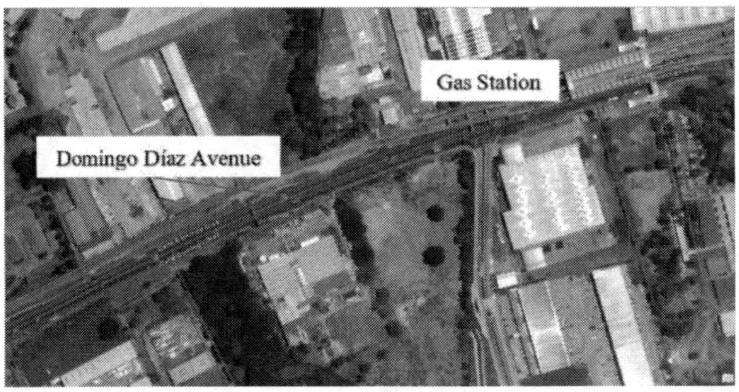

Fig. 5. Street view map of location Avenue Domingo situated at latitude 9° 4' 5.197" N and longitude 79° 24' 33.336" W. Red pin marks the gas station where data was collected.

In order to define the sample of cars to be surveyed, a simple non-probabilistic sample was established with a confidence level of 80% and a margin of error of 9.5% of the

population of 517217 vehicles in the District of Panama, resulting in a sample of 46 vehicles. Even though this is a low confidence level for representing the district's particular car emissions, this data will provide preliminary numbers for estimating the contribution of vehicles to air pollution in a characteristic urban area of the country.
Questions about the brand, model, year, cylinder capacity in cubic centimeters (cc), VIN, fuel type, age in kilometers (km), emission control device and last maintenance were asked.

Information from forty-eight (48) passenger cars was collected and grouped depending on the category of the IVE model (Table 1). However, during the survey it was realized that people were not familiar with the type of emission control system of their cars and the lack of data on the characteristics of the vehicles imported into the country, especially their emission control system, because of the laxity of the vehicle emission regulations in the country.

For those vehicles whose records obtained by the VIN indicated that they complied with the international Euro 2 emissions regulation and users confirmed that their vehicle has a catalytic converter, it was assumed that the value of the exhaust gas recirculation (EGR) system, which in internal combustion engines (gasoline and diesel) helps to reduce nitrogen oxide emissions [10], is not in this type of vehicle represent 29% of the sample. Meanwhile, the vehicles whose users confirmed that their vehicle has a catalytic converter, and any records deny the presence of the EGR system are assumed to have it, 58% of the sample. The remaining 12% of the people surveyed confirmed that they had been removed or their car did not have a catalytic converter. Therefore, it was also assumed that these vehicles had no EGR system. Vehicles are categorized according to their technologies, and abbreviations are added to simplify the description (Table 1).

Table 1. Abbreviations of vehicle technology categories.

Abbreviation	Meaning
Gas	Gasoline
Ds	Diesel
PC	Passenger car, weighing less than 9000 lb
Lt	Cylinder capacity <1.5 cc
Med	Cylinder capacity between 1.5 – 3.0 cc
Hv	Cylinder capacity > 3.0 cc
MPI	Multi-point fuel injection
NoC	Vehicle without catalyst
3Wy	3-way catalyst
EGR	Exhaust gas recirculation
PCV	Positive crankcase ventilation
NoPCV	No Positive crankcase ventilation
Dir-Inj	Direct injection
MPI	Multipoint injection
EGR+Improv	improved fuel injection in diesel vehicles and EGR
A	Kilometers <79K km
B	Kilometers between 80-161K km
C	Kilometers >161K km

 https://doi.org/10.1051/e3sconf/202453001003

Sixteen (16) categories were identified within the collected sample (Table 2). All vehicles are passenger cars and the majority use gasoline, only 4.17% use diesel. It is also important to note that very few vehicles have a low kilometer; 60% have a high age, more than 161,000 km. Additionally, including the fraction of km driven per vehicle, the value represents the fraction of the fleet with the determined characteristics running on the street. For the calculations, each vehicle travels the route only once. For example, 100% is represented by the 48 vehicles, 8 (eight) of which are in the N°7 category, so it exemplifies 16.67% of the fleet.

$$Fraction\ of\ km\ driven\ per\ vehicles = \frac{Quantity\ of\ vehicles\ per\ category}{Quantity\ of\ vehicles\ in\ the\ fleet} * 100$$

Table 2. Vehicle categories identified from the vehicle sample.

N° Category	Description	Fuel	Weight	Air/Fuel Control	Exhaust Control	Evaporative Control	Age	Number of cars	Fraction of km driven per vehicles
1	PC	Gas	Lt	MPI	NoC	PCV	B	1	2.08
2	PC	Gas	Lt	MPI	NoC	PCV	C	2	4.17
3	PC	Gas	Med	MPI	NoC	PCV	B	1	2.08
4	PC	Gas	Med	MPI	NoC	PCV	C	1	2.08
5	PC	Gas	Hv	MPI	NoC	PCV	C	1	2.08
6	PC	Gas	Lt	MPI	3Wy	PCV	B	3	6.25
7	PC	Gas	Lt	MPI	3Wy	PCV	C	8	16.67
8	PC	Gas	Med	MPI	3Wy	PCV	B	1	2.08
9	PC	Gas	Med	MPI	3Wy	PCV	C	2	4.17
10	PC	Gas	Lt	MPI	3Wy/EGR	PCV	A	4	8.33
11	PC	Gas	Lt	MPI	3Wy/EGR	PCV	B	2	4.17
12	PC	Gas	Lt	MPI	3Wy/EGR	PCV	C	6	12.5
13	PC	Gas	Med	MPI	3Wy/EGR	PCV	A	3	6.25
14	PC	Gas	Med	MPI	3Wy/EGR	PCV	B	4	8.33
15	PC	Gas	Med	MPI	3Wy/EGR	PCV	C	7	14.58
16	PC	Ds	Med	Dir-Inj	EGR + Improv	NoPCV	C	2	4.17

2.2 Ambient conditions and local information

The field study was conducted in early July, and on that specific date, ambient humidity was at 76.7%, with a temperature of 31 °C. One of the model requirements is to identify the usage of vehicle air conditioning. With temperatures at 31 °C, it is assumed that 50% of users will use it throughout their journey, a probability suggested by Model IVE. The fuel characteristics were obtained from the Panamanian Commission of Industrial and Technical Standards (COPANIT), which establishes the composition of the fuels sold in Panama. The values for Benzene exceed those available in the IVE model, so the software indicates the maximum value (3 %). The frequency of vehicle ignition and shutdown is indicated as one time.

Table 3. Fuel specifications

Specifications	Gasoline	Diesel
Sulfur, ppm	15	15
Benzene, % (v/v)	5%	-
Oxygenate, % (p/p)	0	-

2.2.1 Driving Pattern Input Bins

For this section, the vehicle's speed, acceleration, and deceleration during the route should be included. Consequently, a statistical analysis must be conducted [11]. However, a study from Bangkok, Thailand, was employed as a reference for this work.

3 Results and analysis

The IVE model was employed to estimate the primary air pollutants (CO, NOx, SOx, and PMx). Subsequently, the following charts present the emission results (g/km) on Domingo Díaz Avenue under the previously established conditions for the 48 passenger cars during a one-kilometer journey.

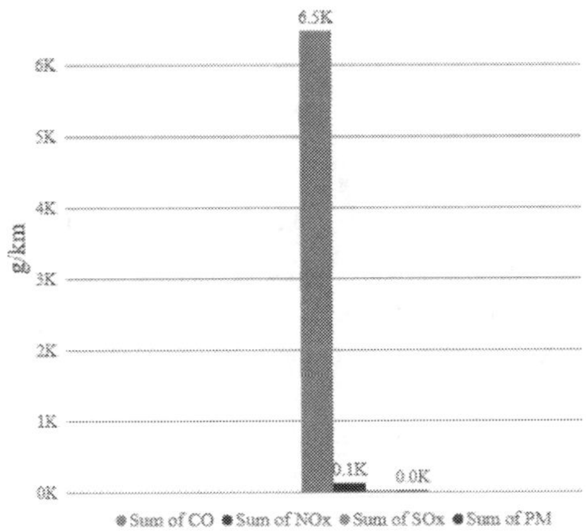

Fig.6. CO, SOx, NOx, and PMx Emissions by the fleet

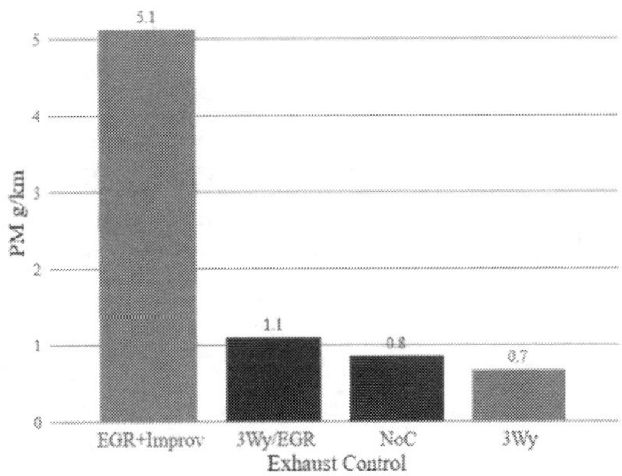

Fig.7. PMx emissions by fleet emission control technology.

Figure 6 shows the results of the emission of CO, Sox, NOx, and PM (g/km) for the entire fleet. Carbon Monoxide (CO) has the highest contribution (98%) compared to NOx, Sox and PMx.

96.78% of this pollutant is primarily emitted in similar quantities by light and medium vehicles, 3.18K and 3.09K CO g/km, respectively. This may be attributed to the fact that these vehicles constitute a significant portion of the fleet, accounting for 93.8%. This aligns with Figure 2, where CO is the primary air pollutant predominantly emitted by mobile sources.

Figure 7 compares emission control systems against PM emission levels. Cars fueled/powered by diesel (EGR.-Improve) are mainly responsible for particulate matter emissions (PMx). Despite being only 4.17 % of the fleet, they have the 66% of the total PMx emissions. For those cars fueled by gasoline, there is no difference if they have a catalyzer or not.

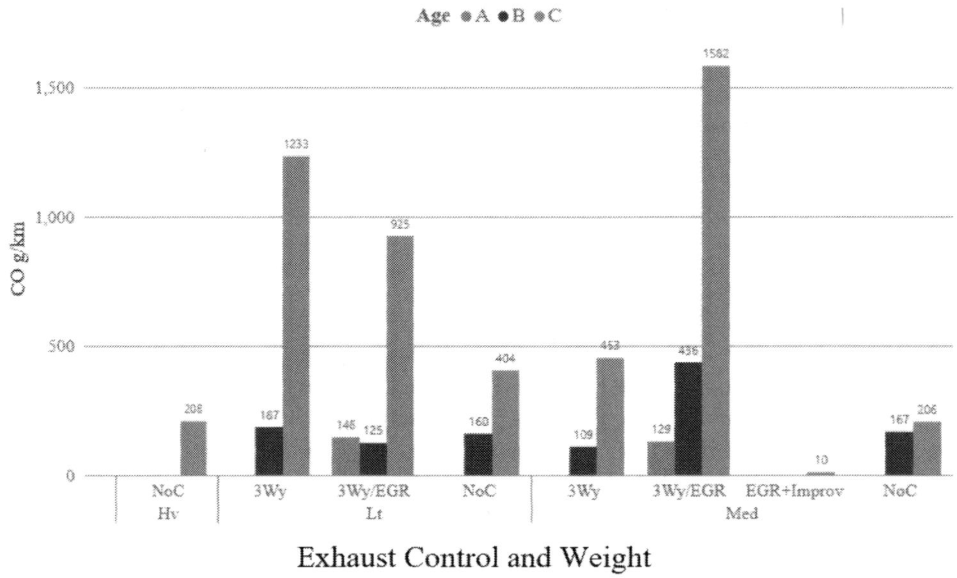

Fig. 8. CO Emissions by Emissions Control, Weight, and Age.

Figure 8 illustrates CO emissions versus vehicle weight and its emission control system, enabling the correlation between emission control technology and vehicle weight to understand their impact. In each section where the exhaust control type is compared, values are higher for vehicles with mileage exceeding 161,000 km. This could be associated with 50% of the fleet falling within this mileage range. In the case of heavy vehicles, only one car was surveyed. Medium-sized gasoline vehicles with 3Wy/EGR equipment and a 161,000km mileage emit the highest values at 1582 CO g/km, representing 24% of emissions and generated by 14.6% of the vehicles. In contrast, vehicles with the same characteristics but in the light size category emit 925 CO g/km, accounting for 12.5% of the fleet, reflecting a substantial reduction of 58.6%.

4 Conclusions

The vehicle fleet mainly comprised vehicles with mileage larger than 161000 km, fifty (50%) of the fleet, generating 79% of the total calculated emissions. It cannot be asserted that a specific technology type influences CO emissions from a vehicle; the results demonstrated that combining technologies and vehicle characteristics will determine its contribution to air pollution. The fleet's total contribution is 6604.825 g/km in combined CO, NOx, SOx, and PM. Of this, CO represents 98% of the emissions. For this reason, it is crucial to categorize vehicles for the development of mobile emission inventories to establish mitigation plans that can be effectively assessed. Gasoline vehicles are the primary sources of CO emissions, as most of the fleet comprises this type of vehicle. However, despite their low proportion in the fleet, diesel vehicles are the primary contributors to particulate matter (PMx) emissions. An increase in the number of these vehicles would significantly raise PMx emissions. These particles pose a health risk, as they can trigger health issues and adversely affect urban air quality.

We can also emphasize the effect of vehicle size on emissions. It was observed that vehicles with mileage exceeding 161,000 km and an emission control system 3Wy/EGR can increase their CO emissions by up to 71% more in medium-sized vehicles (1582.34 g/km) compared to light-sized ones (924.85 g/km). Local conditions or the effectiveness of the catalyst and EGR system in medium and light passenger cars may influence this increase, considering the similar number of vehicles in both categories. The research demonstrates that the IVE Model would allow the establishment of a cost-effective method for evaluating vehicular air pollutants to develop a more representative emissions inventory for mobile sources. With a greater variety of data, emissions can be estimated annually, enabling comparisons with international emission standards.

We extend our gratitude to the Technological University of Panama (UTP) for fostering scientific research. In addition, special thanks to the National Secretariat for Science, Technology, and Innovation (SENACYT) for funding the IDDS22-43 and FIEDS22-12 projects, which were crucial for developing this research. We appreciate the management of the Center for Multidisciplinary Studies in Science, Engineering, and Technology AIP (CEMCIT AIP). Finally, we acknowledge the invaluable support of all collaborators for the success of this project.

References

1. World Health Organization, "Ambient (outdoor) air pollution," https://www.who.int/news-room/fact-sheets/detail/ambient-(outdoor)-air-quality-and-health.
2. The U.S. Environmental Protection Agency (EPA), "Managing Air Quality - Emissions Inventories." Accessed: Jun. 19, 2023. [Online]. Available: https://www.epa.gov/air-quality-management-process/managing-air-quality-emissions-inventories

3. The Intergovernmental Panel on Climate Change, “Directrices del IPCC de 2006 para los inventarios nacionales de gases de efecto invernadero, Volumen 2,” 2006. Accessed: Jun. 19, 2023. [Online]. Available: https://www.ipcc-nggip.iges.or.jp/public/2006gl/spanish/pdf/2_Volume2/V2_3_Ch3_Mobile_Combustion.pdf
4. T. T. Trang, H. H. Van, and N. T. K. Oanh, “Traffic emission inventory for estimation of air quality and climate co-benefits of faster vehicle technology intrusion in Hanoi, Vietnam,” *Carbon Manag*, vol. 6, no. 3–4, pp. 117–128, Jul. 2015, doi: 10.1080/17583004.2015.1093694.
5. P. Outapa and S. Thepanondh, “Development of air toxic emission factor and inventory of on-road mobile sources,” *Air, Soil and Water Research*, vol. 7, Jan. 2014, doi: 10.4137/ASWr.S13320.
6. E. V. Rojas Reinoso, V. J. Romero Hidalgo, and J. M. Pancha Ramos, “Analysis of behavior of CO2 emissions, CO and the lambda factor of a vehicle with a conventional injection system with catalyst and without catalyst,” *Ingenius*, vol. 2020, no. 23, pp. 23–29, Jul. 2020, doi: 10.17163/ings.n23.2020.02.
7. P. Lyu, P. (Slade) Wang, Y. Liu, and Y. Wang, “Review of the studies on emission evaluation approaches for operating vehicles,” *Journal of Traffic and Transportation Engineering (English Edition)*, vol. 8, no. 4, pp. 493–509, 2021, doi: 10.1016/j.jtte.2021.07.004.
8. International Sustainable Systems Research Center, “Manual del Usuario del Modelo IVE Versión 2.0,” 2008. [Online]. Available: www.issrc.org/ive.
9. National Institute of Statistics and Census (INEC), “Automobiles in circulation destined for passenger transportation in the Republic, by license plate class and capacity, by district and type: year 2019 (P),” Panama, Nov. 2020.
10. L. Tanicuchi, “Análisis De Los Sistemas De Ventilación Positiva Del Cárter Y Recirculación De Gases De Escape En Un Mep En Función De Las Emisiones Generadas,” 2021.
11. P. Outapa, S. Thepanondh, A. Kondo, and N. Pala-En, “Development of air pollutant emission factors under real-world truck driving cycle,” *Int J Sustain Transp*, vol. 12, no. 6, pp. 432–440, Jul. 2018, doi: 10.1080/15568318.2017.1384968.

E3S Web of Conferences **530**, 01004 (2024) https://doi.org/10.1051/e3sconf/202453001004
ICFEE 2024

Improving Machine Learning Based $PM_{2.5}$ Prediction by Segregating Biomass Emission Factor from Chemical Transport Model

Supitcha Sukprasert[1*], *Hikari* Shimadera[1], *Shin* Araki[1], *Tin* Thongthammachart[1], *Pongpisit* Thanasutives[2], *Luong* Viet Mui[1], *Tomohito* Matsuo[1], and *Akira* Kondo[1]

[1] Graduate School of Engineering, Osaka University, Osaka, Japan
[2] Graduate School of Information Science and Technology, Osaka University, Osaka, Japan

Abstract. Located in the heart of Mainland Southeast Asia, Thailand is associated with high biomass burning (BB) activities from local and neighbouring countries. The seasonal pattern of BB manifests itself as a potential predictor for $PM_{2.5}$ concentration. Consequently, we enhanced machine learning based $PM_{2.5}$ prediction by segregating BB factor from the Community Multiscale Air Quality (CMAQ). Two Light Gradient Boosting Machine (LightGBM) models with different CMAQ predictors were developed: the BB-integrated model, which incorporated CMAQ-simulated $PM_{2.5}$ from all emission sources and the BB-segregated model, which incorporated CMAQ-simulated $PM_{2.5}$ from sources other than BB (CMAQ_PM25_Othr) and CMAQ-simulated $PM_{2.5}$ from BB emissions (CMAQ_PM25_BB). The two models had shared control predictors, which included simulated meteorological variables from WRF model, population, elevation, and land-use variables, and they were evaluated using a cross-validation (CV). The BB-segregated model outperformed the BB-integrated model, achieving overall-CV R^2 values of 0.86 and 0.82, respectively. The analysis of feature importance for the BB-segregated model indicates that CMAQ_PM25_Othr and CMAQ_PM25_BB are the two most significant predictors. These findings emphasize the importance of considering BB emissions when predicting $PM_{2.5}$ concentrations, particularly in regions with high BB activities.

1 Introduction

Biomass Burning (BB) is a prevalent occurrence within the Southeast Asia (SEA) region, primarily arises from slash-and-burn agriculture practices and land clearing [1-3]. Mainland SEA (MSEA) typically experiences elevated levels of BB activities, predominantly occurring throughout the dry season, which is linked to reduced precipitation [3-4]. BB releases a significant number of airborne particles into the atmosphere, mainly consisting of Black Carbon and Organic Carbon [4-5]. This, in turn, leads to an increase in the levels of $PM_{2.5}$ (particulate matter with a diameter of ≤ 2.5 μm), reaching their highest point in March [4]. It

* Corresponding author: supitcha.suks@ea.see.eng.osaka-u.ac.jp

was estimated that around 57% of the increase in $PM_{2.5}$ levels across MSEA was attributed to BB, with significant BB influence observed in Laos and Myanmar [6].

Air pollutants originating from BB have the capacity to be transported across extensive distances, playing a role in the occurrence of transboundary pollution within downwind regions [7]. Located in the centre of MSEA, transboundary BB contributes to approximately 67% of the heightened $PM_{2.5}$ levels in Thailand, while 33 % from local BB [8]. As a result, this has led to Thailand becoming one of the highly polluted nations in SEA [9]. In 2020, the annual average of $PM_{2.5}$ reached 23.8 $\mu m/m^3$, which is almost five times greater than the limit set by the World Health Organization (WHO) [9].

Extensive evidence exists regarding the negative impact of $PM_{2.5}$ on human health. PM2.5 is not solely responsible for cardiopulmonary conditions such as heart disease, respiratory infections, and lung diseases [6, 8, 10-11], but also escalates the likelihood of various disorders such as diabetes mellitus [11-13] and adverse birth outcomes [11, 14]. In 2016, $PM_{2.5}$ resulted in roughly 18,003 cases of premature mortality throughout Thailand, with Central Thailand bearing the highest brunt of these cases [8]. Consequently, numerous studies aim to generate high spatiotemporal $PM_{2.5}$ data, particularly in regions with limited monitoring stations, to facilitate essential epidemiological and toxicological studies crucial for developing health-protective measures against $PM_{2.5}$ pollution [15-18, 20-21].

In recent years, there has been growing interest in applying machine learning (ML) models to predict $PM_{2.5}$ concentrations at a fine-grained spatiotemporal resolution [15-21], e.g., 1 km × 1 km daily $PM_{2.5}$ prediction. Several studies have attempted to integrate the ML model with chemical transport models (CTMs) since this helps capture atmospheric dynamics, physics, and chemical processes, leading to further improvement in the performance of the ML model [15, 20]. Nonetheless, in areas where BB activities are prevalent, accounting for its major contribution rather than those of the anthropogenic factors alone is important to prevent $PM_{2.5}$ underestimation by CTMs [22].

In a recent study conducted by Thongthammachart et al. [21], a Light Gradient Boosting Machine (LightGBM) model was developed to predict $PM_{2.5}$ concentrations at 1-km resolution in the Central Thailand [21]. The model incorporated simulated $PM_{2.5}$ concentration from the Community Multiscale Air Quality (CMAQ) model as a significant predictor, encompassing a wide array of emission sources, including BB emissions. While the model demonstrates impressive capabilities in predicting $PM_{2.5}$ concentrations, it falls short of fully exploiting the BB component in $PM_{2.5}$ predictions, which potentially limit the model's performance. Therefore, we aim to elevate the LightGBM model's performance by decoupling BB emissions from CMAQ model. We assess the performance and feature importance analysis of two LightGBM models. Each model utilizes two different sets of CMAQ-simulated $PM_{2.5}$ predictors, both of which consider BB component but in different ways—one integrates the BB component, and the other isolates the BB component.

2 Method

This study extended the methodology established by Thongthammachart et al. [21] and maintain consistency in the study area, dataset, and key hyperparameter configurations for the LightGBM models. Python version 3.11.2 is utilized for both data pre-processing and the development of the LightGBM models.

2.1 Study area and air monitoring stations

The study area is constrained to an expanse that largely covers the Central Thailand provinces, including the Bangkok Metropolitan Region (BMR), which is located at the centre of the study area (Fig. 1). The area is mostly flat and has fertile alluvial soil, which was deposited by the Chao Phraya River and its tributaries over centuries. These factors contribute to creating a highly favourable environment for agricultural pursuits, with a particular emphasis on rice cultivation [23].

In this study, we used the $PM_{2.5}$ measurements obtained from the ground-based air monitoring stations which are mostly located in BMR (Fig. 1). The stations are under the management of Thailand's Pollution Control Department (PCD) and the Bangkok Metropolitan Administration (BMA). Owing to the data's availability, we utilized the $PM_{2.5}$ observations recorded in the year 2019. Subsequent to this, a thorough data quality assessment was performed and only 55 monitoring stations satisfied the criteria of possessing a minimum of 70% valid data for the specified year.

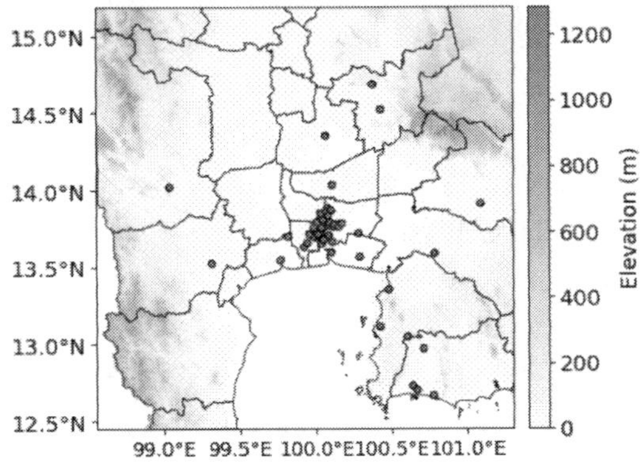

Fig. 1. The study area. The red dots are the ground-based air monitoring stations.

2.2 Model predictors

To assess the impact of BB emissions on the model's performance, we organized two groups of predictors: control predictors and CMAQ predictors, as outlined in Table 1. The set of control predictors encompasses simulated meteorological variables from the Weather Research Forecasting (WRF), population, elevation, and land-use data. CMAQ predictors consist of simulated $PM_{2.5}$ concentrations, accounting for diverse conditions linked to BB emission inputs. The specific methodologies for preparing each predictor are elaborated as follows.

2.2.1 WRF and CMAQ predictors

For the simulation of meteorological variables and $PM_{2.5}$ concentrations, we utilized WRF version 3.8 and CMAQ version 5.2.1, respectively. These simulations were carried out at a spatial resolution of 5 kilometres. More detailed information on WRF and CMAQ configurations can be found in the study of Thongthammachart et al. [21].

To facilitate the construction of CMAQ predictors, two distinct scenarios were set in CMAQ simulation. One scenario included BB emissions, whereas the other excluded them. The outcomes from these scenarios are respectively denoted as CMAQ-simulated $PM_{2.5}$ from all emission sources (CMAQ_PM25) and CMAQ-simulated $PM_{2.5}$ from other emission sources (CMAQ_PM25_Othr). To quantify the contribution of BB emissions, we isolated the $PM_{2.5}$ concentrations attributed to BB by calculating the difference between the simulated $PM_{2.5}$ concentrations from the two scenarios denoted as CMAQ_PM25_BB. Subsequently, the results of the simulations conducted using WRF and CMAQ were resampled to a 1-km grid cell resolution using bilinear interpolation methods. Figure 2 shows the time series plot of each CMAQ predictor and the $PM_{2.5}$ observations throughout the entire study year.

Table 1. Extracted model predictors.

Predictors	**Unit**
Control predictors	
Major road density	km/km^2
Minor road density	km/km^2
Built-up area ratio	Unitless
Forest area ratio	Unitless
Rice-harvested area ratio	Unitless
Population density ratio	Unitless
Elevation	m
WRF-simulated planetary boundary layer (WRF_PBL)	m
WRF-simulated wind speed (WRF_Wind)	m/s
WRF-simulated relative humidity (WRF_RH)	%
WRF-simulated ambient temperature (WRF_Temp)	K
CMAQ predictors	
CMAQ-simulated $PM_{2.5}$ from all emission sources (CMAQ_PM25)	µg/m^3
CMAQ-simulated $PM_{2.5}$ from other emission sources (CMAQ_PM25_Othr)	µg/m^3
CMAQ-simulated $PM_{2.5}$ from BB emissions (CMAQ_PM25_BB)	µg/m^3

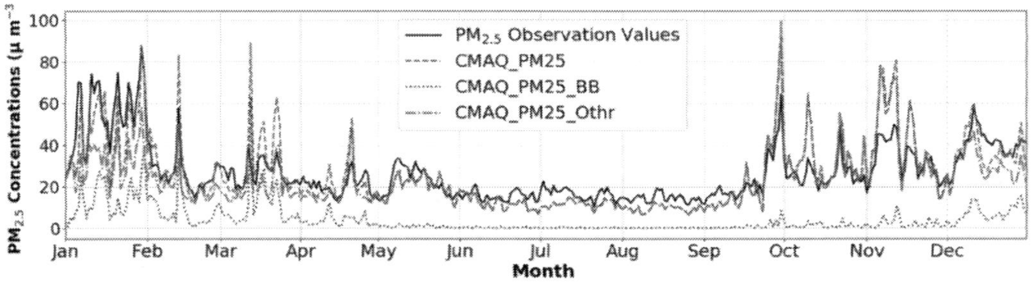

Fig. 2. Time series plot of $PM_{2.5}$ observation values and CMAQ predictors.

2.2.2 Population, elevation, and land-use data

Land-use data utilized in this study are built up, forest area, rice-harvested area and road data. The data sources have been presented in the study of Thongthammachart et al. [21]. Detailed information regarding data formats and original resolutions is provided in Table 2. All datasets primarily exist in raster format, except for the road data. Initially, the road data was categorized into five groups: trunk, motorway, primary, secondary, and tertiary roads. Subsequently, we refined this classification to comprise two main groups: major roads (including trunk, motorway, and primary roads) and minor roads (including secondary and tertiary roads). The vector-based road data was then converted to raster format, with each grid cell representing road density.

All the raster-based land use variables were resampled using bilinear interpolation methods to a 1-km grid cell, ensuring consistent resolution across all input predictors. A focal sum operation with an inverse distance weight matrix was later applied to land-use variables to account for the distance-decay effect, which efficiently helps construct a better model [24].

Table 2. Details of the population, elevation, and land-use data.

Data	Format	Original resolution (m)
Population	Raster	~100
Elevation	Raster	~100
Built up	Raster	10
Forest area	Raster	30
Rice-harvested area	Raster	~150
Road	Vector	-

2.2.3 The LightGBM model

We developed two distinct LightGBM models: the BB-integrated model and the BB-segregated model. These models share common control predictors while differing in their combination of CMAQ predictors. The BB-integrated model incorporates CMAQ_PM25, while the BB-segregated model incorporates the decomposed version of CMAQ_PM25 which are CMAQ_PM25_Othr and CMAQ_PM25_BB. Three evaluation metrics—R^2, RMSE, and MAE—were utilized to quantify the model's performance across three dimensions: overall, spatial, and temporal. For the evaluation algorithm, we adopted the five-fold cross-validation (CV), which randomly partitioned the data into five groups based on data sampling indices, monitoring stations, and dates for the respective overall, spatial, and temporal dimensions. Each CV dimension was executed 50 times using different random seed numbers. For each evaluation metric, we computed the average of the 50 metric values obtained from 50 CV iterations. Subsequently, we conducted a one-tailed paired t-test to statistically compare the evaluation metrics of the two models.

To comprehensively investigate the isolation effect of BB emission within the model's performance, we carried out feature importance analysis for both models. The importance score was calculated based on the gain of a feature, which represents the improvement in the model's performance that is achieved by splitting the data based on a certain feature. To facilitate clearer visualization, the importance scores were subsequently normalized.

3 Results and discussion

As shown in Table 3, the BB-segregated model exhibits notable improvements across all evaluation metrics. Specifically, the overall-CV R^2, spatial-CV R^2, and temporal-CV R^2 stand at 0.86, 0.73, and 0.75, respectively—each significantly surpassing the corresponding values of the BB-integrated model (p-value < 0.001). Furthermore, the BB-segregated model demonstrates significantly lower CV-RMSE and CV-MAE in comparison to the BB-integrated model (p-value < 0.001). These results collectively indicate the superior performance of the BB-segregated model.

Table 4 presents CV-evaluation metrics for two seasons: the wet season, spanning from April to September and the dry season, spanning from October to March. We compare the performance of BB-integrated model and BB-segregated model across the two seasons. A paired t-test analysis reveals significant enhancements in the overall-CV R^2, spatial-CV R^2, and temporal-CV R^2 during the wet season, with the values increasing from 0.75, 0.46, and 0.63 to 0.77, 0.47, and 0.64, respectively (p-value < 0.001). During the dry season, the significant improvements is even more pronounced, as the overall-CV R^2, spatial-CV R^2, and temporal-CV R^2 increase from 0.76, 0.64, and 0.59 to 0.82, 0.68, and 0.66, respectively (p-value < 0.001). These results indicate that the isolation effect of BB emission significantly contributes to enhancing the model's performance in predicting $PM_{2.5}$ concentrations, particularly during periods of elevated BB activities.

Table 3. CV evaluation metrics for the entire study year.

	R^2	RMSE	MAE
BB-integrated model			
Overall CV	0.82	6.68	4.43
Spatial CV	0.69	8.64	6.24
Temporal CV	0.71	8.38	5.55
BB-segregated model			
Overall CV	0.86	5.93	3.98
Spatial CV	0.73	8.16	5.96
Temporal CV	0.75	7.83	5.19

Visualizing the decision-making pathways of both models, we present the feature importance score plot in Fig 3. The CMAQ predictors emerged as paramount in significance across both models. In the BB-integrated model, the CMAQ_PM25 constituted approximately 45.03% of the overall importance. Within the BB-segregated model, the most crucial predictor was CMAQ_PM25_Othr, followed by CMAQ_PM25_BB, contributing to around 31.82% and 23.56% of the overall importance, respectively. On the other hand, the control predictors exhibited consistency between the two models.

Table 4. CV evaluation metrics for wet and dry seasons.

		R^2	RMSE	MAE
BB-integrated model				
Wet season	Overall CV	0.76	4.58	3.26
	Spatial CV	0.46	6.78	5.19
	Temporal CV	0.63	5.44	3.87
Dry season	Overall CV	0.76	8.12	5.41
	Spatial CV	0.64	9.95	7.29
	Temporal CV	0.59	10.47	7.14
BB-segregated model				
Wet season	Overall CV	0.78	4.44	3.16
	Spatial CV	0.47	6.70	5.15
	Temporal CV	0.64	5.42	3.86
Dry season	Overall CV	0.82	6.99	4.65
	Spatial CV	0.68	9.28	6.84
	Temporal CV	0.66	9.55	6.40

Deconstructing CV results through the lens of feature importance, the high importance scores of CMAQ_PM25_Othr and CMAQ_PM25_BB emphasize their substantial contribution to the enhanced BB-segregated model's performance. The high CMAQ_PM25 levels span approximately from October to March, while the CMAQ_PM25_Othr values distinctively rise from solely October to December (Fig. 2). The isolation helps in revealing seasonal contributions of BB emissions hidden in the CMAQ_PM25, emphasizing the explanatory power of the key predictors. As a result, there is an increase in the temporal-CV

R^2 of the BB-segregated model. Moreover, BB emissions can have both local and transboundary effects on $PM_{2.5}$ concentrations in Central Thailand. The use of the CMAQ_PM25_BB in the BB-segregated model accounts for a wider range of predicted $PM_{2.5}$ concentrations, capturing the spatial distribution influenced by both local and transboundary BB emissions. This enhances the model's ability to represent the complex interplay of $PM_{2.5}$ sources, resulting in an increase in the spatial-CV R^2 of the BB-segregated model.

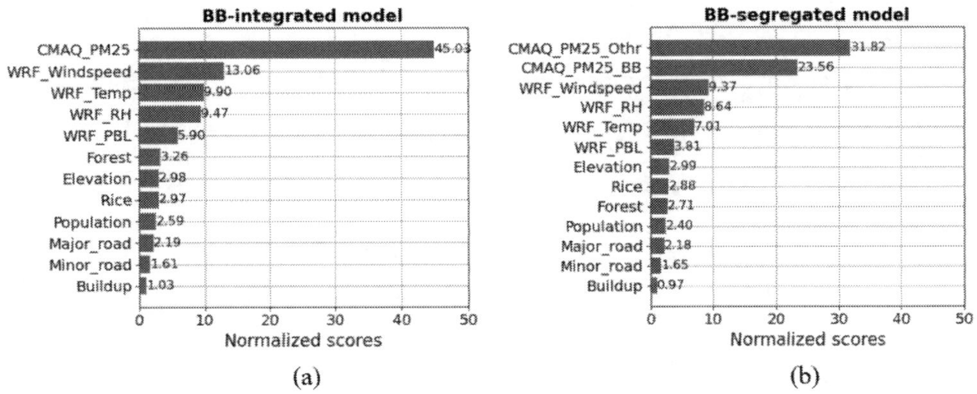

Figure 3. Feature importance of (a) BB-integrated and (b) BB-segregated models.

4 Conclusion

In this study, we successfully enhanced the LightGBM model performance by segregating the BB emission component from CMAQ. We formed two sets of predictors: control predictors (simulated-meteorological variables from WRF, population, elevation, and land-use variables) and CMAQ predictors. Two LightGBM models named the BB-integrated model, and the BB-segregated model were constructed to study the contribution of BB-derived $PM_{2.5}$. Both models share common control predictors but differ in their incorporation of CMAQ predictors. The BB-integrated model integrates CMAQ_PM25, whereas the BB-segregated model combines CMAQ_PM25_Othr and CMAQ_PM25_BB. The results have shown that BB-segregated model exhibited better performance compared to the BB-integrated model in all aspects of CV R^2: overall-CV R^2 of 0.86, spatial-CV R^2 of 0.73, and temporal-CV R^2 of 0.75. The improvement becomes even more pronounced during the dry season, with an overall R^2 increase from 0.76 to 0.82. The analysis of feature importance in the BB-segregated model highlights that BB-derived $PM_{2.5}$ emerges as a noteworthy predictor. It attains the second-highest score among all predictors. Given the recurrent annual cycle of BB events in Thailand, wherein $PM_{2.5}$ concentrations consistently peak from January to March, this pattern contributes valuable information to the model. Consequently, the model can effectively learn these temporal trends, leading to enhanced predictive performance. Hence, particularly in regions with substantial BB emissions, it holds great significance to incorporate the BB components into the model for improved predictive accuracy.

This research was supported by the Environment Research and Technology Development Fund (JPMEERF20195055 and JPMEERF20215005) of the Environmental Restoration and Conservation Agency Provided by the Ministry of Environment of Japan and JSPS KAKENHI Grant Number JP22H03757.

References

1. D. Chang, Y. Song, Atmos. Chem. Phys. **10**, 2335–2351 (2010)
2. Y. Shi, T. Sasai, Y. Yamaguchi, Ecol. Model. **272**, 98–115 (2014)
3. S. Yin, Environ. Int. **145**, 106153 (2020)
4. S. Yin, X. Wang, X. Zhang, M. Guo, M. Miura, Y. Xiao, Environ. Pollut. **254**, 112949 (2019)
5. Y. Shi, Y. Yamaguchi, Atmos. Environ. **98**, 8–16 (2014)
6. N. N. L. Thao, S. Pimonsree, K. Prueksakorn, P. T. B. Thao, P. Vongruang, Atmos. Pollut. Res. **13**, 101418 (2022)
7. T. Amnuaylojaroen, J. Inkom, R. Janta, V. Surapipith, Sustainability **12**, 10049 (2020)
8. P. Punsompong, S. K. Pani, S. -H. Wang, T. T. B. Pham, Atmos. Environ. **247**, 118176 (2021)
9. M. Greenstone, C. Hasenkopf, K. Lee, Air quality life index report (2022)
10. P. Thangavel, D. Park, Y. -C. Lee, Int. J. Environ. Res. **19**, 7511 (2022)
11. S. Feng, D. Gao, F. Liao, F. Zhou, X.Wang, Ecotoxicol. Environ. Saf. **128**, 67–74 (2016)
12. B. Bowe, Y. Xie, T. Li, Y. Yan, H. Xian, Z. Al-Aly, Lancet Planet. Health **2**, E301–312 (2018)
13. W. J. Requia, M. D. Adams, P. Koutrakis, Sci. Total Environ. **584–585**, 1077–1083 (2017)
14. Z. Li, Y. Tang, X. Song, L. Lazar, Z. Li, J. Zhao, Ecotoxicol. Environ. Saf. **169**, 248–254 (2019)
15. C. Huang, J. Hu, T. Xue, H. Xu, M. Wang, Environ. Sci. Technol. **55**, 2152–2162 (2021)
16. R. Schneider, AM. Vicedo-Cabrera, F. Sera, P. Masselot, M. Stafoggia, K. de Hoogh, I. Kloog, S. Reis, M. Vieno, A. Gasparrini, Remote Sens. **12**, 3803 (2020)
17. K. de Hoogh, H. Héritier, M. Stafoggia, N. Künzli, I. Kloog, Environ. Pollut., **233**, 1147–1154 (2018)
18. C. Zhao, Q. Wang, J. Ban, Z. Liu, Y. Zhang, R. Ma, S. Li, T. Li, Environ. Int. **134**, 105297 (2020)
19. Q. Di, H. Amini, L. Shi, I. Kloog, R. Silvern, J. Kelly, M. B. Sabath, C. Choirat, P. Koutrakis, A. Lyapustin, Y. Wang, L. J. Mickley, J. Schwartz, Environ. Int. **130**, 104909 (2019)
20. T. Thongthammachart, S. Araki, H. Shimadera, S. Eto, T. Matsuo, A. Kondo, Atmos. Environ. **262**, 118620 (2021)
21. T. Thongthammachart, H. Shimadera, S. Araki, T. Matsuo, A. Kondo, Atmos. Environ. **297**, 119595 (2023)
22. L. Che, Y. Ga, M. Ma, L. Wan, Q. Wan, S. Guan, X. Yao, H. Gao, Environ. Pollut. **318**, 120835 (2023)
23. Office of Agricultural Economics, Agricultural statistics of Thailand (2021)
24. S. Araki, M. Shima, K. Yamamoto, Sci. Total Environ. **634**, 1269–1277 (2018)

Black Carbon and Traffic-Related PM2.5 in Bangkok Area

Pornsuda Phanukarn[1,2], *Apaporn* Ruchiraset[3], and *Sopa* Chinwetkitvanich[2]

[1]Environment and Sanitation Section, DinDaeng District office, Bangkok Metropolitan Administration (BMA), Bangkok, Thailand
[2]Faculty of Public Health, Mahidol University, Bangkok, Thailand
[3]Faculty of Science and Technology, Rajamangala University of Technology Rattanakosin, Nakhon Pathom, Thailand

Abstract. This study was to investigate level of black carbon (BC) in urban areas of Bangkok, Thailand. Hourly BC concentrations were measured at two stations (operated by PCD), i.e., Bangna station and Thonburi station (roadside) during dry season by using the aethalometer (Model AE-42). The averages of hourly BC concentrations observed were 5.1±2.9 and 4.3±1.9 μg/m3, respectively. The diurnal pattern of BC levels obtained from two stations exhibited two peaks of BC concentrations occurring in the morning (0500 to 0800 LST) and late evening (2100 to 2300 LST), which the late evening peaks were higher than the morning ones. The morning peaks of BC were evidently influenced by traffic condition, while the evening peaks were possibly related to traffic associated with meteorological effects. BC concentrations from roadside stations (Thonburi) were higher than those from Bangna station possibly due to less distance of sampling equipment from roads. There was some difference of BC concentrations observed between during weekday and weekends, particularly, at roadside station (Thonburi station). In addition, increase of diesel vehicles proportion, particularly in the nighttime, contributed to higher BC concentrations. The relationships among BC concentration, air pollutants (PM2.5), vehicle fleets and meteorological variables were found.

1 Introduction

Vehicle emissions are typically the largest contributors to air pollution exposure in urban environments. In Asian developing countries, the increase in vehicles causing traffic congestion, coupled with limited expansion of road surface areas, poor vehicle maintenance, and low fuel quality, poses a significant threat to urban air quality and the well-being of the population. Traffic sources emit the majority of particulate matter, including PM2.5 (particulate matter less than 2.5 micrometres in aerodynamic diameter) and PM10 (particulate matter less than 10 micrometres in aerodynamic diameter), nitrogen oxides (NOx), carbon monoxide (CO) and non-methane hydrocarbons (HC).

Particulate matter (PM) is regarded as the principal air pollutant in urban areas. PM emitted directly from road vehicles has a wide range of particle sizes, including ultrafine, fine, and coarse particles. PM emissions from road vehicles include the exhaust emissions (tailpipe) and non-exhaust emissions (wear and tear of vehicle parts and re-suspension dust). The major PM emitted from non-exhaust emissions is likely to be coarse particulate matter (PM2.5-10) while PM2.5 is the main particulate matter contributed from exhaust emissions. PM is a chemically complex and dynamic mixture of solid and liquid particles. Sources of PM include both natural and anthropogenic processes resulting in various morphologies and compositions. It can contain combinations of inorganic ions, elemental carbons, various trace elements, crustal compounds, organic compounds, and biological matter [1]. In urban areas, the composition of PM2.5 at roadside sites was dominated by the carbonaceous combustion component and the secondary ammonium nitrate/sulphate [2].

Black carbon (BC), representing the fraction of carbonaceous compounds, is the major constituent of PM2.5 mass concentrations. The fraction of BC in PM2.5 mass concentration in urban areas was approximately 38% [3]. In developing countries, BC characterization has become increasingly concern due to its effects on human health and climate. BC lifetimes has typically ranged from 6-10 days in the absence of precipitation and to be easily transported over long distances. BC is often emitted from incomplete combustion of fossil fuels and biomass burning. In urban environments, transportation has been identified as a significant contributor of BC emissions. Ambient BC concentrations were found in the range of 1.0-15.4 µg/m3 elsewhere [3–5]. BC has been reported to be an adequate indicator of diesel soot in urban areas. Due to the being particles with large specific surface area, BC could be an efficient absorber and a carrier of toxic substances, and can penetrate deeply into the human respiratory system [6]. Epidemiological studies have linked particles containing black carbon (BC) with adverse human health effects. Smith et al. [7] reported that, based on a concentration of 1 µg/m3, the percentage increase in all-cause mortality for PM2.5 was 0.58. The effects of sulphate were approximately twice those of PM2.5, while the effects of elemental carbon (EC, an indicator of black carbon mass) were about ten times greater.EC and BC have all been associated with negative cardiovascular health effects [8]. Long-term exposure to traffic-related air pollution, especially to diesel exhaust, has been associated with an increase in cardiopulmonary mortality [9–10] showed that cardiac admissions to hospitals for all ages increased by about 1.1% per 10 µg/m3 increase in black smoke, suggesting a significant role of diesel exhaust.

The increasing severity of road transportation has the potential to greatly increase pollutant emissions, which also have an influence on health, economy, and society. Nowadays, the routine automatic ambient air quality monitoring of fixed roadside stations in Bangkok mainly focuses on parameters of NOx, CO, O3, and PM10, while PM2.5 was only an addition in some monitoring stations. The measurements of BC, which are more health-damaging, are not included in routine monitoring programs. At present, BC concentrations in Thailand are only reported by researchers, thus, more information on traffic-related BC mass is required. The purpose of this study is to monitor and analyse the traffic-related BC concentrations and their relationships with major roadside pollutants in urban areas of Bangkok. Atmospheric dispersion of BC emitted from vehicles will be investigated using a mathematical model. The mitigations for air quality management, especially the black carbon aspect in urban areas, are expected if possible.

2 Material and methods

Bangkok is located at 13.45 latitude and 100.31 longitude and it is situated at elevation 12 meters above sea level. The climate of Bangkok is a tropical monsoon climate and is under the influence of southwest monsoon and northeast monsoon. It has three seasons as follows: rainy (wet season) or southwest monsoon season (mid-May to mid-October), winter (dry season) or northeast monsoon season (mid-October to mid-February), and summer or pre-monsoon season (mid-February to mid-May) [11]. The monitoring of black carbon (BC) in this study was done from November 28th, 2016 – January 4th, 2017 because this period was usually under the influence of the dry season.

2.1 Sampling sites

Samples were collected from automatic continuous air quality monitoring stations located in the Bangkok area. Due to resources and time limitations, only two stations (Bangna station and Thonburi station) were selected for this study period. These two air monitoring stations were managed by the Pollution Control Department (PCD) of Thailand and their locations as shown in Figure 1.

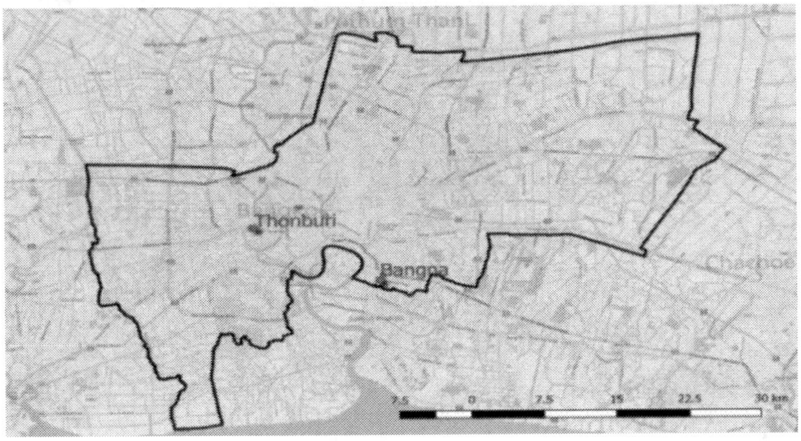

Fig. 1. Location of two monitoring stations in this study.

For the Bangna station, it was installed in the area of Thai Meteorological Department Bangna, Bangkok Thailand (13° 39′ 52′ N, 100° 36′ 35′ E). "Bangna" is also the name of the district of BMA with less density than Thonburi district (about 5,020 populations/km^2). This station is defined as a general station located at the south-southeast side of Sukhumvit Road and about 700 meters far from the Bangna Intersection and Bangna Expressway. However, Sukhumvit Road is one of the main roads connected to Samut Prakarn Province with high traffic density (especially diesel vehicles/trucks) and congestion during rush hours. This Sukhumvit road is surrounded by several commercial buildings and residential areas. Moreover, there is the sky train structure covering some parts of this road, especially near this monitoring station. In this study, BC samples were collected on the rooftop of this Bangna station, which was 150 meters away from the curbside of the Sukhumvit road (Figure 2a). The investigation of black carbon occurrence in this station was expected to be interesting.

For Thonburi station, "Thonburi" is the name of the southern district of Bangkok Metropolitan Administration (BMA), which is considered one of the high-density districts of BMA (13,487 populations/km^2). This Thonburi air quality monitoring station is located near the Metropolitan Electricity Authority (MEA), Thonburi Substation (13°43′30′ N, 100°29′9′ E) and about 100 meters from Bang Yi Ruea Junction. This station is defined as a roadside monitoring station located on the north side of Intharapitak road surrounded by several commercial buildings and residential areas. Intharapitak road is composed of six lanes with a total width of 21 meters, which is one of the main roads with high traffic density and congestion during rush hours. The structure of this Thonburi station was installed very close to the road area (5 meters away from the curb side) as shown in Figure 2b. The aethalometer was equipped on the rooftop of this station to collect black carbon (BC) samples with permission from PCD. Traffic in this area was postulated to play an important role in the air quality monitoring data of this station.

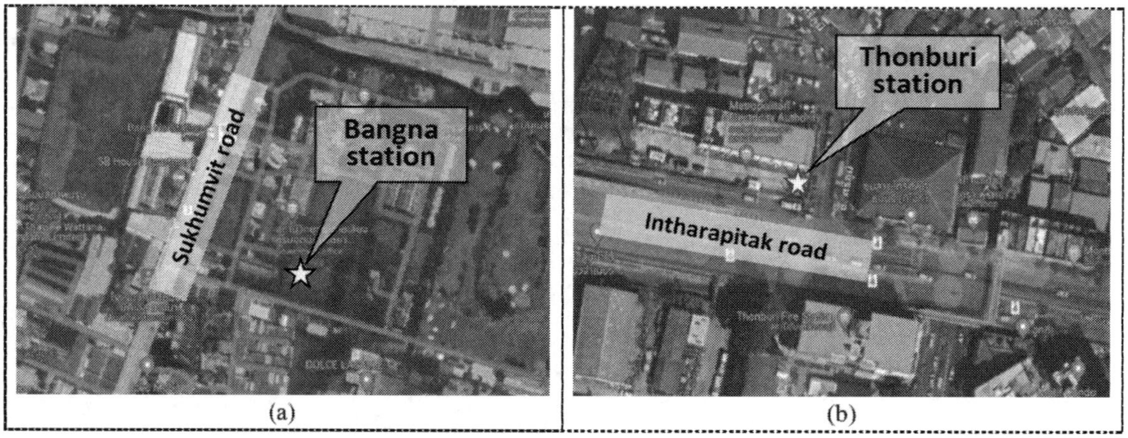

Fig. 2. Maps of Thonburi station and Bangna station.

2.2 Samples collection and data analysis

Black carbon concentrations were measured by using Model AE-42 aethalometer (Magee Scientific, Berkeley, CA, USA). According to Hansen [12], seven wavelengths (UV at 370 nm, Blue at 470 nm, Green at 520 nm, Yellow at 590 nm, Red at 660 nm, IR-1 at 880 nm, and IR-2 at 950 nm) were set up. The sampling part of this aethalometer allowed only particles smaller than 2.5 μm in aerodynamic diameter to be collected on a quartz fiber tape. A sampling flow rate of 2 liters/minute and the 2 min data logging interval were designated.

Black carbon (BC) concentrations were measured during the dry season, covering weekdays (Monday to Friday) and weekends (Saturday, Sunday, and Holidays) as concluded in Table 1. At the Bangna station, the total days of BC data were 18 days, consisting of 12 days of weekdays, 2 days of Saturday, and 4 days of Sunday/Holiday. For Thonburi station, BC data were collected for 21 days (consisting of 13 weekdays, 3 days on a Saturday and 5 days on a Sunday/Holiday).

Particulate matter (PM2.5 and PM10), gaseous pollutants (CO, NO_2, O_3), and meteorological data (wind speed, wind direction, temperature, relative humidity, and precipitation) of these two air monitoring stations were supported by PCD.

Table 1. Sampling sites and number of BC data in this study

Sites	Total (Days)	Weekday		Weekend*	
		(Days)	(Hours)	(Days)	(Hours)
Bangna (Nov 28 – Dec 15, 2016)	18	12	211	6	134
Thonburi (Dec 15, 2016 – Jan 4, 2017)	21	13	278	8	191

Note : *Weekend = Saturday and Sunday/Holiday

2.3 Traffic count

Traffic volumes related to the monitoring stations were analysed by using digital video files of nearby CCTV cameras obtained from the Traffic and Transportation Department, Bangkok Metropolitan Administration. The number of vehicles passing through the camera was manually counted for at least twenty minutes in every hour, then hourly data were interpolated. Vehicles were categorized into eight different types, that is, passenger cars, pick-ups, vans, taxis, motorcycles, three-wheelers (or "Tuk Tuk"), buses, and trucks.

3 Results and Discussion

3.1 Black carbon (BC) and particulate matter (PM)

In the case of the Bangna station, the averages of PM10 and PM2.5 data obtained from the PCD (Pollution Control Department of Thailand) system during the same period of BC sampling were 58±25 and 46±18 μg/m³, respectively. However, their maximum concentrations were observed as high as 164 and 118 μg/m³, respectively. The average black carbon (BC) concentration of this station was 4.3±2.0 μg/m³ while the median and the maximum were 3.8 and 14.2 μg/m³, respectively.

For the Thonburi station, PM10 and PM2.5 were observed with averages of 57±25 and 44±19 μg/m³, respectively, the maximum concentrations of which were 170 and 125 μg/m³, respectively. The average BC concentrations from this station were 5.1±2.9 with a median and a maximum of 4.3 and 20.4 μg/m³, respectively.

From the results, the averages of particulate matter (PM10 and PM2.5) observed from both stations were quite similar, while their black carbon (BC) concentrations showed a slightly higher range in the Thonburi station. This could be possibly due to the shorter distance between the equipment and the road (about 5 m) than the Bangna station (about 150 m.). However, the averages of BC concentrations from the Thonburi station in this study were quite lower than the average of 17.9±6.6 μg/m³ reported in the previous study [13], which they monitored at another roadside monitoring station in the Bangkok area (Dindaeng station) during the dry season. Some other densely populated cities reported higher BC concentrations such as Istanbul in Turkey or London in the United Kingdom — which reported BC concentrations of 8.9±3.6 μg/m³ [5] and 9.7±0.8 μg/m³ [14], respectively.

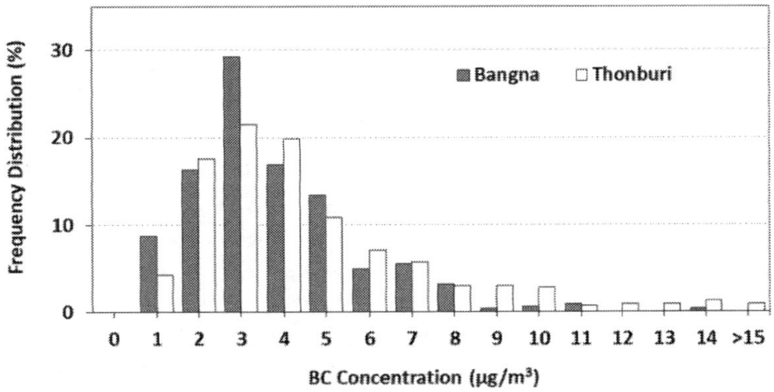

Fig. 3. Percentage occurrence of BC mass concentrations at two sampling stations.

In comparison between these two stations, Thonburi station showed that their hourly BC data higher than 10 μg/m³ were recorded during the sampling period more frequently (4.5%) than the Bangna stations (1.2%). The frequency distribution of hourly BC concentrations in Figure 3 supported that the majority of hourly BC concentration data from both stations were still in the range of less than 10 μg/m³. Other urban cities that also reported their BC concentrations in a similar range, e.g., Chiang Mai province in northern Thailand observed BC data less than 3.3 μg/m³, and biomass-burning (both forest fire and agricultural burning) was mentioned as the cause [15]. Shanghai in China reported that 91% of their BC data were less than 6 μg/m^3 [16]. The BC data in the range of 1 – 5 μg/m^3 was recorded with a frequency of 62% in Ostrava, Czech Republic [17]. Differently, Istanbul in Turkey showed a little higher BC concentration, with their frequency of BC data higher than 10 μg/m^3 was as high as 40% of total data [3].

3.2 Diurnal variation of BC concentrations

This study was to investigate black carbon occurrence in the urban areas of Bangkok Metropolitan. Vehicles emission was expected to be an important source of black carbon in this urban area. Any activities of biomass burning and grilled street foods were observed very little in the surrounding area of these two stations, however, these effects were not verified in this study. Traffic characteristics should be related to BC accumulation in the surrounding ambient air. Therefore, sampling periods were assigned to cover both weekdays–weekends pattern and daytime–nighttime pattern. Weekdays mean Monday to Friday, while weekends mean Saturday to Sunday. Daytime was specified as 6:00 – 18:00 and nighttime was 18.00 – 06.00. Statistically, hourly BC concentrations obtained from these two stations were calculated for discussion both weekdays–weekends and daytime–nighttime patterns as presented in Table 2. During weekdays, the averages of daytime data from the Bangna station (3.7 μg/m^3) and Thonburi station (5.2 μg/m^3), were less than those of nighttime data (4.8 and 5.6 μg/m^3, respectively). Similarly, the averages of daytime data during weekends (3.9 and 4.4 μg/m^3, respectively) were less than those of nighttime data (4.5 and 5.0 μg/m^3, respectively) as well.

Table 2. Statistics of hourly BC concentrations in this study

Sites	Parameter	Weekday		Weekend	
		Daytime	Nighttime	Daytime	Nighttime
Bangna	range	1.2 – 12	2.1 – 14	1.4 – 11	2.7 – 8.8
(μg/m^3)	average ± sd	3.7 ± 2.1	4.8 ± 2.0	3.9 ± 1.9	4.5 ± 1.6
	median	3.1	4.6	3.6	4.8
Thonburi	range	1.7 – 15	1.1 – 20	1.2 – 9.7	1.9 – 18
(μg/m^3)	average ± sd	5.2 ± 3.1	5.6 ± 3.5	4.4 ± 1.6	5.0 ± 2.7
	median	4.4	4.5	4.1	4.3

For the Thonburi station, averages of daytime (5.2 μg/m^3) and nighttime (5.6 μg/m^3) during weekdays appeared to be higher than their weekends (4.4 and 5.0 μg/m^3, respectively). This could be due to less traffic congestion during weekends on this Intharapitak road (Thonburi station). For the Bangna station, there seemed not to be a substantial difference between weekdays and weekends. This Sukhumvit road is a main road connecting to Sumutprakarn Province, where transportation for industrial and commercial purposes with heavy-duty vehicles are found. The density of vehicles increased particularly on weekends due to these heavy-duty vehicles being allowed to access the inner part of the Bangkok Metropolitan area. Hence, the diurnal variation of BC concentrations on weekends at this station was a little higher than those on weekdays.

3.3 Correlation between black carbon (BC) and PM2.5

Due to black carbon (BC) representing a significant fraction of PM2.5 concentrations in the urban atmosphere, Figure 4 shows the hourly variation of BC and PM2.5 concentrations at the Bangna station. The diurnal variation of BC and PM2.5 was quite consistent similarly to those of Thonburi station (Figure 5). However, the hourly concentrations of BC and PM2.5

(at this Bangna Station) showed a significant morning peak at approximately 7:00 to 8:00. Then, BC and PM2.5 decreased toward midday and increased in the evening. The second peaks of BC and PM2.5 were observed at around 21:00 (a little earlier than Thonburi station). The measurement results showed these two peaks of BC concentrations happening in the morning and the late evening, similar to other studies [3, 18–20]. According to the explanations of the previous studies, this could be postulated that BC concentrations in this study should be correlated to traffic emissions and the formation of boundary layer height.

Figure 5 illustrates the hourly variation of BC and PM2.5 concentrations observed from the Thonburi station. The diurnal variation of these BC and PM2.5 concentrations was consistent. The hourly concentrations of BC and PM2.5 showed a significant morning peak at approximately 5:00 to 6:00, which further decreased toward midday and increased again during the evening. Both pollutants showed a second peak at around 21:00 to 23:00. This variation suggested that traffic congestion, particularly during morning (5:00 to 7:00) and evening (16:00 to 17:00) rush hours, led to higher concentrations of pollutants.

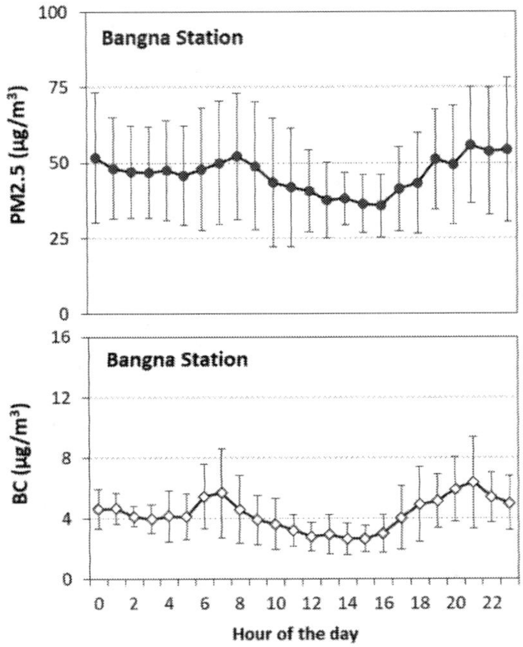

Figure 4. Hourly variation of BC and PM2.5 concentrations of Bangna station.

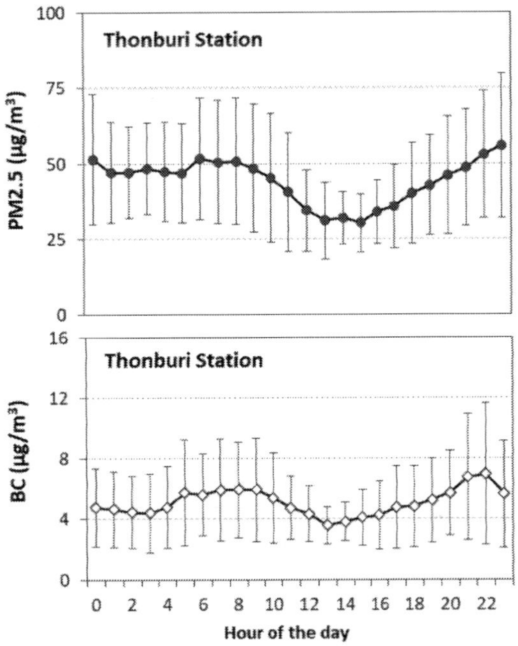

Figure 5. Hourly variation of BC and PM2.5 concentrations of Thonburi station.

3.4 Traffic characteristics

Due to the assumption that BC concentrations in urban areas were somehow influenced by traffic characteristics, traffic volumes on the roads near the monitoring station during the sampling period were also observed. At the Bangna station, total vehicle volumes on Sukhumvit road during the sampling period were counted in the range of 189 – 11,424 vehicles/hour, of which the average was 3,549 vehicles/hour. The average vehicle volume on weekdays and weekends was 3,577 and 3,625 vehicles/hour, respectively (table 3). Motorcycle (29 – 30%) was the highest composition of this Sukhumvit road similar to the Intharapitak road. The others were private cars (26 – 27%), pickups (21 – 22%) and taxis (14 – 15%). For vans, buses, and trucks, each held its share of around 2 – 4%, and tuk-tuks held the smallest share of below 1%. The dominant shares of motorcycles, private cars, pickups, and taxis were observed similarly both during weekdays and weekends.

At the Thonburi station, the total vehicle volume on Intharapitak road during the sampling period was counted in the range of 648 – 6,900 vehicles/hour. The averages during weekdays and weekend were 3,993 and 3,576 vehicles/hour, respectively (Table 3). The fleet composition was dominated by motorcycles (37 – 38%), private cars (27%), taxis (18-19%), and pickups (11-12%), while the others held a share of 1 – 4%. The smallest composition of this Intharapitak road belonged to the truck fleet (less than 1%). The higher shares of motorcycles, private cars, taxis, and pickups were observed similarly both during weekdays and weekends. However, the compositions of private cars, taxis, and Tuk–Tuks tended to increase during weekends.

Table 3. Percentage of traffic fleets composition (%).

Vehicle type	Intharapitak road		Sukhumvit road	
	Weekdays	Weekends	Weekdays	Weekends
Private car	26.6 ± 1.0	27.2 ± 0.8	26.5 ± 0.7	26.3 ± 0.3
Pickup	12.0 ± 0.3	11.4 ± 1.1	20.6 ± 1.2	21.8 ± 1.0
Van	1.0 ± 0.1	0.7 ± 0.0	3.0 ± 0.2	3.1 ± 0.2
Taxi	17.5 ± 0.5	18.7 ± 3.0	15.2 ± 0.7	14.3 ± 0.4
Motorcycle	37.5 ± 1.4	36.7 ± 2.5	28.9 ± 0.7	28.5 ± 0.3
Tuk-Tuk	3.1 ± 0.2	3.7 ± 0.0	0.2 ± 0.1	0.2 ± 0.0
Bus	1.7 ± 0.1	1.4 ± 0.0	3.5 ± 0.1	3.7 ± 0.2
Truck	0.6 ± 0.1	0.5 ± 0.2	2.0 ± 0.2	2.1 ± 0.4
Average (vehicles/hour)	3,993 ± 108	3,576 ± 369	3,577 ± 1,013	3,625 ± 208

4 Conclusions

Hourly BC concentrations in this study were measured with a single-wavelength Aethalometer at two air monitoring stations, one roadside station (Thonburi station), and a general station (Bangna station) in Bangkok during the dry season of the year 2016-2017. The results showed the level of BC concentrations from the roadside station was a little higher than that from the general station, possibly due to the distance from the main roads. Two peaks of BC concentrations in this study were more likely due to rush hour traffic concurrently with meteorological conditions. The main emission sources of BC concentrations should be emitted from local vehicular emissions due to BC peak concentrations being correlated to traffic rush hours. Heavy-duty diesel vehicles should be an important fleet affecting BC concentrations, especially during nighttime and weekends. However, analytical results imply that BC concentrations in urban Bangkok were not only released from fossil fuel emissions, but also emitted from biomass burning emissions. From the dataset elsewhere [21], BC concentrations showed a significantly positive correlation with air pollutions, particularly, PM10, PM2.5, CO, and NO_2. This finding indicated that the main sources of these pollutants in study areas were emitted from vehicle emissions, residential combustion, and industrial combustion. In addition, a recent study of black carbon across urban Europe [22] showed a decrease in equivalent BC emissions from traffic over the past decade, but those from residential and commercial combustion were almost the same. Recently, the PM2.5 crisis in the Bangkok area has been discussed, and its causes are not only from traffic, but also from biomass burning (both local and regional areas). Further study and monitoring of PM2.5 with black carbon (or elemental carbon) should be strongly recommended.

Authors would like to express our gratitude to Dr.Hathairatana Garivait for her valuable support and assistance for BC analyser from the Environmental Quality Promotion Department. Also, Special thanks to the Pollution Control Department for ambient air quality data and sampling locations accessibility.

References

1. K. Yoo, J. Ji, A. Aufdenkampe, J. Klaminder, J. Geophys. Res. **116**, G01014 (2011)
2. R.M. Harrison, A.M. Jones, R.G. Lawrence, Atmos. Environ. **38**, 4531–38 (2004)
3. H. Ozdemir, L. Pozzoli, T. Kindap, G. Demir, B. Mertoglu, N. Mihalopoulos, C. Theodosi, M. Kanakidou , U. Ime, A. Unal, Sci. Total. Environ. **473–474**, 451–8 (2014)
4. Y.H. Cheng, B.S. Shiu, M.H. Lin, J.W. Yan, Environ. Sci. Pollut. Res. **20**, 1537–45 (2013)
5. J. Feng, M. Zhong, B. Xu, Y. Du, M. Wua, H. Wang, C. Chen, Atmos. Res. **147–148**, 1–9 (2014)
6. G. Pavese, M. Calvello, F. Esposito, Aerosol. Air. Qual. Res. **12**, 1146–56 (2012)
7. K.R. Smith, M. Jerrett, H.R. Anderson, R.T. Burnett, V. Stone, R. Derwent, R.W. Atkinson, A. Cohen, S.B. Shonkoff, D. Krewski, C.A. Pope 3rd, M.J. Thun, G. Thurston, Lancet. **374**, 2091–103 (2009)
8. K.I. Jansen, T.V. Larson, J.Q. Koenig, T.F. Mar, C. Fields, J. Stewart, L. Morton, Environ. Health. Perspect. **113**, 1741–46 (2005)
9. R.D. Brook, S. Rajagopalan, C.A. Pope 3rd, et al., Circulation. **121**, 2331–78 (2010)
10. A. Tertre, P. Quenel, D. Eilstein, et al., Arch. Environ. Health. **57**, 311–9 (2002)
11. Thai Meteorological Department (TMD), Weather of Bangkok (2017)
12. A.D.A. Hansen, *The aethalometer*™ Magee Scientific Company Berkeley, California, USA. Available from: http://mageesci.com/support/downloads/Aethalometer_book_2005.07.03.pdf. (2005)
13. N.T.Q. Hung, S.B. Lee, N.T. Hang, J. Kongpran, N.T. Kim Oanh, S.G. Shim, G.N. Bae, Atmos. Environ. **92**, 231–9 (2014)
14. V. Singh, K. Ravindra, L. Sahu, R. Sokhi, Atmos. Environ. **178**, 148–57 (2018)
15. S.K. Pani, S. Wang, N. Lin, S. Chantara, C. Lee, D. Thepnuan, Environ. Pollut. **259**, 112871 (2020)
16. S. Zha, T. Cheng, J. Tao, et al., Atmos. Res. **135-136**, 159–71 (2014)
17. M. Kucbel, A. Corsaro, B. Svedova, H. Raclavska, K. Raclavsky, D. Juchelkova, J. Environ. Manage. **203**, 1178–89 (2017)
18. L. Shen, L. Li, S. Lu, X. Zhang, J. Liu, J. An, G. Zhang, B. Wu, F. Wang, Particuology **20**, 10–15 (2015)
19. X. Zhang, R. Rao, Y. Huang, M. Mao, M.J. Berg, W. Sun, J. Quant. Spectrosc. Radiat. Transfer. **150**, 3–11 (2015)
20. A.C. Targino, P. Krecl, Aerosol. Air. Qual. Res. **16**, 125–37 (2016)
21. P. Phanukarn, Analysis of Black Carbon and Traffic-related Air Pollutants in Bangkok Area [PhD Dissertation in Environmental Technology]. Bangkok: Department of Sanitary Engineering, Faculty of Public Health, Mahidol University (2019)
22. M. Savadkoohi, M. Pandolfi, C. Reche, J.V. Niemi, D. Mooibroek, et al., Environ. Int. **178**, https://doi.org/10.1016/j.envint.2023.108081 (2023)

 https://doi.org/10.1051/e3sconf/202453001006

The health Loss Distribution of Coal-fired Power Plants and the relationship to Air Pollution Charges

Sih-Ping Chen[1] , *Xiang Hou* Wang[1*]*and Chao Heng* Tseng[1]

[1] Institute of Environmental Engineering and Management National Taipei University of Technology, Taipei, Taiwan

Abstract. In this study, the regional health loss due to the air pollution of a coal-fired power plant is analyzed by using the atmospheric dispersion model. AERMOD was used to simulate the increment concentration map of point source emission, and the GIS geographic information system was used to displace the distribution of air pollution concentration and regional health loss in grid receptors. The data of the continuous automatic monitoring facilities and the source of pollution sources were analyzed. Moreover, the effects of emissions, seasons, population, and distance on the simulated concentration were evaluated. The various influencing factors of air pollution concentration were evaluated.the R2 value of the empty product prediction map in summer has a better correlation with the concentration from monitor station,between 0.38 and 0.83, and the sulphur dioxide correlation has the best R2 value of 0.83. AERMOD was used to combined with pollutant emissions, weather, and terrain data, to simulate the distribution of rising concentration values due to coal-fired power plant emissions within the research area, and the study used the simulated rising concentration values to calculate the health loss. [1] The study used health loss of 25,647,291,161 NTD/year and medical loss of 319,153,198 NTD/year.

Keywords: AERMOD,ArcGIS, Airquality, Thermal power generation,health benefits, air pollution fees

1. Introduction

1.1 Research background and motivation

Coal-fired power plants emit a variety of pollutants, including particulate matter (PM), sulfur oxides (SOx), nitrogen oxides (NOx), carbon dioxide (CO2), ozone (O3), and volatile organic compounds[2]. There have been many studies in Taiwan demonstrating the relationship

* Corresponding author: wan60310@gmail.com

between these pollutants and mortality from cardiovascular disease, respiratory disease, and central nervous system disease. Taiwan's power industry is dominated by thermal power generation, which ranks first among all sectors in carbon emissions, accounting for approximately 60% of Taiwan's total emissions. In addition to causing greenhouse gas emissions, the air pollutants it produces can also have negative impacts on health. The power plant uses 12 million tons of bituminous coal and 2.5 million tons of sub-bituminous coal every year. It is the second largest coal-fired power plant in the world and the largest power plant in Taiwan.

Air pollution fees refer to special revenue funds that have specific sources of income and are used for special purposes. They are earmarked for special purposes. The current reporting objects for stationary pollution sources in Taiwan's air pollution fees are emissions of SOx, NOX, Volatile organic compounds (VOCS), particulate pollutants, lead, cadmium, mercury, arsenic, hexavalent chromium, and dioxin are subject to self-calculated charges based on the types, emissions, and operation records of air pollutants emitted each quarter. Fee amount. Since the current air pollution prevention fees for stationary pollution sources are levied based on the type and quantity of air pollutants they emit, without taking into account the health damage to local residents, this study takes the largest power plant in Taiwan's power industry as an example to calculate the local Health losses of residents, analyze the distribution of health losses of coal-fired power plants and local residents.

2. Framework of the Study

The study only used the correlation analysis between pollutant concentration and emissions from a certain power plant, and did not consider the impact of other pollution sources, which may cause bias.

The building height is not set during the simulation, so the concentration diffusion changes in the area will be different from the actual ones.

The study uses air pollutant concentration and meteorological condition data from the Air Quality Monitoring Network of the Environmental Protection Department of the Executive Yuan. The data represents the atmospheric concentration data in the region and is easily affected by other pollution sources. It may be difficult to compare with simulated concentrations.

The study only uses the total outpatient and outpatient (including emergency room) data collected by the Ministry of Health and Welfare. It is unable to collect detailed personal information such as past medical history, occupation, and lifestyle of individual people seeking medical treatment. Therefore, it is impossible to determine whether people are seeking medical treatment. No personal factors.

Epidemiological study results show. that chronic diseases caused by long-term or short-term exposure are more important than short-term exposure(Ruijun Xu.2022) [3]. Therefore, scholars recommend using the Cohort Study exposure response of the American Cancer Society (ACS). coefficient, but generational research is expensive and difficult, and only Western developed countries have conducted long-term research. Therefore, this study uses retrospective data to explore the exposure dose-response coefficients(Leslie Stayner PhD.,1992) [4] of diseases caused by short-term exposure to air pollutants in Taiwanese people, which will be inconsistent with the results of generational studies.

Although the interaction between pollutants has been discussed in domestic and foreign literature, its impact on human health is still unclear. Therefore, this study only considers the single impact of air pollutants on human health, and the increase or decrease in the concentration of air pollutants. Health loss, regardless of effects due to interactions.

There is a linear relationship between emissions and pollutant concentration in the region, and air pollutant emissions (P) and the concentration-emission linear coefficient (m) changed by climate change are independent of each other and do not affect each other.

pointed out that ozone is a secondary pollutant in the atmosphere, and its concentration (Su Huizhen2008) [5]will be affected by its precursors (such as NOX, VOCS and NMHC). According to the TEDS database, the coastal counties and cities in western Taiwan The ratio of NMHC/NOx is mostly between 1 and 15, and most urban areas in Taiwan belong to NMHC-Limited areas. Therefore, to control ozone pollution, we should first control the emission of NMHC to achieve better control effects. . Therefore, this study assumes that NHMC is the main precursor for simulating ozone concentration in this study.

3.Methodolgy

The scope of this study is preliminarily calculated to be a rectangular area of 30 kilometers upward, 40 kilometers downward, 70 kilometers to the right, and 40 kilometers to the left centered on the power plant.

Focus on changes in the concentration of six major air pollutants: suspended particulates particulate matter10(PM10), fine suspended particulates (PM2.5), SO2, NOX, CO and O3.

Stationary source: a coal-fired power plant

By multiplying the Coefficient of Response Functions (CRFs) of various pollutants with the Rollback Model, the emission change (ΔC) is derived. Subsequently, the following formula is employed to estimate the change in pollutant concentration and calculate the corresponding RR variation. Finally, the Attributable Proportion (AP) is determined using Equation to assess the impact on mortality or morbidity:

The World Health Organization (WHO) recommends using the Attributable Proportion (AP) to calculate the effect of RR changes on disease or mortality rates, as depicted in Equation

The calculated results from Equations and yield the disease or mortality attributable proportion. Combining this information with simplified life tables, mortality rates, and healthcare expenditure data allows for the computation of changes in life expectancy and healthcare expenses attributable to exposure to individual pollutant concentrations.

4.Results and discussion

The construction and use of pollution maps require the integration of all relevant information through the platform. It is expected that an integrated model can be developed for use and query in the future.

This study uses a linear model to simulate the relationship between various air pollutants and atmospheric conditions. However, O3 is a secondary air pollutant with nonlinear response in the atmospheric environment. It is recommended that a nonlinear model can be used to replace this study in the future, which can be more realistic. Realize the influence of secondary reactions, thereby improving the applicability and accuracy of the model.

This study uses AERMOD to simulate PM2.5 pollution diffusion and emission maps. Since AERMOD only simulates and calculates the pollution caused by a power plant, the simulation results are underestimated. It is recommended that other models can be used to replace it in the future.

The diffusion base map generated by AERMOD in this study only has one-year average and four-season diffusion maps. In the future, we hope to draw one diffusion concentration increment map every month or conduct long-term tracking statistics to increase the pollution increment. Maps can be more valuable for reference.

Utilizing the relative risk (RR) values from the literature, exposure reflection regression coefficients are computed. The Exposure-Response Regression Coefficients, also known as Concentration-Response Factors (CRFs), refer to the percentage increase in disease or mortality rates for every 1 μg/m3 rise in air pollutant concentration. It is also referred to as the Exponential Slope Factor, denoted as β. The calculation formula is illustrated in Equation *3.3*: (Equation 3.3)

5.Conclusions

In this study, the emission data of the fixed source thermal power plant was input into the AERMOD, and the concentration diffusion map of six pollutants PM10, PM2.5, SO2, NOX, CO and O3 was established, and combined with GIS spatial analysis to build Spatial model of air quality to calculate the health benefits of air pollutants.

The concentrations monitored by the air pollutant measuring station were integrated into the map to verify the air pollutants map drawn by this institute. After verification of the four seasons and annual average concentrations, it was found that the accuracy of the air pollutant map predictions and actual values varies with different pollutants and seasons. On the other hand, the R2 value of the empty product prediction map with better correlation in summer falls between 0.38-0.83, among which the best R2 value of SO2 correlation is 0.83.

After combining the population density map, the pollution concentration increment map and the health loss map, the health benefit statistics resulted in a medical expenditure of 319,153,198 yuan/year, a reduction in life span of 23,840,645,242 yuan/year, and a total health benefit of 25,647,291,161 yuan/year. The average annual medical expenditure per person is 65.3031 yuan/person-year, the amount of life reduced is 4,878 yuan/person-year, and the total health benefit is 5,248 yuan/person-year.

It can be seen from the population grid density map (Figure.1.) that Taichung City, Changhua County and Nantou Puli Basin have higher population density and pollution concentration (Figure .2.) Therefore, after combining the population density map and the health loss map, it can be seen that population density and health loss are positively correlated. The average annual medical expenditure per person in Taichung City is 58.1467 yuan/person-year, the amount of life reduced is 4,278 yuan/person-year, and the total health benefit is 4,355 yuan/person-year. The average annual medical expenditure per person in Changhua County is 84.7730 yuan/person-year, the amount of life reduced is 6,530 yuan/person-year, and the total health benefit is 7,327 yuan/person-year. The average annual medical expenditure per person in the Puli Basin is 89.9633 yuan/person-year, the amount of life reduced is 7,160 yuan/person-year, and the total health benefit is 7,249 yuan/person-year.

Using the health loss to estimate the air pollution fee of 25,647,291,161 yuan/year. Using the medical loss to estimate the air pollution fee of 319,153,198 yuan/year, the difference is smaller, because the amount of life lost per person is relatively high when calculating health losses.

When examining medical losses by county, Taichung City has the highest rate at 51%. Regarding seasons, the highest medical losses occur in autumn, reaching 29%.

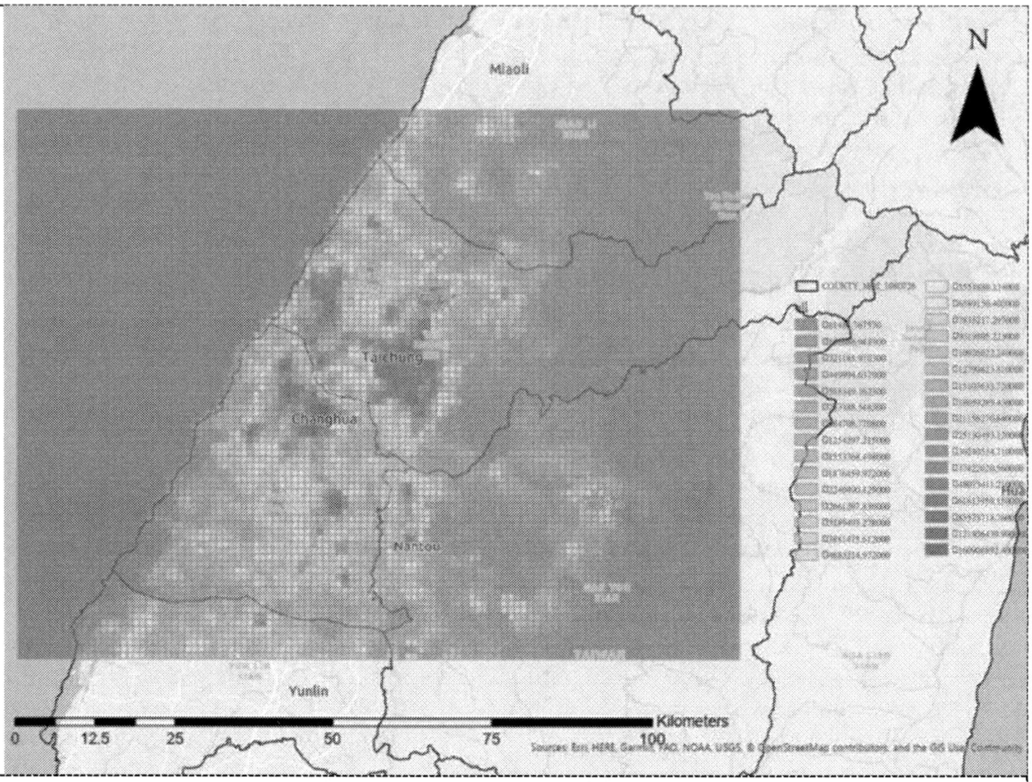

Fig. 1. Population grid map of study scope

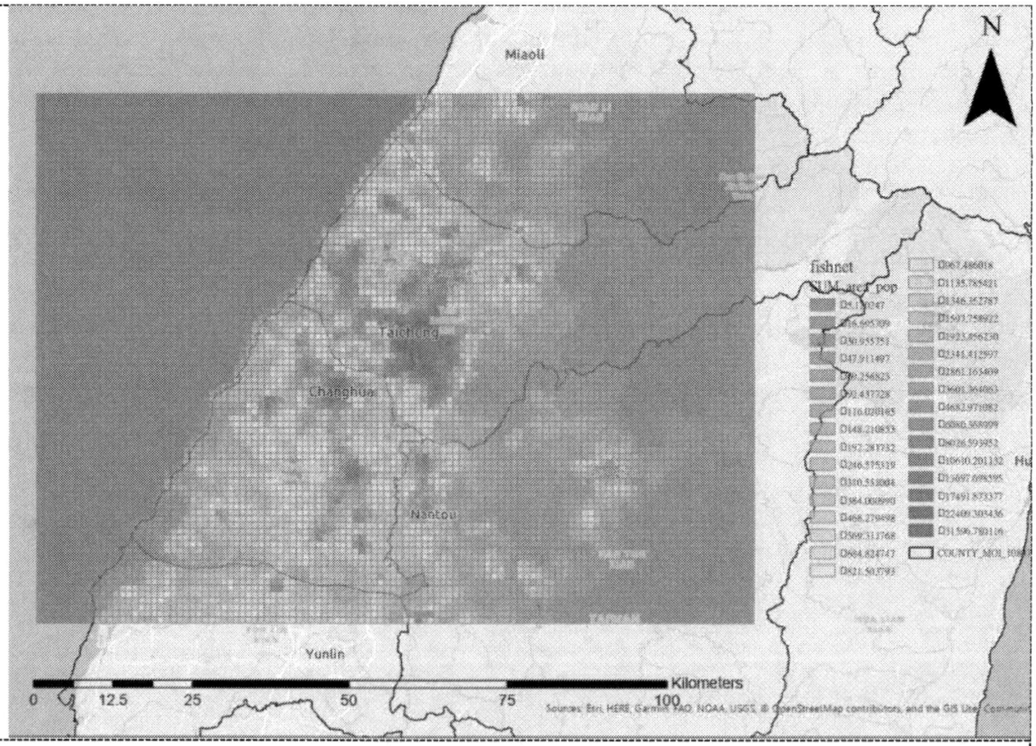

Fig. 2. Health Loss Grid Map

https://doi.org/10.1051/e3sconf/202453001006

References

1.Sih-ping chen, 2020 The Health Loss Distribution of Coal-fired Power Plants and the relationship to Air Pollution Charges.

2. Machol, B., & Rizk, S. (2013). Economic value of U.S. fossil fuel electricity health impacts. Environment International, 52, pp 75-80.

3. Ruijun Xu 2022 Association of short-term exposure to ambient PM1 with total and cause-specific cardiovascular disease mortality

4. Leslie Stayner PhD [1], Randall Smith MA[1] 1992 A dose-response analysis and quantitative assessment of lung cancer risk and occupational cadmium exposure

5. Su Huizhen, Lin Qiankun, Chen Peishi, Impact of climate change on public health, Scientific Development, Issue 421: 12~17, 2008.

Biodegradation of Prazosin by *Bacillus* sp. Isolated from PPCPs Contaminated River in Malaysia

NurAqilah Farhanah binti Mohd Mohsi[1], *Atiqqoh binti* Apandi[2], *Hirofumi* Hara[2,3], and *Motoo* Utsumi[1*]

[1]Graduate School of Life and Environmental Sciences, University of Tsukuba 1-1-1 Tennoudai, Tsukuba, Ibaraki, 305-8572, Japan
[2]Department of Environmental Engineering and Green Technology, Malaysia-Japan International Institute of Technology, Univerisiti Teknologi Malaysia, Jalan Sultan Yahya Petra, 54100, Kuala Lumpur, Malaysia
[3]Department of Biotechnology, Graduate School of Agricultural and Life Sciences, The University of Tokyo, 113-8657, Tokyo, Japan

Abstract. Prazosin (PRZ) is a type of drug used to treat hypertension. It is found to contaminate the aquatic environment, therefore has substantial threats to the ecosystem and human health. PRZ has been found to contaminate rivers in Malaysia which is presumed due to excretion and inappropriate disposal. A tropical bacterium, PR5 has been isolated from a PRZ contaminated site in Kuala Lumpur, Malaysia. It has demonstrated its capabilities to utilize PRZ as a nutrient source and efficiently degrade PRZ. UPLC-TUV is used to measure PRZ concentration during growing cell assay and the results showed that isolate PR5 able to degrade 78% of 0.01 mM PRZ within 7 days of incubation. Species identification based on the 16S rRNA gene sequence showed isolate PR5 has a 99.66% similarity with *Bacillus velezensis*. This study reports on the first finding of PRZ biodegradation by *Bacillus velezensis*.

1. Introduction

Our environment and nature especially water resources play a vital and important role in the life of all living things including human beings. Inevitably, water is being used daily and very essential for fruits/vegetables farming, husbandry, human utilization, and environmental ecosystem. As stipulated by World Health Organization (WHO), the most primary necessity for freshwater for a person is ranging from 20 to 50 litres per day [1]. However, due to rapid growth of human population around the globe, water pollution is parallelly increasing which results in shortage of clean freshwater for domestic use. There are wide range of compounds detected in the water system which result in the water pollution.

Many common anthropogenic disposals into the water system have been detected, for example, inorganic substances, heavy metals, pesticides, and untreated sewage water [2]. One of the newly emerging pollutants detected polluting the water system is pharmaceutical and personal care products (PPCPs). The existent of PPCPs in the natural environment is very concerning as

*Corresponding author: utsumi.motoo.ge@u.tsukuba.ac.jp

the continuous usage makes them continuously being discharge and recalcitrating in the environment especially the water system [3].

One of the PPCPs which was detected as a pollutant is PRZ. Prazosin (IUPAC name of 4-(4-amino-6, 7-dimethoxyquinazolin-2-yl) piperazin-1-yl] -(furan-2-yl) methanone (PubChem Compound Database; CID=4893)), a pharmaceutical substance which is primarily used for antihypertensive remedy [4]. PRZ was measured to contaminate the water system in Malaysia at the highest concentration of 525 ng/L [5]. According to the Ministry of Health Malaysia, PRZ is among the top 50 utilized drugs in Malaysia and one of the two drugs for alpha-adrenergic blocker given for treatment in Malaysia [6]. Therefore, it is projected that the recalcitrant of this substance in the water system will be continuously increased.

Previous researches have shown many technologies for PPCPs degradation, for example, adsorption, advanced oxidation processes (AOPs), membrane separation, hybrid treatment and biodegradation. Among the technologies studied, adsorption and AOPs showed the highest efficiency in degradation [7]. However, adsorption requires different adsorbent material for different pollutant [8] and the pollutants are not degraded but adsorbed onto the adsorber at their still harmful state. In contrast, AOPs which is consider highly efficient to mineralize pollutants with no toxic end products [7] has its own disadvantages. Cost efficiency is not an attribute for AOPs technologies due to its costly reagents and huge energy consumption [7]. Hence, biodegradation is one of the top alternatives in removing xenobiotic pollutants. PPCPs degradation by bacteria have shown 47% to 100% removal by various bacteria [9]. The microbial communities can remove, degrade, change, or detoxify various chemical pollutants in the environment, through their enzymatic pathways [10]. However, to date very minimal study was reported on the biodegradation of PRZ [11], even though it is continuously accumulating in the environment. Therefore, more studies are needed to uncover the potential of PRZ biodegradation by bacteria.

The aims of this study were: (1) to isolate a tropical bacterial strain from PPCPs contaminated river in Malaysia that can degrade PRZ; (2) to determine the isolate growth curve and PRZ degradation during growing cell assay; and (3) to identify the species of PRZ degrading bacteria isolate by 16S rRNA gene sequence.

2. Materials and methods

2.1 Sampling site

Water samples were collected from a PRZ contaminated Pantai River, Kuala Lumpur, Malaysia (3°07'01.5"N 101°39'51.4"E) [12]. Pantai River is located near to two main hospitals, Pantai Hospital and University Malaysia Medical Centre, and also surrounded by a central business district.

The Minimal Salt Media (MSM) was prepared with the following composition in gL-1, 1.816 g Na_2HPO_4. $12H_2O$; 0.6 g K_2HPO; 3 mg MgSO4 and 3.9 mg $CaCl_2$. $2H_2O$ [13][14]. The water samples were filtered and the filtrate water was diluted to obtain the heterogenous solution containing the microorganisms [13]. The solution was then poured and spread onto a solid screening MSM, with the inclusion of 0.1 M of PRZ [13][14]. This step was repeated for few times until single colonies are observed and isolated. Then, isolate PR5 which has the most efficiency in degrading PRZ was chosen for further study.

2.2 Growing cell assay of PRZ degradation

Luria-Bertani (LB) agar was prepared with the following composition in gL^{-1}, 10 g tryptone, 5 g yeast extract, 10 g NaCl, 7.5 g agar [15]. MSM was also prepared with the addition of 1.5 g of peptone. Both type of MSM were autoclaved for sterilization at 121°C for 20 minutes. Then, after the media cooled down, 0.01 mM PRZ was added into the liquid media.

The single isolate, PR5, was streaked on LB agar. Then, the single colony grown on the plate was picked and transferred into MSM. The growing cell assay was done in duplicate and a sterile control was also included. The cultures were incubated at 30°C and set to swirl at 160 rpm for 7 days. 1 mL of culture of each sample was taken daily for the measurement of the isolate's growth. The isolate's growth was measured by using spectrophotometer with the wavelength of 600 nm and the average optical density value was calculated.

2.3 PRZ concentration measurement by UPLC-TUV

Another 1 mL of culture of each sample was also taken daily for the measurement of the PRZ concentration by using UPLC-TUV system. The samples were centrifuged to separate the supernatant from any bacterial cells or residues. Then, the supernatant was filtered by using Merck Millex Milipore 0.22 μm PTFE Hydrophilic filter attached to a 1 mL syringe. All the filtered samples were kept in a -20°C refrigerator.

PRZ standard was also prepared for the UPLC analysis at the concentration of 0.238 mM, 0.15 mM, 0.10 mM, 0.08 mM and 0.05 mM to construct the calibration curve. PRZ peak was identified by referring to the standard sample peak at retention time 6.9 min. PRZ standards were prepared by dissolving the PRZ Hydrochloride (Cayman Chemical, USA) powder with HPLC grade methanol (Wako, Japan) and vortexed to mix. The standards were also filtered and kept in a -20°C refrigerator until analysis.

Before analysing, all the samples including all of the standards underwent 1:10 dilution with methanol. The analysis was done by using Waters Acquity UPLC-TUV system. C18 BEH (1.7 μm, 2.1 x 100 mm) column was used as a stationary phase. 2 mobile phases, A; 10 mM Ammonium acetate (Wako, Japan) + 5% methanol (Wako, Japan) + miliQ H_2O and B; 100% Acetonitrile (Wako, Japan) were used. The mobile phase flow rate was maintained at 0.2 mL/min and the UV detection wavelength was set at 254 nm [16].

2.4 Species identification by 16S rRNA gene sequencing

Bacterial isolate PR5 was streaked onto a LB agar plate. The streaking was diluted from each quadrant of the media in order to strictly obtaining single colonies. The cultured plate was incubated in 30°C, overnight.

After the formation of the single colony, the colony was picked and genomic DNA extraction was performed. Then, Polymerase Chain Reaction (PCR) was done with the universal primer 27F: 5'-AGAGTTTGATCCTGGCTCAG-3' and 1492R: 3'-CTACGGCTACCTTGTTACGA-5'. After that, the PCR amplicon underwent library preparation and proceeded for sequencing. The sequencing data was aligned onto the National Center for Biotechnology Information (NCBI) database by using blastn (https://blast.ncbi.nlm.nih.gov/Blast.cgi) to find the homologue species.

8 highly similar species with 2 outlier species were selected and their 16S rDNA sequences were pooled into one FASTA file together with PR5 16S rDNA sequence. Pairwise alignment was done by using ClustalW and the gaps was deleted. Then, the alignment result was used to construct a neighbour joining phylogenetic trees to show species similarities based on statistical analysis on MEGA 7. The 16S rDNA sequence was deposited on DNA Data Bank of Japan (DDBJ) under the accession number, LC795929.

3. Results and discussion

3.1 Isolate PR5 growth curve

Growth curve of the isolate PR5 was measured during the growing cell assay in MSM added with 0.01 mM PRZ. As measured, the measurement started with optical density (OD) value at 0.004 on Day 0. Then, it was slowly increasing until Day 1 at OD value 0.009. OD value was rapidly increased to 0.107 upon Day 2. After that, the growing reached the highest point on Day 3 at 0.175 and started to decline on Day 4 onwards.

From Figure 1, it is shown clearly that lag phase was on Day 0 until Day 1. During the lag phase the bacteria cells underwent physiological adaptation which is adapting to the new conditions or environment, therefore minimal growth is observed [17]. In previous research, growth characteristic of *Bacillus velezensis* strain MHNK1 in MSM showed the same pattern of lag phase which lasted for 1 day during atrazine biodegradation [18]. Then the growing of isolate PR5 moved into exponential phase with the mid-exponential phase occurred on Day 2. Isolate PR5 has a very short stationary phase on Day 3. A short stationary phase demonstrated by isolate PR5 is an advantage as it can reduce stationary-phase adaptive mutation to the cells [19] which is very common to bacteria cells with long-term stationary phase [20]. This could ensure robustness in future application. After that the growing starting to decline into the death phase from Day 4 onwards.

OD value of isolate PR5 cultured in MSM added with 0.01 mM PRZ but without peptone was also measured and included in Figure 1. Isolate PR5 recorded no growth under the stringent condition of PRZ as the sole carbon or nitrogen source. Hence, peptone was added to boost the growth isolate PR5. This demonstrates the mixotrophic attribute of isolate PR5. Based on the mixotrophic characteristics, isolate PR5 needed low nitrogen source from peptone for the activation step to grow [21]. The presence of peptone induced isolate PR5 consumption on carbon or nitrogen in PRZ. This is because mixotrophic bacteria is functioning well with low nutrient source and simultaneously speed up the metabolism of carbon, nitrogen, sulfur, and heavy metal [22].

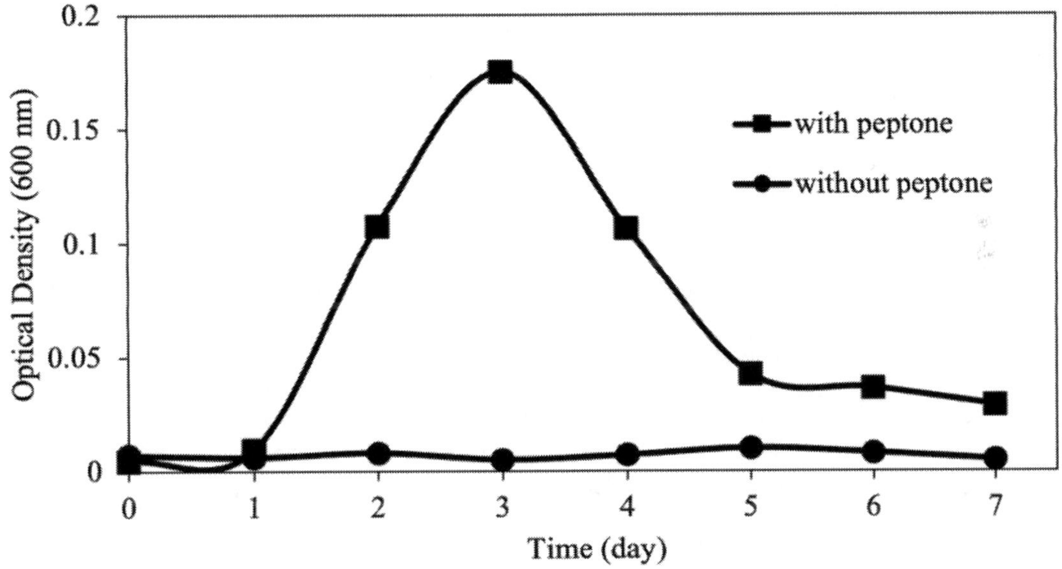

Figure 1. Growth curve of isolate PR5 in MSM added with 0.01 mM PRZ, with and without peptone.

3.2 PRZ degradation by isolated bacteria PR5

PRZ concentration in growing cell assay samples were measured by UPLC-TUV then the PRZ biodegradation by isolate PR5 was calculated (Figure 2). The initial concentration of PRZ was measured at 0.012 mM. The concentration remained consistent from Day 0 until Day 1. Then, rapid degradation occurred on Day 2, where the PRZ concentration decreased into 0.0040 mM, which indicates 68% of PRZ degradation (Figure 2). After that, PRZ concentration continued to remain stagnant from Day 3 until Day 7 and the final measured concentration was recorded at 0.0027 mM. In total, 78% of PRZ was degraded within 7 days of growing cell assay. Meanwhile, PRZ concentration in sterile control remained constant throughout the assay (Figure 2).

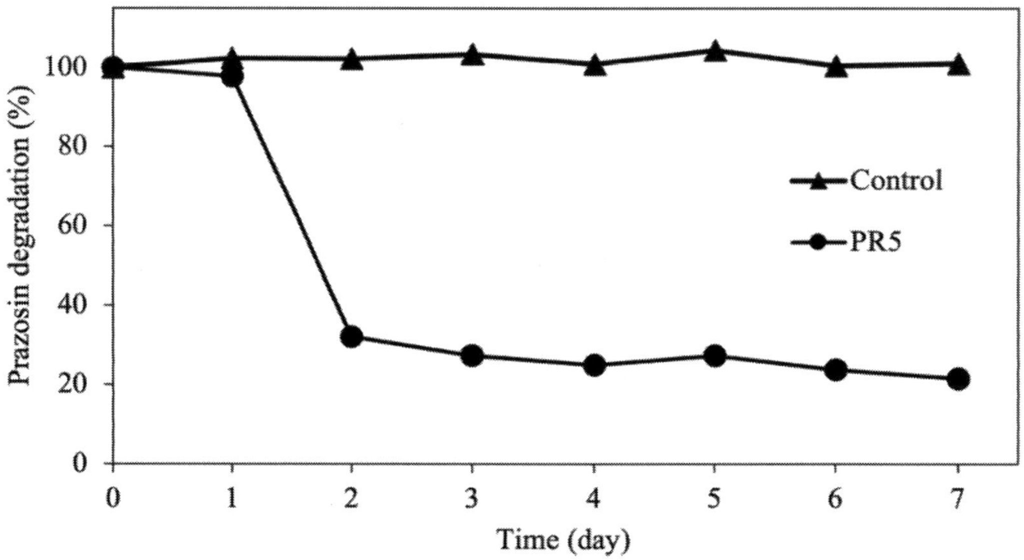

Figure 2. Degradation percentage of PRZ by isolate PR5 from day 0 until day 7.

Comparing Figure 1 and Figure 2, it is seen clearly that the highest degradation activity occurred during the mid -exponential phase which was on the second day of growing cell assay. This is explained by the active mRNA expression during mid-exponential phase of bacterial growth curve [23], that might lead to the certain protein expression which highly to be responsible on the PRZ biodegradation. Then, the degradation activity started to decrease during the stationary phase on Day 3 and PRZ continued to degrade slowly during the death phase onwards until Day 7. Although there was still some growth occurring from Day 2 until Day 3, the degradation of PRZ had already begun to slow down. This could be attributed to a phenomenon called the transition phase, where the cells continue to grow steadily until they reach the starvation point, where the growth stops abruptly [24]. It was reported previously that genes in *Bacillus* sp. that accounted for metabolic processes were observed to become upregulated during the mid-exponential growth phase to facilitate metabolic activity. Conversely, during the transition phase or late exponential phase, these same genes were found to be downregulated [25]. This phenomenon may account for the slower degradation of PRZ from Day 2 to Day 3. Simultaneously, the sterile control sample remained at a consistent concentration, which indicates the stability and recalcitrant attributes of PRZ.

3.3 Species identification by 16S rRNA gene sequencing

The 16S rRNA gene sequence read was aligned against the 16S ribosomal RNA sequences (bacteria and archaea) database on blastn platform to identify the homologous species based on percentage of identity. The full sequence of 16S rRNA gene of isolate PR5 has the 99.66% percentage of identity and 100% query cover (1469/1469 bp) with *Bacillus velezensis* FZB42 (NR_075005.2).

Isolate PR5 colony formation has cloudy white color. It has irregular circular shape, flat and viscous texture with the approximate size of 1-4 mm. The observation on morphology of isolate PR5 has similarity with explanation on other strain of *Bacillus velezensis* [26].

8 highest hits based on the blastn alignment were chosen to construct a phylogenetic tree. Additionally, 2 outlier hit were also selected to show significance. As shown in Figure 3, the phylogeny relationship of 11 16S rRNA gene sequences were deduced using the neighbour-joining method [27]. The constructed phylogenetic tree on Figure 3 clearly show a high similarity of isolate PR5 with *Bacillus velezensis* FZB42.

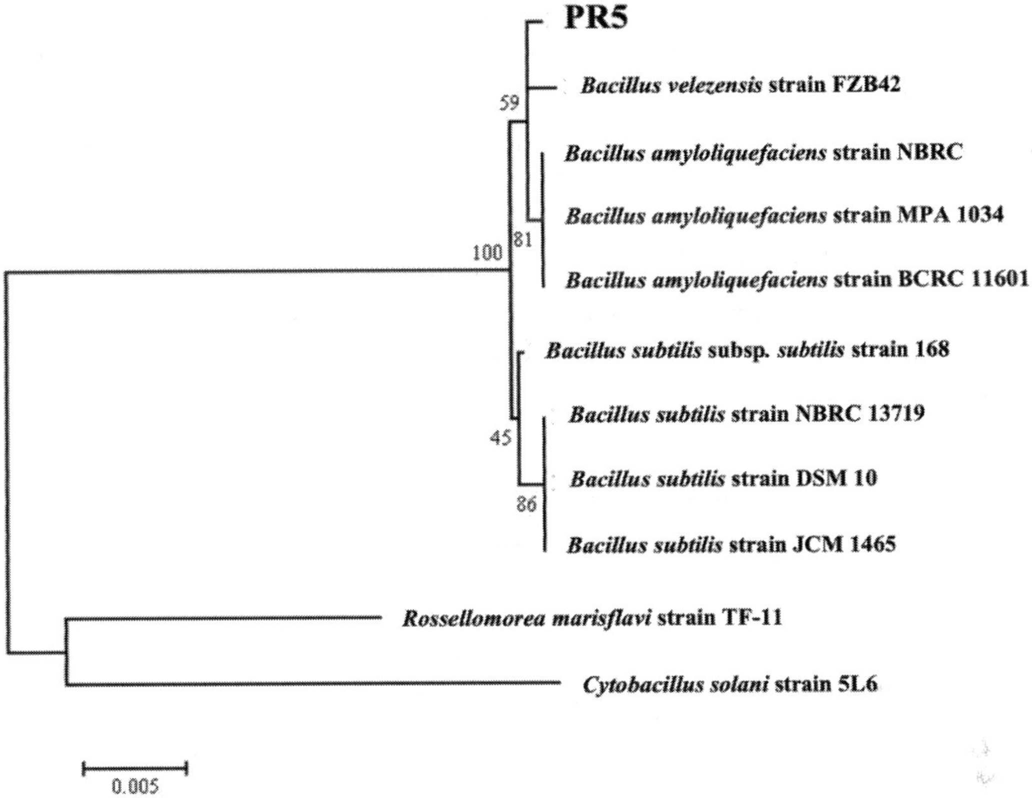

Figure 3. A neighbour joining phylogenetic tree consisting of evolutionary relationship between 11 16S rRNA gene sequences of different bacteria species and strains including isolate PR5. The percentage of replicate trees in which the associated taxa clustered together in the bootstrap test (1000 replicates) are shown next to the branches.

Bacillus velezensis is reputed to have abilities to degrade many xenobiotic compounds [28]. However, this is the first report on *Bacillus velezensis* ability to degrade PRZ. A strain of *Bacillus velezensis* isolated from a tropical climate was reported able to degrade dichlorodiphenyltrichloroethane (DDT), a type of pesticide [29], and it has similar structure with PRZ with the presence of benzene ring. This shows *Bacillus velezensis* is a type of bacteria species with many potentials to mitigate pollutants problems. Besides, it is an advantage to utilize bacteria isolate from a tropical environment because bacteria metabolism is reported higher in tropical area in comparison with temperate regions [30].

On the other hand, based on 16S rRNA gene sequence, isolate PR5 also has high similarities with few strains of *Bacillus amyloliquefaciens*. This is understandable as *Bacillus velezensis* is a heterotypic synonyms with *Bacillus amyloliquefaciens* [31]. Therefore, this information infer that *Bacillus amyloliquefaciens* have the potential for PRZ biodegradation.

4. Conclusion

In summary, isolate PR5 has demonstrated its ability to utilize PRZ as a nutrient source and able to degrade PRZ during the mid-exponential phase of its growing. 78% of PRZ was successfully degraded after 7 days which shows isolate PR5 efficiency to degrade the recalcitrating PRZ. Through 16S rRNA gene sequence similarity searching, isolate PR5 has 99.66% similarity with *Bacillus velezensis* FZB42. Thus, this study highlights on *Bacillus velezensis* ability to degrade PRZ. The result of this study exhibiting isolate PR5 to be a valuable candidate for the removal of PRZ as contaminants in the environment.

The first author is sponsored financially by Ministry of Education, Culture, Sports, and Technology Japan (MEXT) scholarship. This study is funded and supported by University of Tsukuba, Japan.

5. References

1. M.L. Crouch, H.E. Jacobs, V.L. Speight, J. Water Supply Res. T **70**, 1002-11 (2021)
2. N. Akhtar, M.I. Syakir Ishak, S.A. Bhawani, K. Umar, Water-Sui **13**, 2660 (2021)
3. R.Y. Krishnan, S. Manikandan, R. Subbaiya, M. Biruntha, M. Govarthanan, N. Karmegam, Environ. Technol. **23**, 101757 (2021)
4. R. Fuchs, G. Schwach, A. Stracke, N. Meier-Allard, M. Absenger, E. Ingolic, H.S. Haas, R. Pfragner, A. Sadjak, Anticancer Res. **35**, 31-8 (2015)
5. F.F. Al-Qaim, M.P. Abdullah, M.R. Othman, Z.H. Mussa, Z. Zakaria, J. Latip, W.M. Afiq, J. Brazil Chem. Soc. **26** 1124-35 (2015)
6. M.C.C. Fong, A.I.A. Shanizza, N.A. Kadir, Malaysian Statistics on Medicines 2015-2016 Kuala Lumpur (2020)
7. T. Liu, C.O. Aniagor, M.I. Ejimofor, M.C. Menkiti, K.H. Tang, B.L. Chin, Y.H. Chan, C.L. Yiin, K.W. Cheah, Y.H. Chai, S.S. Lock, J Mol. Liq. **374**, 121144 (2023)
8. M. Mallek, M. Chtourou, M. Portillo, H. Monclus, K. Walha, A. ben Salah, V. Salvado, J. Environ. Manage. **223**, 576-85 (2018)
9. M. Narayanan, M. El-Sheekh, Y. Ma, A. Pugazhendhi, D, G. Kandasamy, R. Raja, R.S. Kumar, S. Kumarasamy, G. Sathiyan, R. Geetha, Environ. Pollut. **300**, 118922 (2022)
10. M. Bilal, S.S. Ashraf, D. Barceló, H.M. Iqbal, Sci. Total Environ. **691**, 1190-211 (2019)
11. A. Moghaddam, D. Khayatan, F.B.P. Esmaeili, R. Ranjbar, M. Yazdanian, E. Tahmasebi, M. Alam, K, Abbasi, G.G.H. Esmaeili, H. Tebyaniyan, Int. J. Environ. Sci. Te. **5**, 5659-96 (2023)
12. N.F.M. Mohsi , A. Apandi, M.J. Noor, F.N. Akhir, N. Sugiura, M. Utsumi, A.O. Nor, Z. Zakaria, H. Hara, J. Gen. Appl. Microbiol. **66**, 8-14 (2020)
13. J. Sumitha, T. Sivakumar, Int. J. Curr. Microbiol. Appl. Sci. **8**, 338-46 (2013)
14. R. Chandra, A. Ghosh, R.K. Jain, S. Singh, J. Gen. Appl. Microbiol. **52**, 125-30 (2006)
15. M.P. MacWilliams, M.K. Liao, mBio. **9**, 2006 (2006)
16. A. Shrivastava, V.B. Gupta, Sci. Pharm. **80**, 619-32 (2012)
17. R.M. Maier, I.L. Pepper, Bacterial growth (Academic Press, 2009) pp 36-56

18. P. Jakinala, N. Lingampally, A. Kyama, B. Hameeda, Ecotox. Environ. Safe. **182**, 109372 (2019)
19. C. Ross, C. Pybus, M. Pedraza-Reyes, H.M. Sung, R.E. Yasbin, E. Robleto, J. Bacteriol. **21**, 7512-20 (2006)
20. J. Gross, S. Avrani, S. Katz, S. Hilau, R. Hershberg, Genome Biol. Evol. **12**, 2292-301 (2020)
21. R.M. Madani, J. Liang, L. Cui, R.H. Elsalahi, T.A. Otitoju, D. Zhang, X. Song, M. Yongguang, S. Liu, Water Air Soil Poll. **6**, 185 (2022)
22. L. Huang, X. Xing, P. Zhou, G. Li Puma, Appl. Microbiol. Biot. **18**, 6627-48 (2021)
23. R.G. Cutler, J.E. Evans, J Bacteriol. **91**, 469-76 (1966)
24. H. Funakoshi, A. Yamada, J Math. Biol. **9**, 369-87 (1980)
25. J.S. Jun, H.E. Jeong, S.Y. Moon, S.H. Shin, K.W. Hong, Microorganisms **8**, 1928 (2023)
26. X. Li, X. Gao, S. Zhang, Z. Jiang, H. Yang, X. Liu, Q. Jiang, X. Zhang, Aquaculture **523**, 735165 (2020)
27. N. Saitou, M. Nei, Mol. Biol. Evol. **4**, 406-25 (1987)
28. A.A. Adeniji, D.T. Loots, O.O. Babalola, Appl. Microbiol. Biot. **103**, 3669-82 (2019)
29. N.B. Chima, B.A. Idris, L. Uchenna, S. Ezeanyika, Acta. Biol. Turc. **2**, 76-85 (2021)
30. M.L. Melo, H. Sarmento, Acta. Limno. Brasil. **14**, 31 (2019)
31. C.A. Dunlap, S.J. Kim, S.W. Kwon, A.P. Rooney, Int. J. Syst. Evol. Micr. **3**, 1212-7 (2016)

Exploring PFOA adsorption isotherm in the presence of NOM using DBD plasma-modified GAC

Thera Sahara[1], *Doonyapong* Wongsawaeng[1,*], *Kanokwan* Ngaosuwan[2], *Worapon* Kiatkittipong[3], *Peter* Hosemann[4], *Suttichai* Assabumrungrat[5,6]

[1]Research Unit on Plasma Technology for High-Performance Materials Development, Department of Nuclear Engineering, Faculty of Engineering, Chulalongkorn University, Bangkok 10330, Thailand
[2]Division of Chemical Engineering, Faculty of Engineering, Rajamangala University of Technology Krungthep, Bangkok 10120, Thailand
[3]Department of Chemical Engineering, Faculty of Engineering and Industrial Technology, Silpakorn University, Nakhon Pathom 73000, Thailand
[4]Department of Nuclear Engineering, Faculty of Engineering, University of California, Berkeley, 94720, U.S.A.
[5]Center of Excellence in Catalysis and Catalytic Reaction Engineering, Department of Chemical Engineering, Faculty of Engineering, Chulalongkorn University, Bangkok 10330, Thailand
[6]Bio-Circular-Green-economy Technology & Engineering Center (BCGeTEC), Faculty of Engineering, Chulalongkorn University, Bangkok, Thailand 10330

Abstract. Perfluorooctanoic acid (PFOA), a widely used perfluorinated alkyl substance (PFAS), poses significant environmental and health risks. This study investigates PFOA adsorption in the existence of natural organic matter (NOM) using granular activated carbon (GAC) modified with dielectric barrier discharge (DBD) plasma. PFOA's amphiphilic structure, characterized by a hydrophilic carboxyl group head and a hydrophobic perfluorinated tail, contributes to its versatility and persistence. The selection of the Chao Phraya River (CPR) water as the NOM source, captures the complexity of a major water body subjected to diverse pollution sources. The results were analyzed through the Toth and Temkin isotherm models. Application of the Toth isotherm model reveals enhanced PFOA adsorption capacity in CPR water compared to DI water, emphasizing the influence of NOM. The Temkin isotherm analysis further characterizes the strength and efficiency of the adsorption process, highlighting a stronger interaction between PFOA and plasma-modified GAC in CPR water. The study indicates that the adsorption process in CPR water may be more influenced by PFOA surface coverage on the GAC surface in the presence of NOM. Overall, this research contributes valuable insights into pollutant removal strategies, highlighting the potential of DBD plasma-modified GAC in addressing PFOA contamination challenges in water systems.
Keywords: Perfluorooctanoic acid (PFOA), dielectric barrier discharge (DBD) plasma, Toth isotherm, Temkin isotherm.

* Corresponding author: Doonyapong.W@chula.ac.th

1. Introduction

Perfluorooctanoic acid (PFOA), classified as an anthropogenic perfluorinated alkyl substance (PFAS), has gained extensive application as a surfactant in fire retardants, polymer synthesis, paper coatings, and fabric safeguarding. Its distinctive amphiphilic structure encompasses a hydrophilic carboxyl group head and a hydrophobic perfluorinated tail, rendering it versatile [1, 2]. With a substantial C–F bond energy of approximately 440 $kJ \cdot mol^{-1}$, PFOA exhibits remarkable chemical stability, while its solubility in water (3400 $mg \cdot L^{-1}$) is facilitated by the presence of the carboxyl group [3, 4]. The implications of PFOA exposure are profound, encompassing adverse health effects such as an elevated risk of kidney and testicular cancer, reproductive toxicity, immunosuppression, and neurological disorders [5, 6]. The urgency to address the environmental and health consequences of PFOA exposure is underscored by its persistent presence in various ecosystems and its connection to serious human health impacts.

Natural organic matter (NOM) constitutes a significant portion of the dissolved organic carbon present in water environments [7]. The way NOM behaves when it interacts with the boundary between solids and water holds crucial significance in various physicochemical processes [8]. For instance, NOM that adheres to surfaces at the water-solid interface can exert a stabilizing effect on colloidal particles [9]. Moreover, this adsorbed NOM can hinder the aggregation and precipitation of these particles [10]. Therefore, gaining insights into the arrangement of NOM at the interface between solids and water becomes imperative to comprehending the specific role of NOM in processes involving the aggregation of colloids and the binding of pollutants.

The plasma treatment process utilizes gases like Ar, He, O_2, N_2, or CO_2 to modify material surfaces, introducing chemical functionalities that bind polymers or other molecules to achieve desired surface properties [11]. Depending on the gas and treatment conditions, effects such as surface activation, abrasion, etching, cross-linking, or a combination can be achieved [12]. Plasma treatment leads to the creation of chemically active species, including hydroxyl, carbonyl, and carboxylic acids, which are generated during plasma treatment, interacting with specific chemical functions on the adsorbent surface [13].

The modification and enhancement of granular activated carbon (GAC) possess the utmost significance in water treatment processes. This study adopted a novel approach to enhance adsorbability by employing dielectric discharge plasma (DBD) plasma to modify granular activated carbon (GAC). This technique aims to address the limitations of conventional adsorbents by potentially enhancing their adsorption capabilities. The study investigates the adsorption isotherm behavior using Toth and Temkin isotherm models of PFOA in the presence of NOM, utilizing GAC that has been modified using DBD plasma. There is no literature investigating the adsorption of PFOA and NOM from natural water systems. While experimental PFAS adsorption studies have mainly employed synthetic NOM solutions, natural systems encompass diverse organic compounds. Hence, employing actual water samples is essential for a comprehensive study of adsorption behavior. The Chao Phraya River (CPR), being a major water body in Bangkok, Thailand, is subject to various pollution inputs, such as industrial runoff, agricultural runoff, and urban discharges, all of which can contribute to the diverse composition of NOM present in its water [14]. By selecting CPR as the source of NOM, the study aims to examine the interactions between this complex mixture of organic compounds and the adsorption process under adsorption isotherms investigation.

2. Methodology

2.1 Materials

The CPR sample was taken at lat. 13°54'56.9"N and long. 100°29'38.1"E. and the depth of the river was 3.3 m, the characteristics of CPR water are displayed in Table 1. Subsequently, the water underwent filtration using a pre-washed filter featuring a pore size of 0.3 mm. It was then stored in a refrigerator at a temperature of 4°C. PFOA having a molecular weight of 414.07 g/mol was acquired from Sigma Aldrich. An 85% solution of phosphoric acid (H_3PO_4, PA) was obtained from Anapure. A syringe filter created from glass fibers with a pore size of 2.0 μm was bought from Merck Millipore. Deionized water (DI) was produced in the laboratory and utilized as the standard. Ultra-high purity (UHP) helium and oxygen (O_2) gas cylinders were purchased from Alternative Chemical.

Table 1. CPR water parameters [15].

pH	TOC (mg/L)	DOC (mg/L)	UV_{254} (cm^{-1})	$SUVA_{254}$ (L/mg·m)
6.8	30.153	23.491	0.9572	4.075

2.2 Activated GAC with phosphoric acid (PA)

GAC was soaked in DI water for 4 hours to remove impurities. After thorough washing, the GAC was dried at 50°C for 24 hours. It was then immersed in a 1M solution of phosphoric acid (PA) for 24 hours. After filtration, the GAC was washed with DI water. The filtered residue was dried at 105°C until a constant weight was reached. The resulting PA-activated GAC was used consistently in the entire study.

2.3 DBD plasma treatment of PA-activated GAC

The DBD plasma setup (figure. 1) used an adjustable neon transformer, providing voltages from 0 to 15 kV and a max current of 30 mA. Output power was controlled through a variac linked to the neon transformer. The system featured aluminum parallel-plate electrodes, the discharge electrode (9 cm × 13 cm) and the ground electrode (11 cm × 15 cm). The reaction chamber, an 800 mL borosilicate glass container, was positioned over the ground electrode. A glass dielectric (10 cm × 14 cm) was connected to the discharge electrode. The consistent discharge gap (the space between the glass dielectric's bottom and the reactor's base) was set at 3 mm. During experiments, PA-activated GAC samples underwent DBD plasma treatment. A gas mixture of 30% O_2 and 70% He flowed into the chamber at 1.5 L/min, with independent regulation using mass flow controllers (MFCs). Plasma treatment lasted for 20 minutes and the plasma power was set at 100 W. Afterward, treated GAC samples were sealed in glass containers for subsequent adsorption processes.

Fig. 1. DBD plasma experimental setup.

2.4 Adsorption experiment

In each batch adsorption, 20 mg of plasma-treated GAC was placed in a glass beaker. Then, 100 mL of a solution containing PFOA, initially at 100 mg/L concentration, was added. Room temperature was maintained during the adsorption, using a magnetic stirrer bar within a glass beaker placed on a hot plate stirrer set at 300 rpm. The process lasted for 24 hours [16]. Gas chromatography-flame ionization detection (GC-FID) was employed to assess the remaining PFOA concentration.

2.4.1 Toth isotherm model

The Toth model (Eq. (1)) was created to enable the Langmuir model to be used in heterogeneous systems [17]. It assumes that the majority of adsorption sites have adsorption energies that are lower than the average energy [18].

$$q_e = \frac{K_T C_e}{(a_T + C_e^z)^{1/z}} \tag{1}$$

where K_T (mg.g^{-1}) is the constant, a_T (mg^z L^{-z}) is the Toth constant, and the parameter z in the model quantifies the extent of heterogeneity in the adsorption systems, reflecting the degree of variation or diversity in the adsorption sites [18]. z remains constant and is not affected by changes in temperature, indicating that the degree of heterogeneity in the adsorption systems is independent of the temperature [18]. On the other hand, the value of a_T increases as the temperature rises, suggesting a temperature-dependent relationship with the adsorption process [19]. When the parameter z in the Toth model takes a value of 1, the model simplifies to the Langmuir isotherm. However, as the deviation of z from 1 increases, it indicates a greater level of heterogeneity in the adsorption system.

2.4.2 Temkin isotherm

According to the Temkin model, adsorption occurs via a multi-layer process [20]. The impact of both very high and low adsorbate concentrations in the liquid phase is ignored by this model. In order to incorporate the Temkin isotherm into the Clapeyron-Clausius equation, it was developed a statistical mechanical expression for it [21]. This integration verified that there is a linear decrease in the differential heat of adsorption with increasing coverage.

Eq. 2 represents the Temkin model presented by [20]:

$$q_e = \frac{RT}{b} \ln(AC_e) \qquad (2)$$

where the constants are denoted by A ($L \cdot g^{-1}$) and b ($J \cdot mol^{-1}$).

3. Results and Discussion

3.1 Toth isotherm model analysis

The Toth isotherm is a mathematical model used to describe the adsorption behavior of solutes onto solid adsorbents [22]. The q_{max} of PFOA onto GAC according to the Toth isotherm is 406.42 mg/g for CPR and 229.46 mg/g for DI (Table 2) [15]. The higher q_{max} value for CPR suggests that the NOM enhances the adsorption capacity, either by occupying additional adsorption sites or promoting stronger adsorbate-adsorbent interactions. The interactions between PFOA and NOM compounds can enhance the adsorption capacity by facilitating the formation of complexes or promoting cooperative adsorption mechanisms [23].

The adsorption energy (K_T) of PFOA adsorption onto GAC according to the Toth isotherm is 0.243 L/mg for CPR and 0.237 L/mg for DI. Higher adsorption energy in CPR indicates a stronger affinity between the solute (PFOA) and the adsorbent surface (plasma-treated GAC).

In this case, the higher heterogeneity (n) value for CPR (0.954) compared to DI (0.648) suggests that the adsorption system in CPR is more heterogeneous in nature. This indicates that there is a wider distribution of adsorption energies on the GAC surface, potentially due to the presence of NOM compounds [23]. The lower value of n for DI indicates a relatively more homogeneous adsorption system, suggesting more consistent adsorption energy across the GAC surface in the absence of NOM [23].

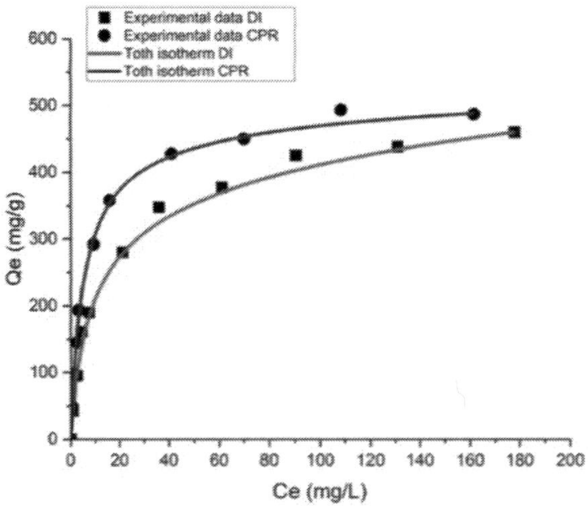

Fig. 2. Toth isotherm graph of PFOA adsorption for both DI and CPR.

 https://doi.org/10.1051/e3sconf/202453002002

Table 2. Toth isotherm model parameters for PFOA adsorption onto plasma-treated GAC for both DI and CPR samples.

Solution	Toth isotherm model			
	q_{max} (mg/g)	K_T	n	R^2
DI	229.46	0.237	0.648	0.994
CPR	406.42	0.243	0.954	0.992

3.2 Temkin isotherm model analysis

According to the Temkin isotherm (figure. 3), if the adsorption energy drops linearly, which implies that the adsorption process follows a heterogeneous distribution of binding energies [24]. The adsorption intensity parameter (b_T) of the Temkin isotherm, which characterizes the strength of the adsorption process, is determined to be 79.49 for CPR and 78.47 for DI in the case of PFOA adsorption onto plasma-treated GAC (Table 3) [15]. A higher b_T value signified a stronger adsorption interaction between PFOA and GAC, indicating a more favorable and efficient adsorption process. Additionally, the rate of decrease in adsorption energy with increasing surface coverage, as indicated by the b_T values, suggests that the adsorption energy decreases at a slightly faster rate in CPR compared to DI [24]. This implies that the adsorption process in CPR may be more influenced by the surface coverage of PFOA molecules on the plasma-treated GAC surface.

The adsorption capacity parameter (A_T) of the Temkin isotherm, which represents the amount of PFOA that can be adsorbed onto GAC, is determined to be 3.84 for CPR and 1.75 for DI. These values represent the maximum adsorption capacity of PFOA onto GAC according to the Temkin isotherm. The higher A_T value for CPR suggests a greater adsorption capacity compared to DI.

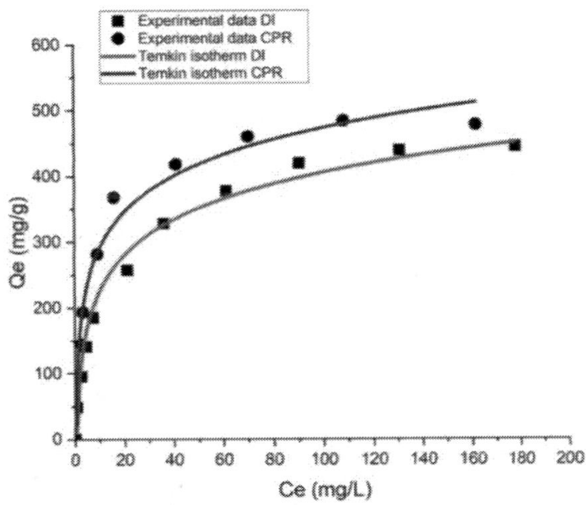

Fig. 3. Temkin isotherm graph of PFOA adsorption for both DI and CPR.

Table 3. Toth isotherm model parameters for PFOA adsorption onto plasma-treated GAC for both DI and CPR samples.

Solution	Temkin isotherm model		
	A_T	b_T	R^2
DI	1.75	78.47	0.986
CPR	3.84	79.49	0.983

4. Conclusion

This study explored the PFOA adsorption in the presence of NOM using GAC modified with DBD plasma. The need to address PFOA's effects on the environment and human health is urgent due to its amphiphilic nature, widespread usage, and enduring persistence in ecosystems.

The application of Toth and Temkin isotherm models yields valuable insights into the enhanced PFOA adsorption capacity in the presence of NOM using plasma-treated GAC. Based on the R^2 values, the Toth isotherm model exhibited better fitting for both DI and CPR water compared to the Temkin isotherm model. The Toth isotherm indicates a higher adsorption capacity (q_{max}) for PFOA in the presence of NOM (CPR), suggesting enhanced adsorbate-adsorbent interactions. The elevated adsorption energy (K_T) and heterogeneity (n) values in CPR further emphasize NOM's impact on strengthening the affinity between PFOA and GAC.

The Temkin isotherm analysis supports these findings, revealing a stronger adsorption interaction (higher b_T) in CPR. The faster decrease in adsorption energy with increasing surface coverage suggests the influence of PFOA surface coverage on GAC in CPR, influencing the adsorption process. The higher adsorption capacity parameter (A_T) for CPR highlights its greater adsorption capacity compared to adsorption in DI water.

References

1. R. Loos, et al. Water Research. **44**(14): p. 4115-4126 (2010).
2. T. Wang, et al. Chemosphere. **129**: p. 87-99 (2015).
3. T. Sidnell, et al. Ultrasonics Sonochemistry. **87**: p. 105944 (2022).
4. Y. Bao, et al. Journal of colloid and interface science. **434**: p. 59-64 (2014).
5. G. W. Olsen, Butenhoff, J. L., and Zobel, L. R. Reproductive Toxicology. **27**(3): p. 212-230 (2009).
6. R. A. Brase, *Legacy and Emerging Per-and Polyfluoroalkyl Substances: Analysis, Environmental Distribution, and Bioconcentration*. 2021: State University of New York at Albany.
7. F. Frimmel. Journal of Contaminant Hydrology. **35**(1-3): p. 201-216 (1998).
8. B. Gu, et al. Geochimica et Cosmochimica Acta. **59**(2): p. 219-229 (1995).
9. M. S. Chandrakanth and Amy, G. L. Water Research. **32**(1): p. 115-124 (1998).
10. T. R. Hundt and O'Melia, C. R. Journal AWWA. **80**(4): p. 176-186 (1988).
11. L. Lin, et al. Journal of CO2 Utilization. **53**: p. 101752 (2021).
12. P. Cools, et al., *Surface Treatment of Polymers by Plasma*, in *Surface Modification of Polymers*. 2019. p. 31-65.
13. Ö. Şahin, Kaya, M., and Saka, C. Applied Clay Science. **116-117**: p. 46-53 (2015).
14. R. Chaysiri and Sukruay, J. J. Res. Unit Sci. Technol. Environ. Learning. **12**: p. 105-121 (2021).
15. T. Sahara, et al. To be published

16. K. Shih and Wang, F. Procedia Environmental Sciences. **18**: p. 106-113 (2013).
17. J. Toth. Acta chim. hung. **69**: p. 311-328 (1971).
18. Y. S. Ho, Porter, J. F., and McKay, G. Water, Air, and Soil Pollution. **141**(1-4): p. 1-33 (2002).
19. W. Rudzinski and Everett, D. H., *Adsorption of gases on heterogeneous surfaces*. 2012: Academic Press.
20. M. Temkin. Acta physiochim. URSS. **12**: p. 327-356 (1940).
21. C. Yang. The Journal of Physical Chemistry. **97**(27): p. 7097-7101 (1993).
22. K. V. Kumar, et al. Physical Chemistry Chemical Physics. **13**(13): p. 5753-5759 (2011).
23. J. G. Jacangelo, et al. Journal-American Water Works Association. **87**(1): p. 64-77 (1995).
24. R. D. Johnson and Arnold, F. H. Biochimica et Biophysica Acta (BBA)-Protein Structure and Molecular Enzymology. **1247**(2): p. 293-297 (1995).

Marine ecological restoration strategies of coastal nuclear power plants in China

Guilan Lin

Third Institute of Oceanography, Ministry of Natural Resources of China, Xiamen 361005, China

Abstract. Low-carbon and green development model is the choice for responding to climate change. Under the dilemma between ever-growing energy demand and the associated environmental issues, nuclear power is one of the low-carbon and high-efficient energy options which compared to wind power and solar power. It is the only stable power source that can replace fossil energy on a large scale at present and even in the future. This paper briefly describes the operational risks of coastal nuclear power plants (CNPPs) and the upgrade process of nuclear reactors, systematically analyzes and summarizes the ecological environment problems, risk problems and ecological restoration and protection strategies in the construction and operation of CNPPs, including coastal wetland ecological environment, marine ecological environment, and the risk prevention measures for cooling water source. The results show that ecosystem restoration by human does not make up for the weakening of the ecological carrying capacity and ecosystem away from its baseline, nonetheless, the restoration of important wetland habitats is conducive to the improvement of the functions of key marine ecosystems and the maintenance of marine biodiversity.

1 Introduction

The growing concern about the crisis of climate change caused by fossil fuels and the approaching deadline of carbon neutrality goals, nuclear power, hydropower, wind-power, solar-power and other renewable energy sources is developing vigorously in many countries. Since hydropower resources are restricted by conditions and wind and solar power are often at low ebb for weeks or even months at a time and current daily-cycle storage is not sufficient to fill in these gaps, nuclear power as one of the low-carbon and high-efficient energy options is needed to achieve deep decarbonization goals and address climate change [1]. It is estimated that another 250 Gwe of nuclear power is needed to optimize the energy structure and ensure energy security for countries worldwide in 2030 [2], figure 1 [3]. Because nuclear power needs a lot of cooling water to operate, many nuclear power plants are located on the coast. The coastal nuclear power plants (CNPPs) must construct ancillary water intake and drainage engineering facilities (canals, culverts or pipelines) and heavy cargo transport terminals, which will more or less changes the coastal topography, landforms, and coastal currents. However, these changes and impacts are generally minimized through the optimization of multiple schemes, but the negative factors of nuclear power operation on coastal and marine environment [4] and the risk of cooling water intake blockage in CNPPs

[5] cannot be ignored. This paper briefly introduces the operation risks of nuclear power plants and the upgrade process of reactors, analyzes the impact and potential risks of the construction and operation of CNPPs on the coastal and marine ecological environment, and discusses the ecological restoration and protection strategies of the sea area around CNPPs.

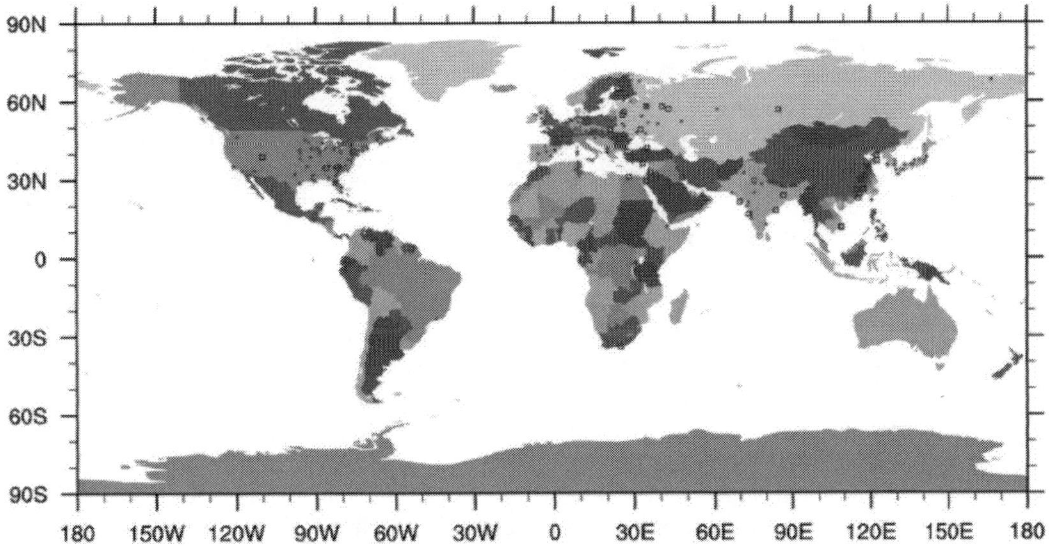

Fig. 1. Global distribution of nuclear power plants 3

2 Operational risks of nuclear power plants

The former Soviet Union built the world's first atomic power plant in 1954, ushering in the era of civilian nuclear energy. However, three horrific nuclear power accidents have caused unbearable pain and economic loss to the local population, land contamination and abandonment, and a poignant reminder that disasters of enormous consequences can occur in the nuclear industry. That is, Nuclear leaks at the Three Mile Island nuclear power plant in the United States of America on 28 March 1979, caused by the failure of water supply pumps and operational errors that led to the continuous release of reactor cooling water and the exposure of fuel rods; at the Chernobyl nuclear power plant in the former Soviet Union on 26 April 1986, caused by operation error that led to the explosion of the reactor due to a sustained increase in reactor temperatures; and at the Fukushima nuclear power plant in Japan, caused by the earthquake and tsunami of 12 March 2011, caused by the failure of the cooling water supply to the reactor to open valves and relieve the pressure, as well as an explosion of the reactor room due to the decomposition of the spent fuel in the presence of high temperatures (higher than 4300 K, i.e., 4027°C), which resulted in the generation of hydrogen gas and led to the nuclear leak. A statistical analysis result of 216 nuclear energy accidents and incidents (events) showed that, although the frequency of events per reactor has become less common, the relative frequency with which events cascade into "dragon king" extremes is large enough that, when multiplied by severity, the aggregate risk to society is still very high [6]. Concerns about nuclear waste disposal and accident risk, several countries had been considering to reduce or cancel the use of nuclear power after the accident at the Fukushima Daiichi Nuclear Power Station in Japan in 2011 [7].

Currently, the main types of nuclear power reactors in operation and under construction in the world are Pressurised Water Reactor (PWR), Boiling Water Reactor (BWR), Pressurized Heavy Water Reactor (PHWR), Gas-cooled Reactor (GR)/Water-cooled Reactor

(WR)/Motlen salt-cooled Reactor (MR), Liquid Metal-cooled Fast Breeder Reactor (LMFBR), as shown in Table 1. The most widely used is the PWR. The development process of these reactors can be divided into four generations [8]: Generation I, 1954-1965, experimental or prototype nuclear power plants, with a total of 38 units in the world; Generation II, 1966-1998, the mainstay of the world's operating nuclear power plants, which can compete economically with thermal and hydropower, and which are technologically mature commercial reactors; Generation III, The upgraded reactors after the Three Mile Island and Chernobyl nuclear power plant accidents, using control rods made of materials that are easy to absorb neutrons, such as boron, boron carbide, silver-indium-cadmium alloy, etc., to control the core fuel fission chain reaction rate or safety rods to quickly stop the fission chain reaction; Generation IV, instead of uranium fuel, plutonium fuel is used, and uranium-238 is placed in the outer area of the core fuel plutonium-239, and the neutrons released by the fission reaction of plutonium-239 are absorbed by uranium-238 and then become plutonium-239, resulting in less nuclear waste. Therefore, the Generation III and IV nuclear power would not lead to melting of the reactor core after an accident, and the spent fuel generated from nuclear power operation is disposed in a closed loop and buried deep in geology, the risk of nuclear leakage has been reduced to a very low level. Nuclear power is the only stable power source that can replace fossil energy on a large scale in the context of wind power, solar power and other renewable energy is difficult to overcome such as static and stable weather, day and night changes caused by intermittent, low power generation efficiency and energy storage conversion bottlenecks at present and even in the future [9]. The assessment report of the Inter-governmental Panel on Climate Change (IPCC) also states that nuclear energy is one of the least life-cycle carbon emitting power generation technologies. Several countries have new nuclear power projects underway, such as China, India, The United Kingdom, and Russia has signed contracts to sell reactors to other countries that are considering starting nuclear power programmes [10].

Table 1 Comparison of nuclear power reactor types

	PWR	BWR	PHWR	GR	WR/MR	FBR
Nuclear fuel	Concentrated uranium	Concentrated uranium	Natural uranium	Concentrated uranium	Concentrated uranium	Plutonium & uranium
Coolant	Water	Water	Heavy water	Helium	Water/ Motlen-salt	Liquid metal
Neutron moderator	Water	Water	Heavy water	Graphite	Graphite	/

3 Influence on marine ecological environment of thermal drainage of CNPPs

Nuclear power needs to be part of any effort to significantly reduce the use of fossil fuels, but it is indicative of the fact that circulating cooling water is a key link to control the operation risk of CNPPs, while the risk of earthquake and tsunami must be fully considered in the site selection and design stage.

The circulating cooling water required for CNPPs operation is taken from the sea (normal temperature water) and discharged into the sea (temperature rising water). Meta-analysis showed an increase of 4.38°C in water temperature near the outfall [4]. The warming zones were depend on CNPPs installed capacity and diffusion environment (terrian, tides, current, wind field and so on) and changes periodically according to tide cycle and sea son. Generally, the affected areas of warming zones, with ebb tides having greater effects, and the total area of all warming zones in summer was approximately twice that in spring. Moreover, with the increase of the operating years of the nuclear power plant, the range of the mixing zone will expand due to the increase of the background temperature of the water body [11]. Warm water discharge rushes straight ahead to form a hot water plume or rises to the surface of water body to form a thermocline, and the thermocline induced by thermal discharge mostly occurred at water depths above 7m [12]. At the same time, the warm drainage contains residual chlorine and nuclear wastewater, which is treated by the waste liquid treatment system at the nuclear power plant to achieve discharge standards to the environment. The authorized discharge limit of radioactive liquid effluents to sea area from nuclear power plant is <1000Bq/L except H-3 and C-14, and may not be discharged across the beach, encouraging offshore emissions. However, the following marine ecological effects of thermal drainage were discovered:

(1) Increased suspended sediment at the outfall and changes in the seafloor topography by scouring. The velocity of the water flow at the outfall is accelerated, and the fine-grained sediment is stirred up by the plume of warm water discharge, causing turbidity in the water near the outfall, coarsening of the sediments, and changes in the seafloor topography by scouring.

(2) Affect the physico-chemical properties of seawater at the outfall, with reduced transparency, lower dissolved oxygen, higher ammonia and nitrogen content, higher total phosphorus and total nitrogen content, and accelerated water eutrophication.

(3) Thermal drainage increases the toxicity of toxic substances in the water. Residual chlorine may react with some organic or inorganic substances in the water to produce toxic compounds; When thermal pollution and toxic substances exist at the same time, the toxic substances will increase the toxicity to aquatic organisms. Studies have shown that when the water temperature increases to 30℃, the toxicity of Cu, Zn and Cd metal ions to zooplankton and benthos increases by 2 to 4 times. Residual chlorine may react with some organic or inorganic substances in the water to produce toxic compounds; When thermal pollution and toxic substances exist at the same time, the toxic substances will increase the toxicity to aquatic organisms. Studies have shown that when the water temperature increases to 30℃, the toxicity of Cu, Zn and Cd to zooplankton and benthos increases by 2~4 times [13].

(4) The change of physical and chemical properties of seawater can result in the change of phytoplankton biomass and population structure. Under the upward effect, the zooplankton population structure may also change in a short time scale, and the fish and crustaceans in the upper trophic level are directly or indirectly affected. The research showed that high temperature stress (resulting in photosynthetic damage) was caused to phytoplankton in the sea area above 33℃, and the diversity and abundance of phytoplankton community decreased [14], but it could recover after being carried away from the high temperature area by the current. It can even promote the rapid growth of phytoplankton [15]. The main effects on organisms were related to changes in the structure and composition of aquatic communities, with species abundance, distribution, dominance, and density [4].

(5) Thermal drainage accelerates the biodegradation of organic matter in the substrate, increases oxygen consumption, and exacerbates water hypoxia, which may pose a threat to fish and benthic organisms. It also increases the ability of organisms to enrich harmful substances. The accumulation of nuclides (Cs, Co, I, Zn) in the bodies of shellfish aquatic organisms increases with increasing temperature. For example, the uptake of cesium and cobalt radioisotopes by Chlorella vulgaris increases with increasing temperature [16].

(6) Impacts on biological communities in coastal wetlands. Changes in the physico-chemical properties of seawater caused by thermal discharges will further lead to changes in wetland biomes and ecosystem vulnerability, especially for important ecosystems such as coral reefs, seagrass beds and mangrove forests in tropical and subtropical area. For example,

for coral reef communities in tropical climates, high summer temperatures coupled with warm drainage can lead to coral bleaching or death. Studies have shown that 32°C is a sub-lethal temperature for corals and 34°C is a lethal temperature for corals [17,18]; However, it is favourable for coral growth and coral reef development in high latitude subtropical area that are stressed by low winter temperatures [19].

4 Restoration and protection strategy of the marine ecosystem around CNPPs

Ecological restoration is the process of assisting the recovery of an ecosystem that has been degraded, damaged, or destroyed [20]. However, Meta-analyses of wetland restoration projects consistently reveal marked underperformance relative to baseline reference states to recover their biological structure and biogeochemical functionality [21]. Ecological restoration and protection strategies for the marine ecosystem around CNPPs should take into account the adaptation of the ecosystem to changes in the coastal topography, landforms, coastal currents, and the seawater physico-chemical environment caused by thermal discharges. Moreover, the stable supply of circulating cooling water is an important guarantee for the safe operation of nuclear power plant, so the ecological safety of cooling water source is also crucial.

4.1 Coastal ecological restoration and protection

The drainage engineering facilities of CNPPs and the construction of heavy cargo transportation terminals may change the topography and coastal currents of the coast, and break the balance of coastal sediment transport, especially for structures located on sandy coasts. For example, a nuclear power plant located on a sandy coast, the construction of a drainage canal (the heavy cargo transport dock was built at the west dyke of the drainage canal) blocked the sediment transport along the coast, causing sedimentation in the upstream and erosion in the downstream, as shown in figure 2. Fortunately, there are rocks below the inshore waters that block some of the waves and currents, so erosion is not severe. In this case, such countermeasures for coastal restoration and protection have to be adopted: the east dike of the canal is extended in an arc to prevent the upstream sediment from silting into the intaking water area, and the downstream of the west dike of the canal is replenishment with sand to prevent coastal erosion. However, it is necessary to consider whether the sediment would affect the water depth and water intaking area at the front of the wharf, and it may be better to adopt ecological bank protection.

Due to the change of physical and chemical properties of seawater caused by the warm drainage of nuclear power plant, the biological population of coastal wetland will change adaptatively, and the ecological restoration and protection strategy should conform to this change. For example, to compensate for coastal wetland biological loss caused by warm drainage, "Coral Base Conservation area" was established on both sides of coastal zone of the Daya Bay CNPPS Base in China, as shown in figure 3, and in 2019, 15 species of stony corals under national second-class protection were found in the coastal zone of 11 kilometers of sinuous coastline, including *beehive coral*, *horn beehive coral*, *porous co-star coral*, *curly black star coral*, *flat head coral*, *spinostar coral*, *sand coral*, and *spinostar coral*.

https://doi.org/10.1051/e3sconf/202453002003

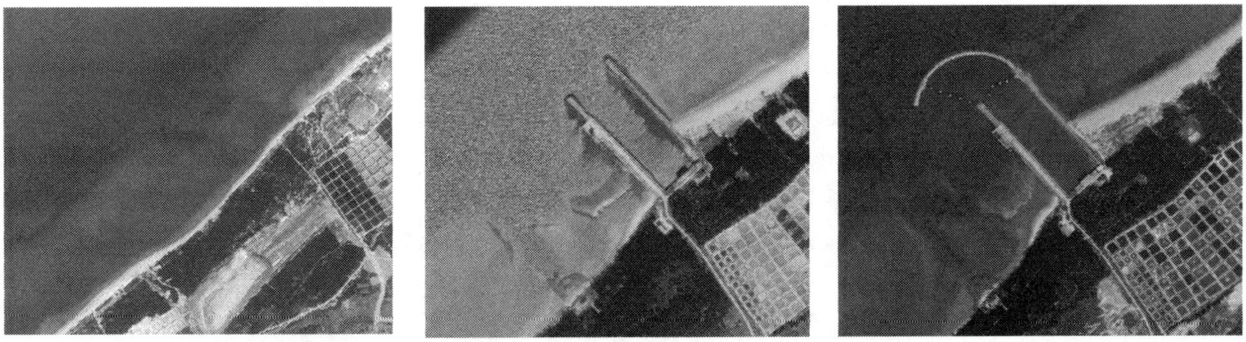

Before the construction of canal | After the construction of canal | The east canal dike is extended

Fig 2 Changes of coastal erosion and siltation before and after the construction of channels

Fig. 3 Daya Bay Nuclear Power Base No. 1 of China

4.2 Marine ecological restoration and protection

The thermal mixed area of the outlet, especially the coastal sea surface temperature in tropical and subtropical areas may be as high as 30 to 32 degrees in summer, and the superimposed temperature rise effect will cause the loss of photosynthesis of phytoplankton, the tolerance temperature of the food chain and Marine organisms, resulting in the scarcity of marine organisms and benthos species. In addition, the entrainment effect of the water intaking process of the nuclear power plant has a high fatality rate for plankton, fish eggs, and fish larvae that can enter the condenser through the filter system (transient high-temperature shock, mechanical damage, chemical hazards of residual chlorine).

According to the temperature tolerance limit of Marine organisms (Table 2), coastal nuclear power plants usually set a 4-degree temperature rise mixing zone (exclusive sea area) at the nuclear power outlet. The temperature rise within this zone can exceed the standard specified value, while the temperature rise outside the mixing zone should be less than 4 degrees. However, with the increase of the operating years of nuclear power plants, the scope of the mixing zone will expand due to the increase of the background temperature of the water body. At the same time, since the life of nuclear power plants is more than 30 years, the effect of global warming should be considered to leave a margin for the environment [11].

Table 2 The temperature limit of different organisms [22]

	Biological group	Temperature
Animals	Fish and other aquatic invertebrates	38°C
	Insect	45~50°C
	Crustaceans	49~50°C
	Protozoa	50°C
Plants	Vascular plants	45°C
	Moss	56°C
	Eukaryotic algae	56°C
	Blue-green algae	60°C
Prokaryotic microorganisms	Fungi	70~73°C
	Photosynthetic bacteria	70~73°C
	Non-Photosynthetic bacteria	99°C

For the biological loss caused by the construction and operation of nuclear power plant, the biological compensation measures are mainly adopted to supplement and restore the biological resources. Aquatic life, including fish, shellfish and algae, can also absorb carbon dioxide from the water, increasing the ocean's carbon sink capacity.
Moreover, Some nuclear power plants have also embarked on aquaculture experiment of fishery species adapted to warm drainage and feasibility studies on winter aquaculture (cage culture, which can be removed at high temperatures in summer) at present, during the experiment process can also monitor whether there will be radionuclides accumulated in organisms..

4.3 Ecological protection measures for the cooling systems in CNPP

In order to ensure the safety of water intake for nuclear power plants, the water intake should be arranged in the water depth that meets the requirements, and the water intake channel should be ensured that it is not blocked by sediment accumulation or a large number of Marine organisms. In view of the influence of sediment coming from sea area on the sedimentation of water intake, the ecological sand fixation prevention measures should be strengthened in the source area of sediment. In view of the large accumulation of Marine organisms at the intake port, the types and growth rules of organisms with large-scale outbreaks near the intake port should be carried out (Marine organisms that affect the safety of water intake include swimming animals such as fur shrimp, plankton such as jellyfish, benthic organisms such as Haiti melon, fouling and attached organisms such as oysters, barnacles, and large algae), and monitoring should be strengthened during the possible outbreak and growth season. Once it appears, fishing or pharmacochemical control should be taken.

4.4 Other strategies

Background monitoring before the operation of the nuclear power, marine environment and tracking monitoring during the operation period, paying special attention to the impact of marine organisms (including plants and animals) and drifting foreign matter gathering at the water intake on the water cooling source channel. The radioactive effluent detection system is set up during the operation of the nuclear power plant to monitor and control the radioactivity level of airborne radioactive effluent and liquid radioactive effluent of the nuclear power plant. When the radioactivity level exceeds a certain set value, the system automatically alarms and automatically closes the corresponding valve. In addition, in order to protect the public, the environment and the safety of the nuclear power plant, the nuclear power plant shall set up a planning restricted area with the reactor as the center and the radius of not less than 5000 meters, and set up signs at the boundaries of the planning restricted area to guide or limit the new construction and expansion projects in the area, so as to consider the possibility of taking appropriate protective measures under the emergency situation.

4.5 Exemplification

The above ecological restoration and protection measures have been fully implemented in the sea area of the Daya Bay Nuclear Power Plant (DNPP), as shown in Table 3. The operation of DNPP was in 1994, and the historical monitoring results of relevant environmental factors are shown in Table 4. The results show that decades of operation of DNPP and exploitation of various human activities (It's not just CNPPs but it's a collection of all human activities and it's inseparable) have pushed Daya Bay's ecosystem away from its baseline and its structure and functions are impaired; ecosystem restoration does not make up for the weakening of the ecological carrying capacity due to sea use [23], nonetheless, the restoration of important wetland habitats is conducive to the improvement of the functions of key marine ecosystems and the maintenance of marine biodiversity. The obvious and stable change is characterized by increasing the species diversity of benthos.

Table 3 Ecosystem restoration actions in Daya Bay [23]

Restoration action	Time	Primary purpose
1. Natural reserve of fishery resource	Since 1983	Improvement of biodiversity
2. Fish fry releases	Since 2002	Improvement of biodiversityand biomass
3. Artificial reefs	2002	Habitat acquisition
4. Natural reserve of mangrove swamps	Since 1999	Habitat acquisition and improvement
5. Natural reserve of sea turtle	Since 1986	Conservation of endangered species and their habitat
6. Mangrove afforestation	Since 2000	Habitat acquisition
7. Coral reefs transplatation	2003	Habitat acquisition
8. Annual moratorium on commercial fishing	July-August since 1999	Restore fish species to self-sustaining levels

Table 4 Biological ecology in Daya Bay from 1983 to 2007 [24]

Year	Phytoplankton Bacillariophyta Genera(Species)	Phytoplankton Pyrrophyta Genera(Species)	Phytoplankton Cyanophyta Genera(Species)	Zooplankton ind ▪·m^{-3}	Benthic animals g ▪·m^{-2}
1982	37(134)	9 (25)	0	316.0	123.1
1983	38 (120)	9 (32)	1 (3)		82.4
1985	38 (127)	8 (30)	1 (3)	352.7	189.3
1987	41 (137)	8 (27)	2 (4)		123.6
1990	37 (140)	17 (61)	2 (5)	727.4	55.7
1994	25 (78)	10 (30)	1 (2)	233.2	
1996					78.6
1998	24 (72)	5 (8)	0	55.4	152.8
2002	25 (96)	9 (27)	2 (4)	283.6	117.7
2005	26 (66)	8 (17)	1 (2)	4110.1	228.2
2007	31 (87)	7 (12)	2 (2)	403.4	308.8

5 Concluction

To achieve deep decarbonization goals and address climate change, nuclear power is the only stable power source that can replace fossil energy on a large scale at present and even in the future. Because nuclear power needs a lot of cooling water to operate, many nuclear power plants are located on the coast. The impact and potential risks of the construction and operation of CNPPs on the coastal and marine ecological environment include coastal topography, landforms, coastal currents, physical and chemical properties of seawater, structures of biological communities, and so on. The ecological restoration and protection strategies in the construction and operation of CNPPs include coastal wetland ecological environment, marine ecological environment, and the risk prevention measures for cooling water source. Although the ecosystem restoration by human does not make up for the weakening of the ecological carrying capacity and ecosystem away from its baseline, nonetheless, the restoration of important wetland habitats is conducive to the improvement of the functions of key marine ecosystems and the maintenance of marine biodiversity.

References

1. Cao J J, Cohen A, Hansen J, Lester R, Peterson P, Qvist S A, Xu H J, 2016. Response to nuclear power: serious risks. *Science*, 354, 6316.
2. Saito S, 2009. Role of nuclear energy to a future society of shortage of energy resources and global warming. Journal of Nuclear Materials, 398(1-3): 1-9.
3. Christoudias T, Proestos Y, Lelieveld J, 2014. Global risk from the atmospheric dispersion of radionuclides by nuclear power plant accidents in the coming decades. Atmos. Chem. Phys., 14 (9): 4607-4616.
4. Guimarães L S F, Carvalho-Junior L, Façanha G L, Resende N S, Neves L M, Cardoso S J, 2023. Meta-analysis of the thermal pollution caused by coastal nuclear power plants and its effects on marine biodiversity. Marine Pollution Bulletin, DOI:10.1016/j.marpolbul.2023.115452.

5. Lin H S, Zhang S Y, Cao R R, Yu S H, Bai W, Zhang R Y, et al., 2023. A review on the risk, prevention and control of cooling water intake blockage in coastal nuclear power plants. Nuclear Engineering and Technology, https://doi.org/10.1016/j.net.2023.10.009.
6. Wheatley S, Sovacool B K, Sornette D, 2016. Reassessing the safety of nuclear power. Energy Research & Social Science, 15. DOI:10.1016/j.erss.2015.12.026.
7. Marcus G H, 2019. Comment: Nuclear power around the world. Nature reviews physics. DOI: 10.1038/s42254-019-0020-z
8. Zhang J L, Wang Q S, 2017. Future research and development of the fourth generation reactor. Science and technology investment in China, 5: 380. (In Chinese)
9. Li L W, Bai Y S, 2023. Strategic thinking on the "three-step" development of China's nuclear energy under the goal of "double carbon". China Nuclear Power Information Network. http://www.heneng.net.cn/index.php/home/hd/infotwo/id/72559/sid/9/catId/169.html.
10. Shellenberger M, 2018. As renewables drive up energy prices, voters in U.S., Asia and Europe are opting for nuclear power. Forbes, https://www.forbes.com/sites/michaelshellenberger/2018/11/08/as-renewables-drive-up-energy-prices-voters-in-u-s-asia-europeare-opting-for-nuclear-power/ #695e2a7f7646.
11. Yu F, Zhang Y X, Cao Y, 2010. Effects of thermal drainage from nuclear power plants on Marine ecosystem under global warming. Transaction of oceanology and limnology, 2: 155-161.DOI:10.3969/j.issn.1003-6482.2010.02.024. (In Chinese)
12. Huang F M, Lin J, Zheng B X, 2019. Effects of thermal discharge from coastal nuclear power plants and thermal power plants on the thermocline characteristics in sea areas with different tidal dynamics. Water, 11, 2577. DOI: 10.3390/w11122577
13. Huang X C, Chen X C, Peng X, Zhang L, Qian X Y, Tang J, 2014. The research progress of the impact of thermal discharge from coastal power plants on marine environment. Marine environmental science, 33(6): 972-976. (In Chinese)
14. Hao Y J, Tang D L, 2010. Changes in phytoplankton community structure in response to water temperature increase in Daya Bay, China. Ecology and Environmental Sciences, 19(8):1994-1800. DOI:10.3969/j.issn.1674-5906.2010.08.006. (In Chinese)
15. Xie Y H, Li T, Jian W J, Hu S M, Tian Y, Liu S, 2015. Influence of ocean warming on the community structure and photosynthetic efficiency of phytoplankton in Daya Bay. Journal of tropical oceanography, 34(2):8.DOI:10.3969/j.issn.1009-5470.2015.02.004. (In Chinese)
16. Yu F, Zhang Y X, 2008. The Review on the effects of thermal effluent from nuclear plants on the marine ecosystem. Radiation Protection Bulletin, 28(1):7.DOI:10.3969/j.issn.1004-6356.2008.01.001. (In Chinese)
17. Fitt W K, Warner M E, 1995. Bleaching patterns of four species of Caribbean reef corals. Biol Bull, 189:298-307.
18. Strychar K B, Coates M, Sammarco P W, 2004. Loss of Symbiodinium from bleached Australian scleractinian corals (Acropora hyacinthus, Favites complanata and Porites solida). Mar Freshwater Res, 55:135-144.
19. Chen T R, Yu K F, Shi Q, Chen T G, Wang R, 2011. Effect of global warming and thermal effluents on calcification of the Porites coral in Daya Bay, northern South China Sea. Journal of tropical oceanography, 2011, 30(2): 1-9. DOI:10.3969/j.issn.1009-5470.2011.02.001. 2011, 30(2): 1-9. (In Chinese)

20. SER, 2004. SER International Primer on Ecological Restoration Society for Ecological Restoration. Tucson, www.ser.org (Accessed 26.01.16).
21. Moreno-Mateos D, Power M E, Comín F A, Yockteng R, 2012. Structural and functional loss in restored wetland ecosystems. PLoS Biol. 10, e1001247.
22. Li Z Y, 1982. Prevention and Control of thermal pollution of nuclear power plants. Environmental pollution & control, 4(4): 35-36. (In Chinese)
23. Chen X Y, Gao H W, Yao X H, Fang H D, Chen Z H, Xu Z Z, 2010. Ecosystem-Based Assessment Indices of Restoration for Daya Bay near a Nuclear Power Plant in South China. Environmental Science & Technology, 44(19):7589-7595. DOI:10.1021/es1008592.
24. Li C H, Xu S N, Du F Y, Lin L, 2015. Responses of the Daya Bay ecosystem to human activities and health assessment. *Chinese Fishery Quality and Standards*, 5 (1): 1-10. (In Chinese)

Design proposal for a ballast treatment system to reduce the invasion of species in the Galapagos.

Patrick Townsend[1*], *María Sallo*[1] *and Juan Carlos Suárez*[2]

[1]Escuela Superior Politécnica del Litoral, ESPOL, Guayaquil, Ecuador
[2]Universidad Politécnica de Madrid, UPM, Madrid, España

Abstract. Ballast water is necessary to maintain the stability of the vessels when loading and unloading their merchandise. However, it is also one of the main causes of species invasions in many ports around the world. The Galapagos Islands are no exception to this problem, since according to investigations several non-native species have been found due to ballast transport. Therefore, this project focused on the development of an alternative model for a ballast water treatment system using the chemical method to prevent the invasion of species in the Islands. As a result, two models were obtained. In the first, one NaClO dosing pump is used and in the second, two pumps are used. Each one of them was carried out from the Isla de la Plata ballast system and using RhinoCeros and AutoCAD software. In conclusion, after the modeling of both proposals and the technical analysis of the current ballast system of the vessel, it was determined that they comply with the provisions of the IMO and that they are the basis for the subsequent implementation of the alternatives.

1 Introduction

Ballast tanks are necessary to maintain the stability and operation of ships when loading and unloading the goods they transport. According to [1], ships move approximately 80% of the world's cargo, causing 3 to 5 billion tons of ballast water to be transferred each year. This has caused an increase in the transfer of invasive species in ballast tanks and therefore invasions in the ports that arrive.

Around the world, the response of the International Maritime Organization was to implement the Ballast Water Management Agreement, where different regulations are presented for the management of the ballast corresponding to the Ballast Water Exchange and Ballast Water Treatment. In Ecuador, the situation is just as worrying as in the rest of the countries of the world, considering that, in the first quarter of 2022, more than 1,200 national and international vessels were mobilized between the different ports of the country [2]. Even though the Galapagos do not directly receive merchant ships with international traffic, it should be considered that the ships that supply merchandise from the continent do so from the port of Guayaquil, which is the area with the highest international traffic [3]. According

* Corresponding author: ptownsen@espol.edu.ec

 https://doi.org/10.1051/e3sconf/202453002004

to the investigations carried out in [4] during 2015 and 2016, 53 species of non-native marine invertebrates arrived on two of the islands of the archipelago and 33 species of algae and cryptogenic species were discovered, most of them introduced by maritime transport.

The importance of designing a proposal for a system that prevents the spread of these invasions lies in the fact that these foreign species are the cause of secondary problems that threaten biological diversity, human health, and economic activity [5]. For this reason, to carry out this project, a prototype ship will be selected to which a treatment system will be adapted to be applied when unloading the ballast. These models will be designed using software such as AutoCad and Rhinoceros to demonstrate how the system works.

2 Theoretical Introduction

2.1 Methods applied for the elimination of ballast water in ships.

All currently developed systems are based on IMO approved methods corresponding to mechanical, physical, and chemical. According to [6], the different methods are mentioned in Table 1.

Table 1. IMO-approved treatment methods.

Mechanical	Physical	Chemical
Filtration	Heat treatment	Hydrogen peroxide
Hydrocyclone	Ultraviolet irradiation	Biocides
Magnetic Separation	Ultrasound	Electro-chlorination
	Cavitation	Ozone
	Deoxygenation	Electrocatalysis
	Vacuum	Chlorine dioxide
		Advanced Oxidation
		Sodium hypochlorite
		Bromine hypobromite

Some examples of these systems are the JFE Ballast Ace- JFE Engineering Corporation and the Pure Ballast 3.1-Alfa Laval.

2.2 Limitation of current treatment systems

Those mechanical methods, such as the hydrocyclone, despite being the simplest to use, are not effective in reducing bacteria, viruses, or phytoplankton, since they are very small organisms and are not affected [7].

Physical methods such as Alfa Laval's PureBallast 3.1 [8] offer optimal results in clean water, but in turbid water, the low transmittance of UV radiation requires a higher dose that reduces the ballast water flow, so aspects external reduce the efficiency of the equipment. On the other hand, chemical methods are efficient as the use of substances has been used for years for water sterilization and it is an attractive treatment technology due to the ease of application [9]. However, it could have toxic effects on humans and marine life if the correct dosage is not used.

That is why all the systems that use biocides are regulated by another entity called GESAMP. Another limitation is that most of these systems are applied at the time of ballasting the vessel, so there is still the possibility that microorganisms that were not eliminated during treatment reach the tanks and cause massive reproduction from of their eggs.

2.2.1 Ecuador's current response to invasions due to ballast water.

Ecuador's response to this problem was to implement articles to manage the ballast. In accordance with resolution No. MTOP-SPTM-2019-0029-R [10] and articles 2 and 3, all foreign and national flag vessels with international traffic and those vessels without a flag that are in national territory must Carry out ballast water exchange outside 50 nautical miles of the Ecuadorian coast and in depths of at least 200 m.

On the other hand, articles 8 and 9 refer to the BWT, where it specifies that when appropriate, a random sampling of the ship will be carried out in the ballast tanks or in the outlets of the discharge pipes to verify compliance with the regulations [10]

3 Development of the proposed solution.

3.1 Selection and description of the prototype ship.

For this work, the Motor Ship Isla de la Plata was selected, which is one of the vessels that makes trips to the Galapagos Islands to drop off cargo. One of its routes is: Guayaquil- San Cristóbal- Itabaca Canal- Isabela-Baltra-San Cristóbal- Guayaquil, covering approximately 1,340 km, equivalent to 720 nm, one way. Table 2 presents the main characteristics of the ship.

Table 2. Main characteristics of the M/V Isla de la Plata (Source: TRANSNAVE, 2023)

LOA (m)	91.30
B (m)	15.80
D (m)	8.20
T (m)	5.30
Displacement (tons)	5172.00
V (knots)	14.00
Ballast Capacity (m3)	1893.60
Year of construction	2012

(a) 3.1.1 Technical analysis of the current ballast system of the vessel.

The selected vessel must comply with the D-1 and D-2 regulations specified by the IMO. Therefore, it must not only have the treatment system as an alternative, but it must also be able to carry out ballast water exchange without it. For this, an analysis of the existing circuit of the motor ship was carried out, so that it can be verified if it complies with the regulations in the classification houses. In addition, this analysis served to establish if the current equipment of the boat can be reused.

(b) 3.1.2 Ballast pump

The sizing and the number of pumps available were verified in accordance with [11], where it is stated that at least two mechanically driven ballast pumps must be installed for the system to comply with its regulations.

(c) 3.1.3 Ballast pipes

The technical analysis of the pipes was carried out considering the rules of [11], where equation 1 was considered to obtain the minimum internal diameter of suction.

$$d = 25 + 1.68\sqrt{L(B + D)}mm \quad (1)$$

The main pipe was determined from the pump flow, while the suction pipes of each tank were established according to DNV 4-6-4/105 [12].

(d) 3.1.4 Major and minor losses

To calculate the pump pressure needed to ballast and unballast the tanks, it was necessary to determine the major and minor losses in the piping network, since these values are directly related to the power required for the fan or pump to keep it flowing [13]. Equation 2 was used for the calculation of major losses from [13].

$$hL = \frac{\Delta P_L}{\rho g} = f\frac{L}{D}\frac{V^2}{2g} \quad (2)$$

Meanwhile, for minor losses, equation 3 [13] was used.

$$hL = K_L\frac{V^2}{2g} \quad (3)$$

4 Design of the solution proposal

Two models were considered for the development of the proposal; however, the procedure is similar for both cases. In accordance with what was mentioned in 2.2, the chemical method was chosen along with the mechanical method due to ease of installation and cost. Therefore, the system will have two main phases: filtering and treatment by means of active substance.

The filtering stage will be applied at the time of ballasting the vessel. While in the second stage the water from the tanks will be sucked to pass through the injector where the biocide will be applied before unballasting, eliminating the remaining microorganisms from the tanks before being expelled into the sea.

4.1 Chosen active substance.

The selected active substance was sodium hypochlorite due to its disinfectant and treatment properties. However, it should be noted that when the substance is chlorine based, approximately 1-10 ppm must be added to kill organisms and pathogens that have passed through the separation filter [14].

4.2 Filter

At the time of ballasting the vessel, a filter capable of removing large microorganisms from seawater was selected, so that it acts as a ballast water pre-treatment. For this type of system, the ideal is a filter that eliminates species and sediments between 40 and 50 micrometers and that has a self-cleaning process.

4.3 Chemical dosing pump

Dosing pumps have the advantage that they inject a precise dose of the liquid or substance. This type of dispenser was selected due to its ease of operation and applicability in the field of water treatment, considering that it will work with chemical products derived from chlorine, oxygen, and other elements. To choose the appropriate pump, the minimum flow was calculated using equation 4 [13] and the variables were determined from the dosage of the active substance and the number of hours of ballast and unballasted equivalent to 7.5 h. For the dosage, a range of between 1-10 ppm was used.

$$Q = \frac{V}{h} \quad \left[\frac{L}{h}\right] \tag{4}$$

5 Results and analysis

5.1 Verification of the existing ballast system of the M/V Isla de la Plata

In the first part, the ballast system without treatment was considered. This was done to verify compliance with rule D-1 and subsequently the design of the treatment system proposal related to regulation D-2. From the equations of 3.2, the results that are summarized in Table 3 were obtained.

Table 3. Results of the dimensions of the current ballast system.

2 Ballast pumps	250 m^3/h
Suction piping per compartment	125 mm
Main pipe	Entre 200 mm y 210 mm
Suction pipe	100 mm

With these results, the current system data is verified. In addition, the major and minor losses for ballasting and unballasting+unloading were calculated, obtaining values of 0.16 MPa and 0.228 MPa, respectively. Considering that the current pump works with a pressure of 0.25 MPa and the total higher pressure required in the calculations is 0.227 MPa without considering the losses from the treatment plant, it is concluded that there is a margin of error available.

5.2 Proposal for a ballast system solution including treatment.

Due to the current system has equipment available that is in good working order and is compliant with regulations, the main features of the current system were maintained, and two models were developed. Both will work with the same equipment, however, the distribution of these in the plan will be different.

5.2.1 Model 1

It has the following elements: a filter for each pipe network at the sea entrance, a chemical dosing pump, a storage tank for the active substance, an injector, and a flow controller.

5.2.2 Filter

Due to the treatment by chemical method will be applied when unballasting the vessel, the filter had to have a filtration degree of 25 μm, so that fewer microorganisms penetrate the ballast tanks. This filter will have the capacity to pass a flow of at least 250 m3/h, which is the total of each pump. Each one will be located after the sea intake and after the pump of each port and starboard pipe branch.

Considering the catalog [15], one of the alternatives is a filter with a maximum flow rate of 294 $^{m^3}/_{h}$, a loss of 0.04 MPa and with diameters of 200 mm for the entrance and exit of the water.

5.2.3 Dosing pump

For the dosing pump you can choose a dosage within the range established by the IMO which is between 1-10 ppm. Fig. 1 shows a curve with the necessary flow for each of the values.

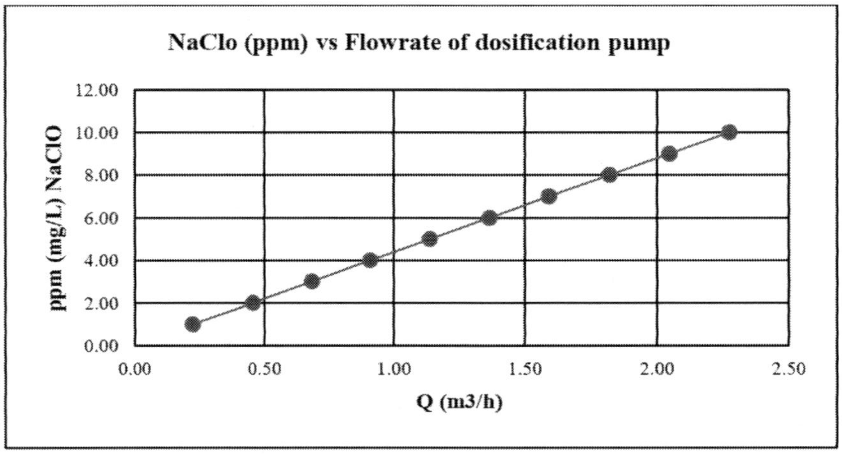

Fig. 1. Dosage of NaClO vs. Required flow rate of the pump.

5.2.4 Storage tank and dosing pump suction and discharge pipes.

The dimensions of the storage tank were based on the volumetric amount of substance that will be needed in each treatment, having an approximate value of 12 L [17]. Therefore, according to [17], two 200 L NaClO tanks could be stored. each one, with measurements of (0.80 x 0.60 x 0.85) m.

The dimensions of the injector and the suction and discharge pipes of the treatment plant will be arranged according to the diameter of the dosing pump. Because the flowrate is small, the dimensions follow the same behavior.

As mentioned above, both models will have the same equipment with the difference that their distribution will be different. Therefore, the first model is characterized by having only one dosing pump plus the added pipes. While the second model has two dosing pumps without the need to add pipes. The 3D result of both developed models is presented below in Fig. 2 and Fig. 3..

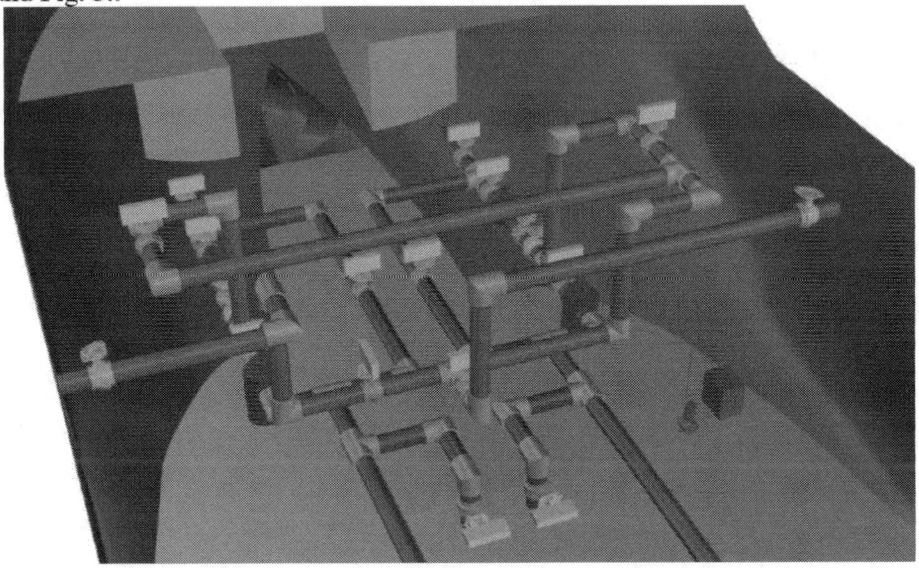

Fig. 2. 3D schematic of the first model of the ballast system.

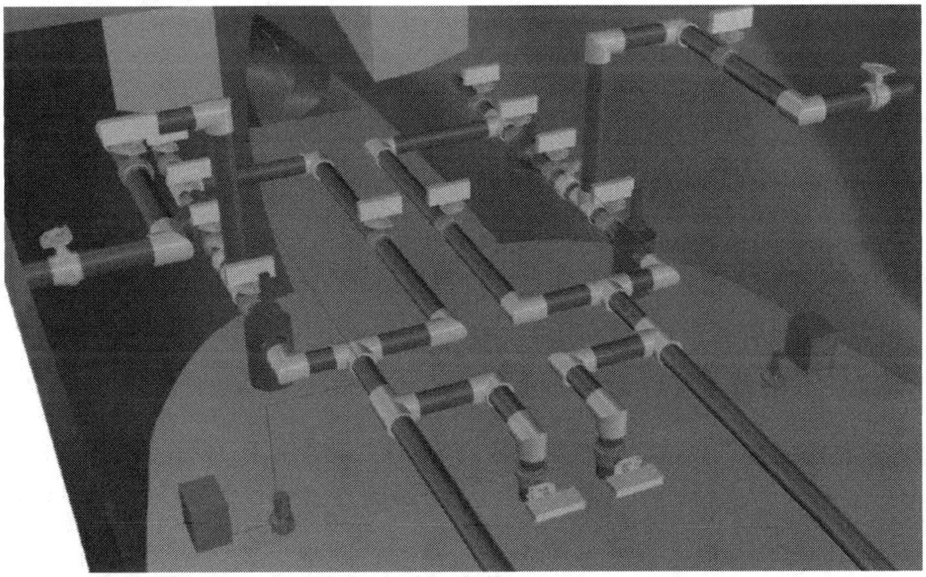

Fig. 3. 3D schematic of the second model of the ballast system.

6 Analysis of results

Based on the models developed during this project, it can be seen that the preliminary equipment that was selected to make up both treatment plants does not require a major change in the structure of the vessel or in the existing circuit, and this can be seen in the Fig. 2 and 3. In addition, the dosing pump needs 0.9 kW of electrical power and weighs approximately 7 kg, according to the reference on which the choice of equipment was based. Therefore, it can be seen that these are not significant values considering the tonnage of the vessel and its cargo capacity.

As seen in Fig. 2, for the first model, pipes had to be added to connect the port and starboard network (represented by red color), so that only one dosing pump could be used. However, it is expected that the modifications related to the pipes will not affect the ballasting and unballasting of the tanks, much less the other surrounding circuits, although it will mean an increase in the cost of purchasing equipment and labor.

In the case of the second model, installation is much easier. As can be seen in Figure 3, the dosing pumps together with all their equipment will be located in each pipe network, so it is expected that there will not be any type of problem in the ballast loading and unloading operation or the stability of the vessel. Likewise, this model does not greatly influence the pressure losses since no equipment or piping of the original ballast system is being modified, unlike model 1 that was proposed. It is important to highlight that, as can be seen in the 3D of both models, the treatment plant was located in a closed compartment in another room to avoid any type of direct contamination with people in the event of a leak or system problems.

Regarding the operation of these alternatives, there is the advantage that the dosing pump performs the main work by applying the dosage per hour required to treat the water when unballasted the vessel. In addition, it is an easy model to use because if the conditions for not using the treatment are met, the system will be able to continue working and applying the BWE, thus complying with what is regulated by the IMO. However, there will also be an alternative that can better control the discharge of organisms, which is more environmentally friendly. It should be noted that, since there are two pipe networks, it would be more advantageous to use the second model, for greater ease of adaptation and considering the economic aspect. However, this would require a deeper economic analysis of the most viable options.

Finally, with this treatment it is expected that the invasion of species in the Galapagos Islands will be reduced. Even though this vessel does not usually unballasted 100% of the time in the Islands, when it does not comply with the total load, it must ballast when leaving the Port of Guayaquil to maintain the initial stability of the vessel. This means that she will have to unballasted on one of the islands in her navigation route, since storing ballast water for so many days would generate more microorganisms. In this way, it is expected to comply with the regulations in IMO D-1 and D-2 and, in addition, to be one of the few vessels that contribute to the care of the marine environment and that help to conserve and maintain the oceans, complying as well as the Sustainable Development Goal #14 about underwater life.

7 Conclusions and recommendations

In conclusion, based on the two proposed models, the vessel would not undergo any type of drastic modification in its installed circuit, nor will it affect the operation and stability of the vessel. This is because, according to the technical analysis of the current system, the equipment is in good condition to continue being used, in addition to the fact that there is space available for the installation of the treatment plant.

The aforementioned goes hand in hand with the design of the proposal, since as can be seen in Figures 2 and 3 and in the preliminary results obtained, they are not large equipment that requires a large space or great power to work.

Finally, it is also important to highlight that this project opens the doors for future work that focuses on the implementation of sustainable systems with the marine environment. Although it is true, this research was theoretical, it is recommended to develop experiments that support the use of these alternatives to treat ballast water, so that there is a possibility that this proposal becomes a fact.

https://doi.org/10.1051/e3sconf/202453002004

References

1. A. Godey, Misra y Sha, Development of A Ballast Free Ship Design, International Journal of Innovative Research and Development, vol. Vol. 1, (2012)
2. J. Villegas, Ministerio de Transporte y Obras Públicas. [On line]. Available: https://www.obraspublicas.gob.ec/mas-de-1200-buques-ingresaron-por-los-puertos-del-pais-en-el-primer-trimestre-del-2022/. (2022)
3. Hurtado & Hurtado Asociados, Estrategia Nacional para la Gestión de Agua de Lastre en Ecuador, Guayaquil, (2016)
4. J. Carlton, I. Keith y G. Ruiz, Assessing marine bioinvasions in the Galápagos Islands: implications for conservation biology and marine protected areas, Aquatic Invasions, vol. 14, p. 20, (2019)
5. I. Keith, W. Bensted-Smith, R. Calderón , W. Iñiguez y F. Terán , Fundación Charles Darwin, [On line]. Available: https://www.darwinfoundation.org/es/investigacion/proyectos/progama-de-especies-invasoras-marinas.
6. American Bureau of Shipping (ABS), GUIDE FOR BALLAST WATER TREATMENT,. [En línea]. Available: https://safety4sea.com/wp-content/uploads/2016/09/ABS-Updated-Ballast-Water-Treatment-Guide-2016_09.pdf. [Last access: May 13th, 2023]. (2016)
7. L. Romero, Universidad Politecnica de Catalunya Barcelonatech, Available: https://upcommons.upc.edu/bitstream/handle/2099.1/18035/PFC.%20Estudio%20del%20dispositivo%20para%20el%20tratamiento%20de%20aguas%20de%20lastre..pdf?sequence=1&isAllowed=y. [Last access: May 13th, 2023]. (2013)
8. P. Vega, Tratamiento de agua de lastre mediante luz ultravioleta en aguas con altas concentraciones de sólidos en suspensión., Oviedo, (2021)
9. E. Tsolaki y E. Diamadopoulos, Technologies for ballast water treatment: a review, Wiley Interscience, (2009)
10. Ministerio de Transporte y Obras Públicas, Nro. MTOP-SPTM-2019-0029-R, Guayaquil, (2019)
11. ABS, Part 4 Vessel Systems and Machinery, de Rules for Building and classing STEEL VESSELS, p. 947. (2019)
12. DNV, Piping systems, deThe RULES FOR CLASSIFICATION OF Ships, p. 89. (2016)
13. Y. Cengel y J. Cimbala, Mecánica de Fluidos Fundamentos y Aplicaciones, 1ra ed., McGraw-HII Interamericana, (2006)
14. World chlorine council, Ballast Water Treatment with Chlorine-based Disinfectants, 2019.
15. Filtersafe, BallastSafe e-Series, (2023)
16. Grundfos, BOMBA DOSIFICADORA DMX y DMH, (2023)
17. JFE Ballast Ace, JFE Ballast Ace Ballast Water Management System, Available: https://jfe-ballast-ace.com/product/. [Last access: May 13th, 2023]. (2016)

Evaluation of the impact of the ^{137}Cs supply from rivers to coastal waters off Fukushima on the ^{137}Cs behavior in seabed sediment

T Ikenoue[1,2*], *T* Nakanishi[1], *H* Shimadera[2], *H* Kawamura[1], and *A* Kondo[2]

[1]Japan Atomic Energy Agency, 2-4 Shirakata, Tokai-mura, Naka-gun, Ibaraki 319-1195, Japan
[2]Graduate School of Engineering, Osaka University, Suita, Osaka 565-0871, Japan

Abstract. The accident at the Fukushima Daiichi Nuclear Power Plant (FDNPP) caused a radioactive contamination in seabed sediment. The ^{137}Cs supply from rivers could be an important process for the long-term behavior of ^{137}Cs in seabed sediment. In this study, a ten-year simulation of the ^{137}Cs behavior in seabed sediment was conducted using an oceanic dispersion model combined with a prediction model of ^{137}Cs behavior in land and river. In the waters north of FDNPP, the simulation results suggested that the ^{137}Cs supply from rivers had a great impact on the concentrations in coastal sediment due to the initial low concentrations in seabed sediment and the large supply of ^{137}Cs from rivers. In the waters near FDNPP and south of FDNPP, the simulation results suggested that the impact of the ^{137}Cs supply on the temporal variation of ^{137}Cs concentration in coastal sediment was relatively small due to the large initial adsorption from seawater. Overall, these results indicated that ^{137}Cs supply from rivers had an impact on the spatiotemporal distribution of ^{137}Cs concentrations in seabed sediment on a decadal time scale and the impact was especially great in the waters north of FDNPP.

Keywords Fukushima Daiichi Nuclear Power Plant accident · ^{137}Cs · Seabed sediment · ^{137}Cs supply from rivers ·ROMS

1. Introduction

The accident at the Fukushima Daiichi Nuclear Power Plant (FDNPP) in March 2011 caused a radioactive contamination in the marine environment. Since large quantities of radiocesium (^{137}Cs) were released due to the accident and ^{137}Cs has a long half-life (30.1 years), there is concern about the impacts of ^{137}Cs on marine organisms. The release of ^{137}Cs increased the concentrations of ^{137}Cs in seawater and seabed sediment [1]. Although the ^{137}Cs concentrations in seawater after the accident rapidly decreased exponentially due to preventing visible leakage and dilution effect of seawater [1,2], the ^{137}Cs concentrations in seabed sediment decreased much slower than those in seawater, at a rate of less than

* Corresponding author: ikenoue.tsubasa@jaea.go.jp

approximately 30% per year [3,4,5]. Consequently, ^{137}Cs could remain in seabed sediment for a long period of time. The concentrations of ^{137}Cs in demersal fish decreased more slowly than those in other marine organisms [6] and Tateda et al. [7] suggested that demersal fish may take up ^{137}Cs in the seabed sediment. Therefore, predicting the ^{137}Cs behavior in seabed sediment by numerical simulation is important for understanding the impact of ^{137}Cs in seabed sediment on marine organisms.

In the early phase after the accident, the adsorption of dissolved ^{137}Cs on the seabed sediment determined the initial spatial distribution of ^{137}Cs deposited on seabed sediment [8,9]. The temporal changes of ^{137}Cs deposited on seabed sediment were explained by: (1) vertical mixing of sediment, (2) resuspension and horizontal export, and (3) desorption from sediment [4]. However, the ^{137}Cs concentrations in seabed sediment have been decreasing more slowly after 2016 and Otosaka et al. showed that the controlling processes may change over time [5]. Both monitoring [10,11] and numerical studies [12,13,14] indicated that particulate ^{137}Cs has been continuously supplied from rivers to the ocean. Ikenoue et al. [14] indicated that the particulate ^{137}Cs supply from rivers for approximately thirty years corresponded to 11–36% of the total ^{137}Cs in the coastal sediment off Fukushima in the early phase after the accident. Several studies showed that the relative impact of the ^{137}Cs supply from rivers on ^{137}Cs in seabed sediment could increase [4,15]. Suzuki et al. [16] estimated the ^{137}Cs concentrations in seabed sediment off Fukushima using a simple box model, considering the supply of particulate ^{137}Cs from rivers. Their study suggested that the particulate ^{137}Cs supply from rivers could maintain ^{137}Cs concentrations in seabed sediment off Fukushima. These results suggested that the particulate ^{137}Cs supply from rivers could be an important process for the long-term behavior of ^{137}Cs in seabed sediment.

Several numerical studies on the^{137}Cs behavior in seabed sediment have been carried out since the accident [15,16,17,18]. However, none of them conducted a long-term simulation of the ^{137}Cs in seabed sediment using an oceanic dispersion model, considering the ^{137}Cs supply from rivers to the ocean. We aim to evaluate the impact of the ^{137}Cs supply from rivers to coastal waters off Fukushima on the ^{137}Cs behavior in seabed sediment for a period of 10 years. In this study, a numerical simulation of the ^{137}Cs in seabed sediment was conducted using an oceanic dispersion model combined with a prediction model of ^{137}Cs behavior in land and river.

2. Methods

2.1. Model

We applied a 3D oceanic dispersion model using ROMS developed by Kamidaira et al. [15] to coastal ocean off Fukushima. Cs-137 transport is modeled under the assumption that ^{137}Cs exist dissolved in the seawater and adsorbed on the suspended matter and the seabed sediment. The model represents transport processes, such as advection, diffusion, adsorption, desorption, deposition, and resuspension. The suspended matter and seabed sediment were divided into three categories, namely, clay, silt, and sand, to consider the difference in the transport characteristics due to particle size. The parameters for the sediment and ^{137}Cs are shown in Table 1. The desorption rate from seabed sediment (s^{-1}), defined as the product of the kinetic coefficient for desorption from suspended matter (s^{-1}) and the correction factor (-) in Kamidaira et al. [15], was set to the same order as the value obtained by monitoring [19]. The other parameters are based on Kamidaira et al. [15].

Table 1. Sediment and ^{137}Cs parameters in a 3D oceanic dispersion model.

Parameter description	Value
Thickness of the active layer in the seabed sediment	0.2 m
Initial thickness of the substrate in the seabed sediment	10 m
Porosity	0.8
Desorption rate from seabed sediment	1.16×10^{-8} s^{-1}

2.2. Computational conditions

We used the oceanic circulation model ROMS to simulate the circulation off Fukushima on a double-nested configuration. Double-nested ROMS model domains and the numerical configurations are shown in Figure 1 and Table 2, respectively. To enhance the reproducibility of oceanographic condition, the spectral nudging was considered in ROMS-L1. The ^{137}Cs input into the ocean from the atmospheric deposition and directly released from FDNPP was set up the same as in Kamidaira et al. [15]. Similarly, the initial values in suspended matter, ^{137}Cs, and seabed sediment were set up the same as in Kamidaira et al. [15]. The supplies of freshwater, suspended matter, and ^{137}Cs from 14 rivers of Abukuma and Fukushima coastal rivers (Figure 1b) were calculated using a distributed radiocesium prediction model developed by Ikenoue et al. [14]. This model was applied from July 2010 to December 2020 in the same computational conditions as Ikenoue et al. [14]. The transport of freshwater and suspended matter was analyzed from July 1, 2010 and the ^{137}Cs transport was analyzed from July 2, 2011. In rivers except for 14 rivers, the supplies of freshwater and suspended matter were calculated the same as Kamidaira et al. [15] based on the Japan River Association [20]. Four size categories of suspended sediment in a distributed radiocesium prediction model were converted to three size categories in the oceanic dispersion model based on the monitoring size distribution data [21]. This study conducted two calculations to evaluate the impact of ^{137}Cs supply from rivers to the ocean. The calculation in Case 0 considers the supplies of fresh water, suspended matter, and ^{137}Cs from rivers and that in Case 1 considers the supply of only fresh water.

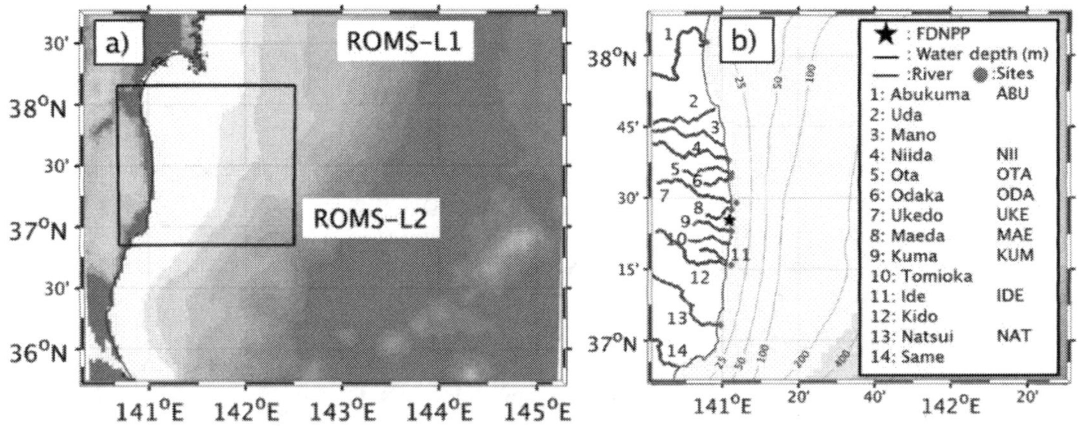

Fig.1. (a) Double-nested ROMS model domains for the parent ROMS-L1 and (b) the model domain for the child ROMS-L2 model.

Table 2. Numerical configurations for the double-nested ROMS models.

Models	ROMS-L1	ROMS-L2
Computational period	01/Jul/2010-31/Dec/2020	01/Jan/2011-31/Dec/2020
Grids (x×y×z)	144×112×32	160×144×32
Horizontal resolution	3 km	1 km
Initial and boundary condition of hydrodynamics	JCOPE2M [22,23]	ROMS-L1 results
Sea surface temperature and salinity	JCOPE2M [22,23]	
Surface wind stress	GPV-MSM [24]	
Sea surface fluxes	COADS [25]	
River input	Japan River Association [20], Ikenoue et al. [14]	Ikenoue et al. [14]
Waves	GPV-CWM [26]	
Tides	-	TPXO7 [27, 28]
Topography	SRTM30 [29] + J-EGG500 [30]	

3. Results and discussions

3.1. Model validation

We evaluated the model performance for simulating ^{137}Cs concentrations in seabed sediment. Since the initial conditions of ^{137}Cs in all phases of the ROMS-L1 and ROMS-L2 models were set to zero in this study, the simulated ^{137}Cs concentrations in seabed sediment plus the background value (1.8 Bq kg^{-1}) [1] were used as the calculated values. The calculated values in subsequent results also include the background value. The calculated ^{137}Cs concentrations were average values for three particle sizes. Figure 2 shows a scatter diagram of the calculated (Case 0) and observed [1,3,31] ^{137}Cs concentrations at the water depths of 25 m to 400 m in seabed sediment between May 2011 and January 2019. Water depths at the monitoring sites ranged from 25 m to 400 m. The model performance for ^{137}Cs concentrations is evaluated by an index, factor 10, which is the ratio of the calculated values to the observed values by a factor of 1/10 to 10 (e.g., Kamidaira et al. [15]). Most of the calculated ^{137}Cs concentrations in seabed sediment were within the range of a factor 10 relative to the observed concentrations with 83.3%. Therefore, the model reproduced ^{137}Cs concentrations at the water depths of 25 m to 400 m well. This result suggested that the total ^{137}Cs input into the ocean was generally reasonable.

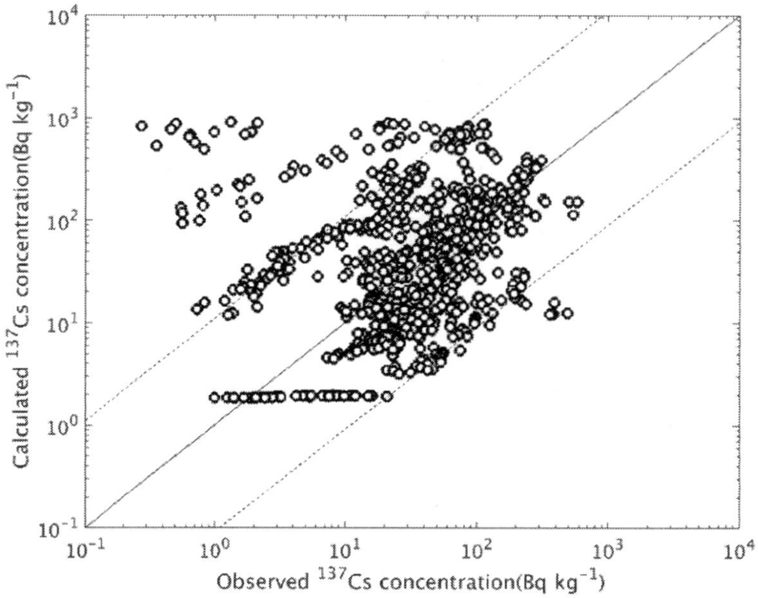

Fig.2. Scatter diagram of the calculated (Case 0) and observed ^{137}Cs concentrations in seabed sediment at the water depths of 25 m to 400 m between May 2011 and January 2019.

3.2. Evaluation of the impact of the ^{137}Cs supply from rivers to the ocean

Uchiyama et al. [18] estimated that 76.6% of the ^{137}Cs supply from the river due to flooding in 2013 was settled on coastal areas shallower than 25 m. This result suggested that focusing on ^{137}Cs in seabed sediment shallower than 25 m is important to evaluate the impact of ^{137}Cs supply from rivers. In this study, therefore, we evaluated the impact of the ^{137}Cs supply from rivers to the ocean by comparing observed and calculated ^{137}Cs concentrations in seabed sediment near estuaries (Figure 1b). Figure 3 shows temporal variations of the observed [32] and calculated ^{137}Cs concentrations in coastal sediment. Considering the ^{137}Cs supply from rivers increased concentrations at five monitoring sites (ABU, NII, OTA, ODA, and UKE) in Case 0 compared to those in Case 1 and improved the reproducibility of concentrations. Based on the distributed radiocesium prediction model, the supply of ^{137}Cs between 2011 and 2020 from Abukuma River, Fukushima coastal rivers north of FDNPP (Uda, Mano, Niida, Ota, Odaka, and Ukedo rivers), Fukushima coastal rivers near FDNPP (Maeada and Kuma rivers), and Fukushima coastal rivers south of FDNPP (Tomioka, Ide, Kido, Natsui, and Same rivers) were estimated to be 15.6, 17.5, 7.72, and 4.70 TBq, respectively. Since high ^{137}Cs deposition in Abukuma River basin and the area northwestward from the FDNPP, the ^{137}Cs supply from rivers north of FDNPP was estimated to be greatly larger than that from south of FDNPP. Both monitoring [33] and numerical studies [15,17] suggested that the initial ^{137}Cs concentrations in coastal sediment near FDNPP and south of FDNPP were greatly higher than those in north of FDNPP. In north of FDNPP, these results suggested that the ^{137}Cs supply from rivers had a great impact on the concentrations in coastal sediment due to the low concentrations in seabed sediment and the large supply of ^{137}Cs from rivers. At four monitoring sites (MAE, KUM, IDE and NAT), the model in Case 0 and Case 1 generally reproduced the temporal variation of concentrations and the ^{137}Cs supply from rivers had a relatively small impact on the temporal variation of concentrations.

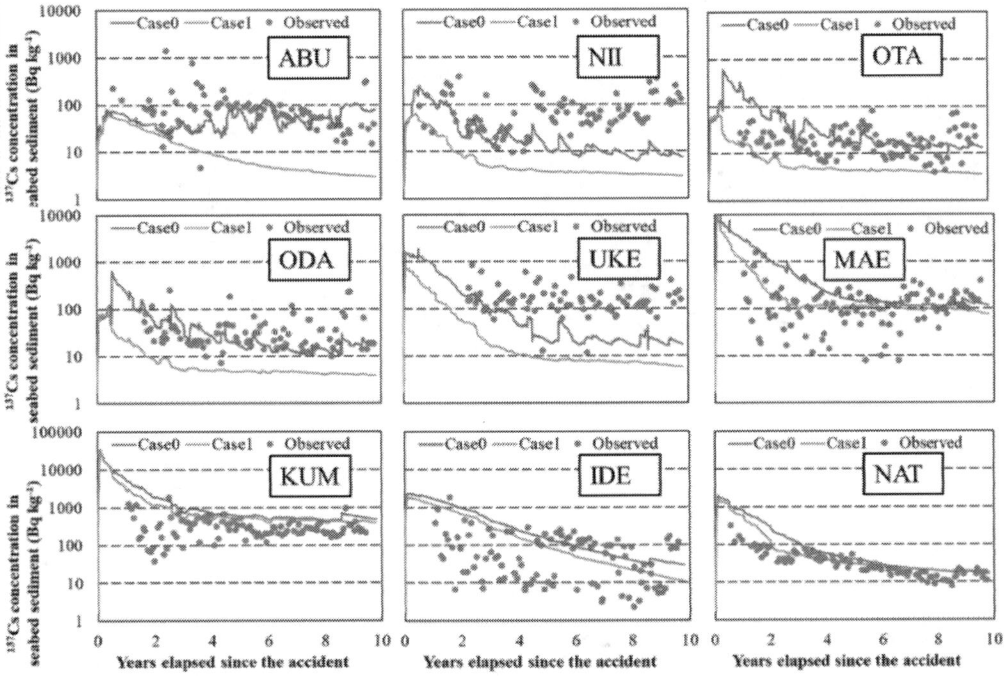

Fig. 3. Temporal variations of the observed and calculated (Case 0 and Case 1) ^{137}Cs concentrations in coastal sediment.

To evaluate the impact of ^{137}Cs supply from rivers on the spatiotemporal distribution of ^{137}Cs concentrations in seabed sediment, we calculated the ratio of the monthly average concentrations in seabed sediment in Case 0 to those in Case 1 (Figure 4). In the waters south of FDNPP with the ratio in concentrations greater than 1.0, although the spatial distributions after June showed a similar trend to the distribution in June, the high ratio gradually approached 1.0. Since the major ^{137}Cs input to the ocean from the atmosphere and FDNPP was set to end by May 30, 2011, these results in the waters south of FDNPP shallower than 25 m suggested that the impact of initial adsorption from seawater to suspended matter and sedimentation of suspended matter was greater than that of ^{137}Cs supply from the river. On the other hand, in the waters with the ratio in concentrations less than 1.0, although the spatial distributions after June showed a similar trend to the distribution in June, the low ratio gradually approached 1.0. These results suggested that ^{137}Cs sedimentation on the seafloor south of FDNPP through June 2011 reduced ^{137}Cs migration to the seafloor in other areas. Although the ^{137}Cs supply from rivers had a slight impact on the spatial distribution of the concentrations in seabed sediment in the waters near FDNPP, the impact of the ^{137}Cs supply on the temporal variation of ^{137}Cs concentration was relatively small due to the large initial adsorption (Figure 3). In the waters north of FDNPP, the ratio in concentrations was greatly high near the estuaries due to the sedimentation of particulate ^{137}Cs supplied from rivers. In these areas, the waters of ratio in concentrations greater than 1.0 gradually expanded and the expanded area was shallower than 25 m. Since the supplies of freshwater, suspended matter, and ^{137}Cs were large due to the large basin area, this trend was great in the waters near the Abukuma River estuary.

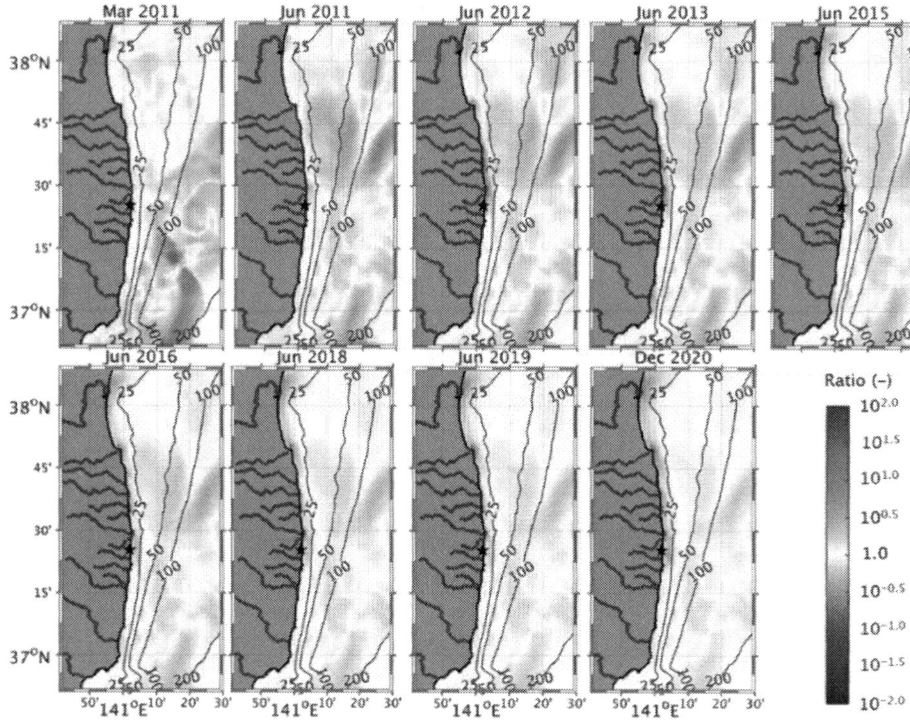

Fig. 4. Spatial distributions of the ratio of the monthly average concentrations in seabed sediment in Case 0 to those in Case 1. The blue lines, black lines, and black pentagram show rivers, water depth (m), and FDNPP, respectively.

4. Conclusions

In this study, a ten-year simulation of the ^{137}Cs behavior in seabed sediment was conducted using an oceanic dispersion model combined with a prediction model of ^{137}Cs behavior in land and river. The model reproduced the ^{137}Cs behavior in seabed sediment shallower than 400 m. In the evaluation of the impact of the ^{137}Cs supply from rivers to the ocean, considering the ^{137}Cs supply from rivers improved the reproducibility of concentrations in coastal areas shallower than 25 m. The waters north of FDNPP, and those near FDNPP and south of FDNPP showed different trends. In the waters north of FDNPP, the simulation results suggested that the ^{137}Cs supply from rivers had a great impact on the concentrations in coastal sediment due to the initial low concentrations in seabed sediment and the large supply of ^{137}Cs from rivers. In the waters near FDNPP and south of FDNPP, the simulation results suggested that the impact of the ^{137}Cs supply on the temporal variation of ^{137}Cs concentration in seabed sediment was relatively small due to the large initial adsorption from seawater. Overall, these results indicated that ^{137}Cs supply from rivers had an impact on the spatiotemporal distribution of ^{137}Cs concentrations in seabed sediment for ten years and the impact is especially great in the waters north of FDNPP.

We evaluated the impact of ^{137}Cs supply from rivers on ^{137}Cs in seabed sediment. This study provided useful information to evaluate the impacts of ^{137}Cs in seabed sediment on marine organisms. Since the particulate ^{137}Cs was estimated to be continuously supplied from rivers to the ocean for decades, quantitatively evaluating its impact on the future behavior of ^{137}Cs in seabed sediments is important. In future studies, future prediction of the ^{137}Cs behavior in seabed sediment will be conducted using the oceanic dispersion model and the predicted results of future ^{137}Cs supply from rivers to the ocean.

References

1. Kusakabe M and Takata H, 2020. Temporal trends of ^{137}Cs concentration in seawaters and bottom sediments in coastal waters around Japan: Implications for the Kd concept in the dynamic marine environment. *J. Radioanal. Nucl. Chem.*, 323, 567–580.
2. Takata H, 2022. Environmental recovery from ^{137}Cs contamination in Japanese coastal waters shown by comparison of temporal distributions with European seas. *J. Environ. Radioact.*, 251–252, 106961.
3. Kusakabe M, Inatomi N, Takata H, Ikenoue, T, 2017. Decline in radiocesium in seafloor sediments off Fukushima and nearby prefectures. *J. Oceanogr.*, 73, 529–545.
4. Otosaka S, 2017. Processes affecting long-term changes in ^{137}Cs concentration in surface sediments off Fukushima. *J. Oceanogr.*, 73, 559–570.
5. Otosaka S, Kamidaira Y, Ikenoue T, Kawamura H, 2021. Distribution, dynamics, and fate of radiocesium derived from FDNPP accident in the ocean. *J. Nucl. Sci. Technol.*, 59, 409–423.
6. Japan Fisheries Agency. Results of the monitoring on radioactivity level in fishery products. Japan: Ministry of Agriculture, Forestry and Fisheries.
   i. https://www.jfa.maff.go.jp/e/inspection/index.html
7. Tateda Y, Tsumune D, Tsubono T, Misumi K, Yamada M, Kanda J, Ishimaru T, 2016. Status of ^{137}Cs contamination in marine biota along the Pacific coast of eastern Japan derived from a dynamic biological model two years simulation following the Fukushima accident. *J Environ Radioact*., 495-501.
8. Otosaka S, Kato Y, 2014. Radiocesium derived from the Fukushima Daiichi Nuclear Power Plant accident in seabed sediments: initial deposition and inventories. *Environ. Sci. Processes Impacts*., 16, 978– 990.
9. Otosaka S, Kobayashi T, 2013. Sedimentation and remobilization of radiocesium in the coastal area of Ibaraki, 70 km south of the Fukushima Dai-ichi Nuclear Power Plant. *Environ. Monit. Assess*., 185, 5419– 5433.
10. Taniguchi K, Onda Y, Smith H.G, Blake W, Yoshimura K, Yamashiki Y, Kuramoto T, Saito K, 2019. Transport and redistribution of radiocesium in Fukushima fallout through rivers. *Environ. Sci. Technol*., 53, 12339–12347.
11. Taniguchi K., Kuramoto T, Onda Y, 2020. Monthly Fluxes of ^{134}Cs and ^{137}Cs in the Rivers Located in the 80 km Radius from TEPCO's Fukushima Daiichi Nuclear Power Plant; Center for Research in Isotopes and Environmental Dynamics, University of Tsukuba: Tsukuba, Japan.
12. Sakuma K., Nakanishi T, Yoshimura K, Kurikami H, Nanba K, Zheleznyak M, 2019. A modeling approach to estimate the ^{137}Cs discharge in rivers from immediately after the Fukushima accident until 2017. *J. Environ. Radioact*., 208, 106041.
13. Ikenoue T, Shimadera H, Nakanishi T, Kondo A, 2023. Thirty-year simulation of environmental fate of ^{137}Cs in the Abukuma River basin considering the characteristics of ^{137}Cs behavior in land uses. *Sci. Total Environ*., 876, 162846.
14. Ikenoue T, Shimadera H, Nakanishi T, Kondo A, 2023. Thirty-Year Prediction of ^{137}Cs Supply from Rivers to Coastal Waters off Fukushima Considering Human Activities. *Water*., 15, 15, 2734.

15. Kamidaira Y, Uchiyama Y, Kawamura H, Kobayashi T, Otosaka S, 2021. A modeling study on the oceanic dispersion and sedimentation of radionuclides off the coast of Fukushima. *J. Environ. Radioact.,* 238–239, 106724.
16. Suzuki S, Amano Y, Enomoto M, Matsumoto A, Morioka Y, Sakuma K, Tsuruta T, Kaeriyama H, Miura H, Tsumune D, Kamiyama K, Wada T, Takata H, 2022. Temporal variability of ^{137}Cs concentrations in coastal sediments off Fukushima. *Sci. Total Environ.*, 831, 154670.
17. Misumi K, Tsumune D, Tsubono T, Tateda Y, Aoyama M, Kobayashi T, Hirose K, 2014. Factors controlling the spatiotemporal variation of ^{137}Cs in seabed sediment off the Fukushima coast: implications from numerical simulations. *J. Environ. Radioact.,* 136, 218-228.
18. Uchiyama Y, Tokunaga N, Aduma K, Kamidaira Y, Tsumune D, Iwasaki T, Yamada M, Tateda Y, Ishimaru T, Ito Y, Watanabe Y.W, Ikehara K, Fukuda M, Onda Y, 2022. A storm-induced flood and associated nearshore dispersal of the river-derived suspended ^{137}Cs. *Sci. Total Environ.*, 816, 151573.
19. Otosaka S, Kambayashi S, Fukuda M, Tsuruta T, Misonou T, Suzuki T, Aono T, 2020. Behavior of Radiocesium in Sediments in Fukushima Coastal Waters: Verification of Desorption Potential through Pore Water. *Environ Sci Technol.* 54, 13778–13785.
20. Japan River Association, 2003. Rainfall and discharge chronology database.
21. The Ministry of Education, Culture, Sports, Science and Technology, and the Secretariat of the Nuclear Regulation Authority, 2012. Nuclide Analysis of Suspended Sediment Sampled from Rivers in Fukushima Prefecture in the Distribution Survey of Radioactive Substances.
22. Miyazawa Y, Kuwano-Yoshida A, Doi T, Nishikawa H, Narazaki T, Fukuoka T, Sato K, 2019. Temperature profiling measurements by sea turtles improe ocean state estimation in the Kuroshio-Oyashio Confluence region, *Ocean Dynamics*, 69, 267-282
23. Miyazawa Y, Varlamov S. M, Miyama T, Guo X, Hihara T, Kiyomatsu K, Kachi M, Kurihara Y, Murakami H, 2017. Assimilation of high-resolution sea surface temperature data into an operational nowcast/forecast system around Japan using a multi-scale three dimensional variational scheme, *Ocean Dynamics*, 67, 713-728.
24. Japan Meteorological Business Support Center, GPV-MSM, http://www.jmbsc.or.jp/jp/online/file/f-online10200.html (in Japanese)
25. Woodruff S.D, Slutz R.J, Jenne R.L, Steurer P.M, 1987. A comprehensive ocean-atmosphere data set. *Bull. Am. Meteorol. Soc.*, 68, 1239-1250.
26. Japan Meteorological Business Support Center, GPV-CWM, http://www.jmbsc.or.jp/jp/online/file/f-online20200.html (in Japanese)
27. Egbert G.D, Erofeeva S.Y, 2002. Efficient inverse modeling of barotropic ocean tides. *J. Atmos. Ocean. Technol.* 19 (2), 183–204
28. Egbert G.D, Bennett A.F, Foreman M.G, 1994. TOPEX/POSEIDON tides estimated using a global inverse model. *J. Geophys. Res. Oceans.*, 99245 (C12), 24821–24852.
29. Rodriguez E, Morris C.S, Belz J.E, Chapin E.C, Martin J.M, Daffer W, Henslet S, 2005. An Assessment of the SRTM Topographic Products Tech. Rep. JPL DO31639 Jet Propulsion Laboratory, Pasadena, CA, p. 143.
30. Japan Oceangraphic Data Center, J-EGG500:JODC-Expert Grid data for Geography.
31. Kusakabe M, Oikawa S, Takata H, Misonoo J, 2013 Spatiotemporal distributions of Fukushima-derived radionuclides in surface sediments in the waters off Miyagi, Fukushima, and Ibaraki prefectures, Japan. *Biogeosciences*, 10,5019-5030

32. Japan Atomic Energy Agency, 2020. Database for Radioactive Substance Monitoring Data.
33. Ambe D, Kaeriyama H, Shigenobu Y, Fujimoto K, Ono T, Sawada H, Saito H, Miki S,Setou T, Morita T, Watanabe T, 2014. Five-minute resolved spatial distribution of radiocesium in sea sediment derived from the Fukushima Dai-ichi Nuclear Power Plant. *J. Environ. Radioact.,* 138, 264-275.

Preliminary investigation of microplastic contamination in river snails (*Filopaludina martensi*) in Eastern Thailand and evaluation of human exposure

Panida Prarat*[1], *Parnuch* Hongsawat[1] and *Bopit* Chouychai[2]

[1] Faculty of Science, Energy and Environment, King Mongkut's University of Technology North Bangkok (Rayong Campus), Rayong, Thailand
[2] Faculty of Engineering and Technology, King Mongkut's University of Technology North Bangkok (Rayong Campus), Rayong, Thailand

Abstract. Microplastic contamination in the aquatic environment is a growing environmental issue with implications for food safety concerns. This study investigated microplastics in river snails (*Filopaludina martensi*) obtained from local markets in the eastern province of Thailand. The study analyzed the abundance, shape, size, color, and polymer type of microplastics extracted from river snail tissue and estimated human intake of microplastics through river snail consumption. The results revealed that the average microplastic abundance was 88.63 ± 51.18 particles/g wet weight. The predominant microplastics were fibers and fragments smaller than 1.0 mm, with colors ranging from transparent-white. Polymer analysis indicated that river snails primarily contained polyethylene (PE), polypropylene (PP), and polystyrene (PS). Microplastics' estimated annual intake (EAI) was significantly higher than the average in other countries. This suggests that humans may ingest microplastics through food consumption due to the biomagnification of microplastic accumulation in the food chain. Given this elevated level of human exposure to microplastics, there is an urgent need to develop policies to regulate the use, management, and disposal of plastic waste in Thailand.

1 Introduction

Microplastics, once primarily associated with environmental pollution, are gaining widespread attention due to their impact on food safety and their potential to harm human health [1]. Microplastics are considered a pathway for seafood-borne risks because they can transport various substances that threaten seafood safety, including heavy metals, persistent

* Corresponding author: panida.p@sciee.kmutnb.ac.th

organic pollutants, microbial infections, and naturally occurring additives, pigments, and dyes found in plastics [2]. One of the most common routes for microplastics to enter the human body is ingestion. Shellfish are a significant component of the human diet. When it comes to human exposure through the consumption of aquatic species, the exposure depends on the specific parts consumed. Research indicates that most microplastics accumulate in the digestive tract and gills [3]. Since shellfish are typically consumed whole, including the gastrointestinal tract [4], contaminated shellfish containing microplastics represent a noteworthy source of human exposure.

Over the past few years, research has been conducted on microplastics in various marine mollusks, including commercially harvested, wild, and aquaculture bivalves like mussels, oysters, and cockles [5-10]. Nonetheless, there have been limited surveys that adequately cover gastropod mollusks. Previous studies have explored the connections between microplastics in water/sediment and mollusks. For instance, microplastics were found in sea snails (*Ellobium chinense*) from the mangrove forests in the northern part of Beibu Gulf, China [11]. Another study identified microplastics in Gonggong snails (*Laevistrombus turturella*) and sediment on Bintan Island, Indonesia [12]. Akindele et al. (2019) investigated microplastic pollution in the West African river system using gastropods as indicators [13]. Only a few studies, as well as independent national surveys regarding mollusk populations, have assessed the dietary intake of microplastics through the consumption of gastropod mollusks [14].

Filopaludina martensi, an herbivorous aquatic gastropod mollusk locally known as "Hoi Khom," is one of the most commonly abundant species consumed in various Southeast Asian regions, including Thailand. It is traditionally collected as a food source due to its edibility [15]. These river snails are typically hand-picked during the rainy season. Furthermore, *F. martensi* holds significant economic value as a commercial species for many river communities.

However, the available literature on microplastic contamination in shellfish in Thailand is limited. Recent research has highlighted the significant impact of microplastic pollution on coastal areas of Thailand [16-19]. Consequently, numerous studies examining microplastic contamination in marine shellfish, including green mussels (*Perna viridis*), blood cockles (*Tegillarca granosa*), clams (*Marcia optima*), and lampshells (*Lingula anatina*), have reported such contamination [8, 20]. Remarkably, despite the crucial role of rivers in microplastic transportation, there needs to be more documentation regarding microplastic contamination in freshwater shellfish. Therefore, the objectives of this research were to evaluate the prevalence and characteristics of microplastics in river snails (*F. martensi*) available in local markets in the eastern province of Thailand and to estimate the extent of microplastic ingestion resulting from the consumption of river snails by the Thai population.

2 Materials and methods

2.1 Study area and sampling

River snail (*Filopaludina martensi*) was obtained from local markets and caught by local fishermen. This study included two local markets: Mae Daeng market (MD: 12° 41' 1.428"N, 101° 16' 29.8554" E) in Rayong province and Khao Chi Chan (KC: 12° 45' 50.958" N, 100° 57' 20.0016" E) in Sattahip district located in Chon Buri province. At each local market, three retail shops were randomly selected. Each local market acquired three kilograms of river snails (2-3 cm shell length) from three different retailers. The samples were then wrapped in aluminum foil, placed in an icebox, and transported to the laboratory, where they were refrigerated at 4°C before examination. These samples were collected in December 2022.

2.2 Microplastic extraction

The microplastic extraction procedure in this study followed an established method [21]. Initially, the soft tissues of 40 individual river snails (n = 40) from each retail shop were separated. Next, samples weighing approximately 35-40 g were placed in a clean glass beaker. Organic matter was digested by immersing the samples in 69% concentrated nitric acid overnight at a ratio of 1.0 g of soft tissue to 5 ml of nitric acid. The mixture was then boiled for 2 h when no biological residue was visible. After that, the solution was diluted with warm distilled water at a 1:10 %v/v ratio. To separate the microplastics, commercial NaCl was added to create a saturated salt solution with a density of 1.2 g/cm^3. After standing at room temperature for 48 h, the solution was vacuum filtered using glass fiber filter paper (Whatman GF/C, pore size 1.2 μm) to separate the microplastics. Finally, the filter membranes were stored in a clean Petri dish with adequate ventilation before further microplastic observation and identification. The same extraction methodology was applied to process samples for polymer analysis. A schematic illustration of microplastic extraction is shown in figure 1.

River snail **Detach and weight** **Soaked in HNO_3** **Digestion** **Density separation** **Filtration**

Fig. 1. Schematic illustration of microplastic extraction from river snail tissue.

2.3 Microplastic observation and identification

Using a Nikon Eclipse E200 stereo microscope from Japan, suspected microplastic particles on the filters were visually examined and photographed at magnifications ranging from 7.5x to 45x. Three types of particle shapes were identified: fibers, fragments, and films. To classify the color of these particles, the following categories were used: transparent-white (colorless/white/silver), black (black/grey), yellow (yellow/orange/brown), blue (blue/green), and red (red/pink/purple). In addition to the longest dimension, the following size categories were assessed: 1 mm, 1-2 mm, 2-3 mm, >3-4 mm, and >4-5 mm [22]. For microplastic measurement, a GF/C filter with a 47 mm diameter was divided into 1 mm^2 sections, and the morphology and sizes of potential microplastic particles were evaluated under a stereo microscope. Micro-Fourier transform infrared spectroscopy (μFTIR) was performed using Thermo Scientific Nicolet iS5 to identify the types of polymers in suspected microplastics. The spectra obtained were compared with the library information from the OMNIC Picta software. Based on these library comparisons, the plastic polymer can be inferred with an accuracy greater than 70% [6].

2.4 Unit and data analysis

Microplastic abundance was assessed by reporting the mean and standard deviation (SD) as the number of microplastics per soft tissue wet weight of shellfish (particles/g wet weight), making it easier to compare the abundance of microplastics with previous studies. Furthermore, the independent t-test analysis was conducted with a 95% confidence level to compare variations in the quantity of microplastics at the sampling sites. The analysis was performed using IBM SPSS Statistics software (version 25.0).

2.5 Dietary intake of microplastics via river snail consumption

The following equation was used to calculate the dietary intake of microplastics (EAI) in river snails, which may result in the transfer of microplastics into the human body:

$$EAI = C \times IR \times 365 \qquad (1)$$

where C represents the mean microplastics concentration in river snail tissues (particles/g wet weight), and the ingestion rate (IR) of river snails for consumers per capita is 0.92 g/person/day according to the National Bureau of Agricultural Commodity and Food Standards survey (Thailand).

3 Results and discussion

3.1 Microplastic abundance and characteristics

Microplastics were detected in all samples of river snails analyzed from the MD and KC markets. Microplastic abundance ranged from 29.80 to 165.50 particles/g wet weight (figure 2a). The highest number of microplastics was found in the MD market, with an average abundance of 114.93 ± 58.42 particles/g wet weight, while a lower abundance, which was two times less than that of MD, was observed in KC, with an average abundance of 62.33 ± 32.57 particles/g wet weight. The independent t-test indicated no statistically significant variations in locations (t = 1.26, p = 0.325). An average of 88.63 ± 51.18 particles/g wet weight was obtained from the sampled river snails. Microplastic abundance in river snails can be attributed to their feeding habits and the microplastic content in their environment. The river snail is a gastropod mollusk that primarily feeds on various food sources, including microalgae, filament algae, and plankton. It predominantly grazes on algae, using its radula to scrape them from rocks or the substrate's biofilm. This substrate is often dominated by fucoid algae, which can potentially trap and retain microplastics [23], making them available as part of the snails' diet. This result demonstrates how microplastics enter the river snail's system through the intermediary of the food it consumes.

River snails' tissue contains microplastics in various morphologies, including fibers, fragments, and films. Figure 2b shows that fibers (>80%) are the most common shape category, followed by fragments (2.39% to 7.10%) and film (1.80%). The diversity of microplastic shapes may have originated from the fracturing and fragmentation of large plastics due to physical processes such as photodegradation, wind, and wave action [24]. Li et al. (2016) reported that the highest occurrence of fibers in mussels' bodies, compared to other shapes, might be due to slower egestion, resulting in long-term accumulation [5]. Additionally, microfibers can be twisted or woven into food, increasing the likelihood that an organism will consume them [7]. Similar results demonstrating the dominance of microfibers have been reported in mussels [9, 25], cockles [8], oysters [7], and marine periwinkles [26, 27].

According to figure 2c, the primary size of microplastics in river snails was 1.0 mm (90%), followed by 1-2 mm (9%), with the least standard size being 2-4 mm (1%). The small-sized microplastics (< 1.0 mm) have been demonstrated in snails in many other studies [11, 28].

https://doi.org/10.1051/e3sconf/202453002006

This result indicates that river snails selectively consume microplastics within specific size ranges. A previous investigation found that mussels preferred to ingest microplastics ranging from 7-35 μm from the water [29]. Smaller microplastics (less than 1000 μm) are comparable in size to food particles such as diatoms, leading to size selectivity in shellfish for their food intake. Furthermore, smaller microplastics may absorb heavy metals and persistent organic pollutants from the environment, posing a threat to biota when ingested [30].

The colors of microplastics isolated from the river snail tissue were identified and are shown in figure 2d. Transparent-white (>80%) was the predominant color of microplastics in river snails, followed by black, blue (comparable to red), and yellow, with averages of 9.33%, 2.26%, 2.20%, and 1%, respectively. The high proportion of transparent-white microplastics reported in our study was similar to those found in China's sea snails (*Ellobium chinense*) in China [11]. This result contrasts with the color patterns of microplastics in bivalves, which exhibit a high proportion of dark colors (blue and black) [6, 7]. Since bivalves are filter feeders, color preference may not significantly influence the uptake of microplastics. In contrast to herbivorous river snails, microplastics can be indirectly absorbed by ingesting food that already contains microplastic particles.

Regarding the polymer types, three types of low-density polymers were identified in river snail tissue: polypropylene (PP), polyethylene (PE), and polystyrene (PS). The most dominant polymer type was PE (30%), followed by PP (20%) and PS (20%) (figure 3a). Given Thailand's production and consumption patterns, the PE, PP, and PS abundance seems reasonable. PP is used in clothing, blankets, fiber-based products, food packaging, and chemical containers, while PS is used in foam products, plastic cups, protective packaging, bottles, and food containers. PE is a raw material used in bottles, homeware, pipes, and toys. Consequently, these polymer types are commonly found in the aquatic environment. Additionally, this analysis found evidence of a non-polymer natural fabric, specifically linen. Under a stereo microscope, it can be difficult to distinguish these non-polymer objects from polymers, and they can be mistaken for microplastics. The FTIR spectra of representative microplastics from the river snails, along with the shapes of selected microplastics under a μFTIR microscope, are shown in figure 3b.

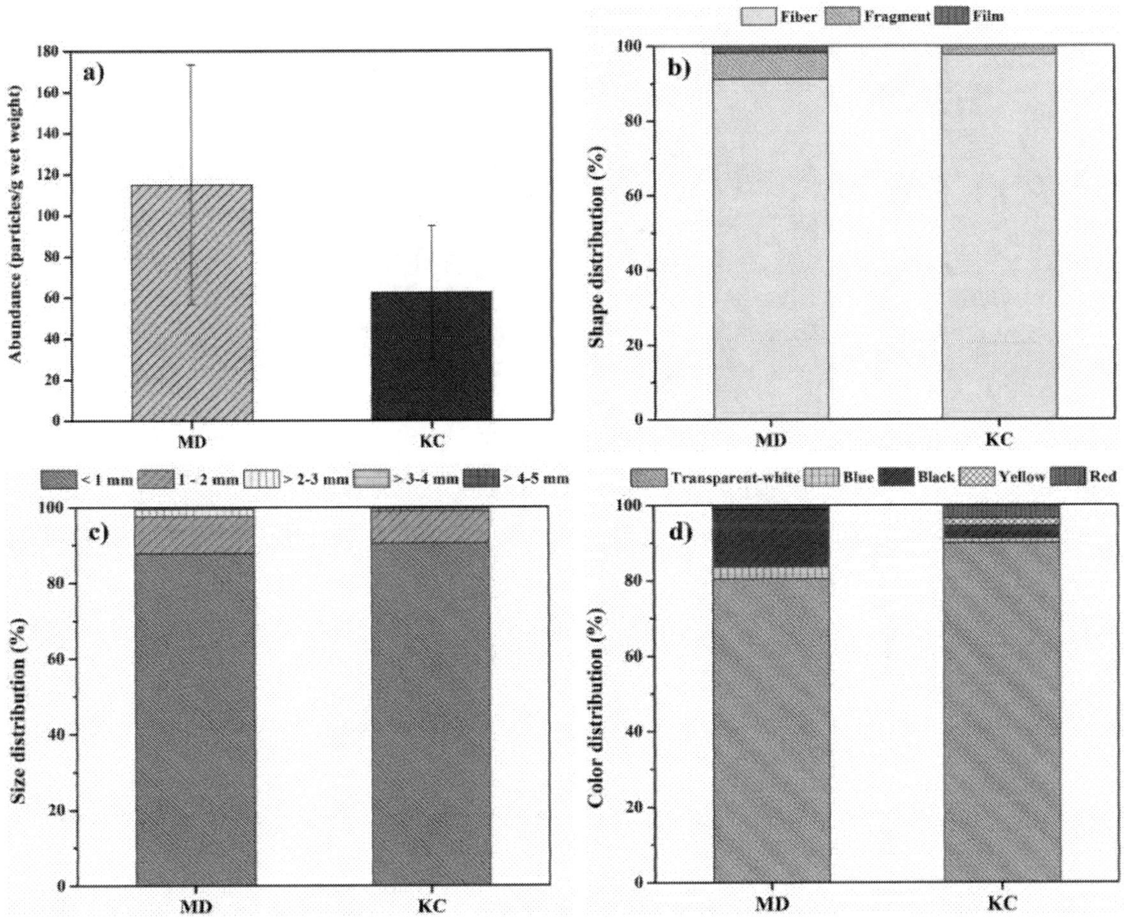

Fig.2. Abundance of microplastics in river snails (*F. martensi*) in the eastern province of Thailand (a), illustrating microplastic distribution by shape (b), size (c), and color (d).

3.2 Human microplastic exposure via river snail consumption

We assessed the Thai population's annual per capita intake (EAI) of microplastics through river snail consumption using Thai food consumption statistics and microplastic concentrations discovered in river snails. The EAI through the consumption of river snails was 29,761.95 particles/person/year. To compare the EAI values with this study, we used shellfish such as mussels, cockles, clams, and oysters, which are consumed whole. As shown in Table 1, the current result is 2.5 to 2.7 times higher than those found in Italy (11,970 particles/person/year) [4], and Germany (11,000 particles/person/year) [31]. The intake of microplastics by the population of Spain [14], Ghana [6], Canada, and South Korea [4] was found to be 8,407, 8,103, 754-4,160, and 243-1,182 particles/person/year, respectively, which is between 3.5 and 122.5 times lower than that of this study.

Our findings suggest that the annual dietary exposure for Thai river snail consumers might be significantly higher, and we advise people to avoid consuming river snails frequently. Although we observed a significant level of human exposure in this study, there are still uncertainties to address. It remains uncertain whether the quantity of ingested microplastics corresponds to the amount absorbed and whether cooking substantially impacts plastic toxicity. While the research offers crucial foundational information regarding the presence and quantity of microplastics in river snails from Thailand's freshwater sources, we acknowledge the limitation of a relatively small sample size. Therefore, we recommend further studies on a larger scale.

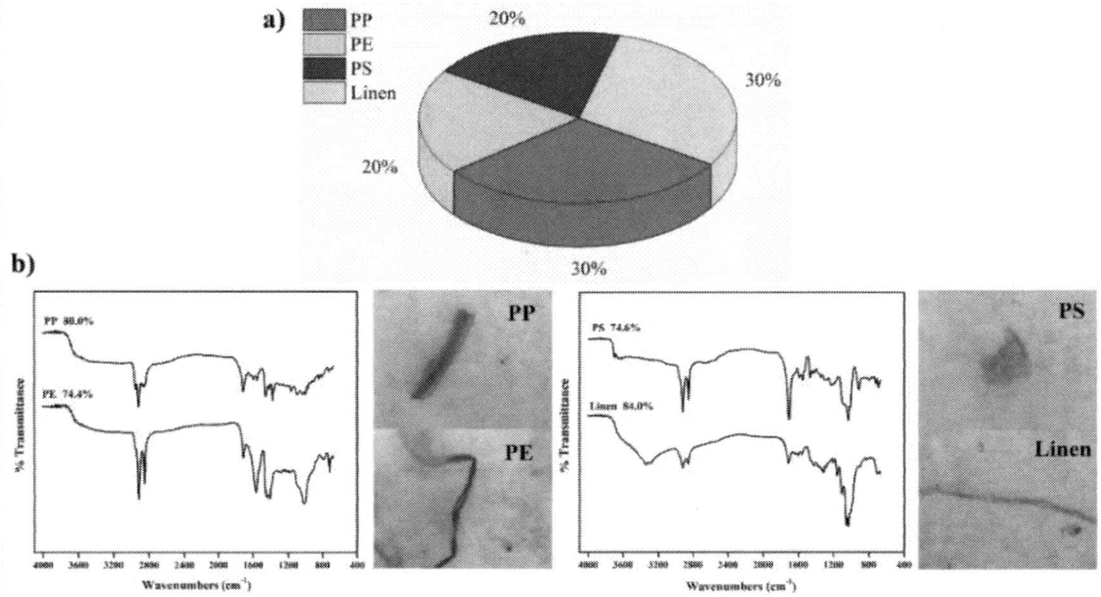

Fig.3. Polymer type distribution (a), and FTIR spectra of representative microplastics from the river snails, along with the shape of selected microplastics under a microscope of μFTIR (b).

Table 1. Estimations of the annual consumption of microplastics through the ingestion of shellfish across various countries worldwide.

Country	Species	Annual microplastic intake (EAI) (particles/person/year)	Reference
Thailand (Eastern region)	River snail (*F. martensi)*	29,761.95	This study
Italy	Mussel	11,970	[4]
Germany	Mussel (*Mytilus edulis*)	1,800 - 11,000	[31]
Canada	Mussel	8,407	[4]
Spain (North-west Mediterranean Sea)	Mollusks (clams, mussel, oysters)	8,103	[14]
Ghana (Gulf of Guinea)	Oyster (*Crassostrea tulipa*)	754 - 4,160	[6]
South Korea	Oyster (*Crassostrea gigas*) Mussel (*Mytilus edulis*) Clam (*Tapes philippinarum*) Scallop (*Patinopecten yessoensis*)	243 - 1,182	[4]

4 Conclusions

This study aimed to examine the occurrence of microplastics in river snails purchased from local markets in Thailand's eastern province and assess the likelihood of human exposure to them. The samples of river snails contained an average of 114.93 ± 58.42 particles/g wet weight. Microplastics primarily appeared in the form of fibers and were transparent-white in color. Most microplastics were smaller than 1.0 mm in size. The sample contained three different types of polymers, including PE, PP, and PS, as well as natural fibers produced from linen. The estimated annual intake (EAI) for river snails was 29,761.95 particles/person/year. This value is significantly higher than the global mean EAI for consuming shellfish. These findings suggest that individuals should consider limiting their consumption of shellfish, especially river snails.

We would like to express our gratitude to the Scientific Instrument and Analytical Testing Center (SIATC) for providing the equipment. This research was supported by the Faculty of Science, Energy, and Environment, King Mongkut's University of Technology North Bangkok (Rayong Campus), Thailand.

References

1. Gamarro E G, Ryder J, Elvevoll E O and Olsen R L 2020 *J. Aquat. Food Prod.* **29** 417-25
2. Barboza L G A, Vethaak A D, Lavorante B R B O, Lundebye A-K and Guilhermino L 2018 *Mar.* Pollut. *Bullet.* **133** 336-48
3. Carbery M, O'Connor W and Palanisami T 2018 *Environ. Int.* **115** 400-09
4. Cho Y, Shim W J, Jang M, Han G M and Hong S H 2019 *Environ. Pollut.* **245** 1107-16
5. Li J, Qu X, Su L, Zhang W, Yang D, Kolandh'asamy P, Li D and Shi H 2016 *Environ. Pollut.* **214** 177-84
6. Addo S, Boateng C M, Diyie R L, Duodu C P, Ferni A K, Williams E A, Amakye AO, Asamoah O, Danso-Abbeam H and Nyarko E 2022 *Heliyon* **8** e12255
7. Patterson J, Jeyasanta K I , Laju R L , Edward J K P 2021 *Mar. Pollut. Bullet.* **171** 112678
8. Ta A, Pupuang P, Babel S and Wang L 2022 *Chemosphere* **303** 134918
9. Ruairuen W, Chanhun K, Chainate W, Ruangpanupan N, Thipbanpot P and Khammanee N 2022 *Trends Sci.* **19** 3073
10. Do V M , Dang T T, Le X T, Nguyen D T, Phung T V, Vu D N and Pham H V 2022 *Mar. Pollut. Bullet.* **180** 113800
11. Li R, Zhang S, Zhang L, Yu K, Wang S and Wang Y 2020 *Environ. Pollut.* **263** 114368
12. Hamra A and Mufti P 2019 *AIP Conference Proceedings* **2202** 020079
13. Akindele E O, Ehlers S M and Koop J H 2019 *Limnologica* **78** 125708
14. Expósito N, Rovira J, Sierra J, Gimenez G, Domingo J L and Schuhmacher M 2022 *Mar. Pollut. Bullet.* **181** 113843
15. Apostolou K, Staikou A, Sotiraki S and Hatziioannou M 2021 *Animals* **11** 272

16. Thepwilai S, Wangritthikraikul K, Chawchai S and Bissen R 2021 *Mar. Pollut.* Bullet. **168** 112467.
17. Prarat P and Hongsawat P 2022 *Mar. Pollut. Bullet.* **174** 113200
18. Jiwarungrueangkul T, Phaksopa J, Sompongchaiyakul P and Tipmanee D 2021 *Mar. Pollut. Bullet.* **168** 112452
19. Bissen R and Chawchai S 2020 *Mar. Pollut. Bullet.* **157** 111345
20. Akkajit P, Khongsang A and Thongnonghin B 2023 *Mar. Pollut. Bullet.* **186** 114404
21. Thushari G, Senevirathna J, Yakupitiyage A and Chavanich S 2017 *Mar. Pollut.* Bullet. **124** 349-55
22. Fang C, Zheng R, Chen H, Hong F, Lin L, Lin H, Guo H, Bailey C, Segner H, Mu J and Bo J 2019 *Aquaculture* **512** 734322
23. Gutow L, Eckerlebe A, Giménez L and Saborowski R 2016 *Environ. Sci.Technol.* **50** 915-23
24. Abidli S, Lahbib Y and Trigui El Menif N 2019 *Mar. Pollut. Bullet.* **142** 243-52
25. Chinfak N, Sompongchaiyakul P, Charoenpong C, Shi H, Yeemin T and Zhang J 2021 *Sci. Total Environ.* **781** 146700
26. Doyle D, Gammell M, Frias J, Griffin G and Nash R 2019 *Mar. Pollut. Bullet.* **149** 110645
27. Mufti P, Santoso C and Hanifah N 2020 *Sains Malays.* **49** 2151-58
28. An L, Cui T, Zhang Y and Liu H 2022 *Sci. Total Environ.* **847** 157461
29. Hermabessiere L, Paul-Pont I, Cassone A.-L, Himber C, Receveur J, Jezequel R, El Rakwe M, Rinnert E, Rivière G, Lambert C, Huvet A, Dehaut A, Duflos G and Soudant P 2019 *Mar. Pollut. Bullet.* **250** 807-19
30. Brennecke D, Duarte B, Paiva F, Caçador I and Canning-Clode J 2016 *Estuar. Coast. Shelf Sci.* **178** 189-95
31. Van Cauwenberghe L, Claessens M, Vandegehuchte M B and Janssen C R 2015 Mar. *Pollut. Bullet.* **199** 10-17

https://doi.org/10.1051/e3sconf/202453002007

Ibuprofen as an Emerging Contaminant of Concern: Occurrence in Southeast Asia Water Environment

Yenni Trianda[1*], *Sandyanto* Adityosulindro[1], and *Setyo Sarwanto* Moersidik[1]

[1]Environmental Engineering Study Program, Department of Civil Engineering, Universitas Indonesia, Depok, Indonesia

Abstract. Ibuprofen is the third most salable pharmaceutical worldwide even being sold in the absence of prescription. This drug is not thoroughly metabolized and excreted into wastewater as an unchanged molecule and its metabolites. Conventional wastewater treatment plants are insufficient to remove ibuprofen, thus this substance ends up in the effluent of wastewater treatment systems and pollutes water bodies, soil, and other ecosystems. As an emerging contaminant, continuous discharge of ibuprofen has become a threat to the aquatic environment. However, studies of ibuprofen are predominantly concentrated in developed regions. The current status of ibuprofen in developing countries such as Southeast Asia region is unclear. This review aims to inform the current status of ibuprofen research in Southeast Asian countries, particularly focusing on the occurrence in the aquatic environment. This study would fill a considerable knowledge gap on ibuprofen in Southeast Asia and contribute to global knowledge development, including the need for future studies on issues unique to the region.

1 Introduction

The occurrence of micropollutants or emerging contaminants (ECs) in the surface and groundwater has developed into a global concern over the past few decades. ECs comprises of natural or synthetic compounds that infiltrate the environment as consequences of varied anthropogenic activities, manufacturing, farming, or health care facilities. Such activities further encourage the previously occuring problems of freshwater resources sustainability [1-2]. Pharmaceuticals and personal care products (PPCPs), agrochemicals, and other industrial chemicals, such as plasticizers, artificial sweeteners, food additives, etc., are among the broad range of these substances [3]. Although the concentrations of the ECs are remarkably low (ng/L to µg/L) in wastewater or water bodies, their persistence and long-term exposure lead to bio-accumulation in ecosystems and living organisms, increasing comprehension of their potential toxicity [4].

* Corresponding author: yenni.trianda@ui.ac.id

The ECs are exceptionally persistent in the aqueous environment and the removal efficiency in conventional wastewater treatment plants (WWTPs) is low, ranging from 30 – 90% depends on wastewater characteristics and WWTPs processes applied [5]. The most generally used treatment system in secondary WWTPs is fundamentally designed for the elimination of organic matter and suspended solids to satisfy the minimum effluent standard [6]. Their complex structure and toxicity make them resistant to microbial activity, hence biological treatment methods have repeatedly exhibited their inadequacy to treat ECs. As the results, various ECs and their transformation products end up in the effluent of treatment systems [7]. Continuous discharge and prolonged exposure to these substances may be hazardous to aquatic environment, even at trace levels [8]. For these reasons, collecting of information on the environmental occurrence, source, and potential toxicity of ECs allows systematic evaluation of the environmental risk related with these chemicals.

One of pharmaceuticals frequently found in wastewater and has become a global concern is ibuprofen (2-(4-isobutylphenyl) propionic acid). This substance is a type of nonsteroidal anti-inflammatory drugs (NSAIDs) that is widely used as an analgesic, antipyretic, and anti-inflammatory agent [9]. It is the world's third most favored, highly prescribed and most profitable over-the-counter medication, even being sold disregarding the prescription [10]. The global production rate of ibuprofen is roughly more than 20,000 tons per year [11]. However, since this drug can be taken over-the-counter or in combination with antibiotics, the amount ingested is overstated. This compound can enter ecosystems through various pathway, including human secretion following the medication use, improper disposal, or even wastewater from the pharmaceutical industry [12]. Continuous discharge of this chemical into the environment accounted for ibuprofen being one of the most frequently detected pharmaceutical in various water bodies.

The therapeutic dose of ibuprofen ranges from 600 to 1200 mg/day and nearly 15% of this value is excreted as unchanged molecules (conjugated with glucuronic acid and thiols) and its metabolites [13]. Meanwhile, the conjugate of this substance will be hydrolyzed in the environment [14]. Furthermore, during the chlorination process, ibuprofen in wastewater will change into intermediate molecules with more toxic properties [15]. Because of its physicochemical characteristics, this molecule is difficult for microbes to degrade in the environment, which might result in bioaccumulation and ecological damages. Therefore, it is necessary to remove ibuprofen from wastewater.

However, the study of ECs quantification including ibuprofen are mostly focused in developed regions, such as Europe and North America, and other developed country in Asia such as China and Japan [16-18]. This is greatly attributed to the enormous resources and financial investment demands required in investigation of ECs. Observing these pollutants in the environment requires the use of scientific and sensitive instrumentation in addition to relevant pretreatment methods. As such, much less is confirmed about emerging contaminant occurrences, i.e. ibuprofen, in developing region such as Southeast Asia.

This literature review aims to present the current status of ibuprofen research in Southeast Asia. The objectives of this study were to identify (i) the occurrence of ibuprofen in wastewater and water environment in Southeast Asian countries, (ii) the source and pathway of ibuprofen in water environment of Southeast Asian countries, and (iii) the potential risks of ibuprofen to the aquatic environment. This study would fill a significant knowledge gap on ibuprofen in Southeast Asia and contribute to global knowledge development, including prospective research topics needed that are definitive to Southeast Asian circumstances.

2 Method

A methodical research was conducted according to Oberg and Leopold [19] and Gaston et al. [20] by using the combination of keywords in the Scopus and Web of Science (Fig. 1).

The criteria for publications to be reviewed consist of: (1) the scope of the research is limited to the issue of emerging contaminants occurrence and quantification in Southeast Asia, particularly ibuprofen; (2) the time span or articles published in the past fifteen years, from 2008 up to November 2023; and (3) they were written in English. By limiting our focus to the past 15 years, the researchers were able to review recent findings on the occurrence of ibuprofen in Southeast Asian water environment.

The phrases "micropollutants" and "emerging contaminants," which are both the most commonly used terminology across the publications, were selected for this review during the preliminary screening. "Pharmaceuticals" was used as the broader categorical term and "ibuprofen" was used as the most comprehensive and relevant database of publications for this review. Management of the compilation of the appointed literature were conducted using EndNote online. Abstracts from more than 3000 publications were meticulously reviewed in the initial screening, and only 534 publications were chosen for the advanced review.

The 534 publications were further reviewed and sorted based on the topic of the journals. Data was compiled by transferring .ris format from EndNote into Ms. Excel spreadsheet, and VOSviewer™ software was used to visualized the inter-topic clustering. To organize and visualize the bibliometric networks between the chosen publications, VOSviewer™ software was utilized. The inclusion criteria for "Grade A" publications resulted from the visualization and used in this study were those that have quantified concentrations of emerging contaminants, particularly ibuprofen, in various environmental matrices, such as wastewater, surface water, groundwater, soil, and sediment. Pilot lab-scale studies, reviews, guidelines, monographs, and research with full text accessibility issues were among the exclusion criteria. Out of 534 publications, only 12 publications that can be categorized as "Grade A" publications. The concluding step was pulling information and synthesizing various things from the chosen literature. Data synthesis's primary goal was to assess different study findings about the prevalence of ibuprofen in Southeast Asian countries. The first section of this paper's framework reports on the prevalence of ibuprofen in various environmental matrices in Southeast Asian countries as reported by information from "Grade A" publications. This is followed by an evaluation of the source of ibuprofen and potential risks it may pose to aquatic environments.

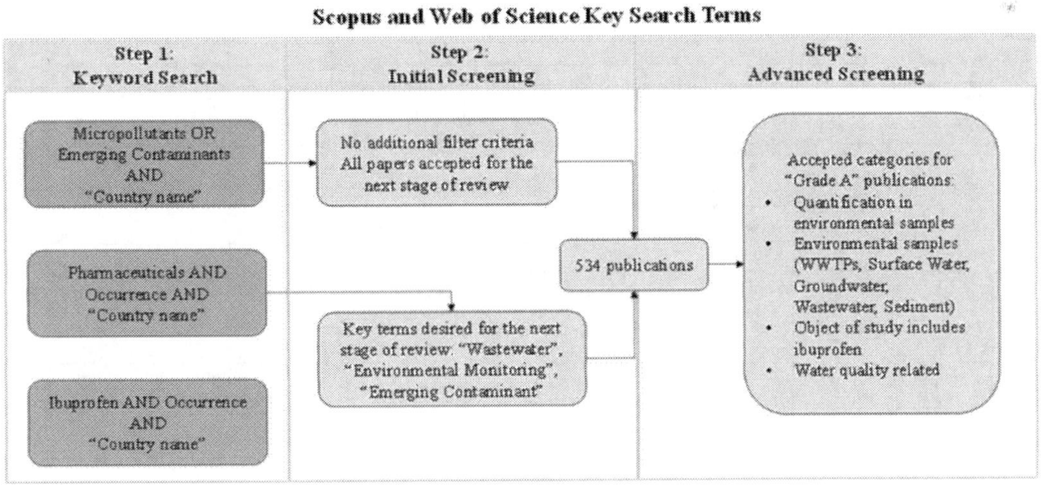

Fig. 1. Flowchart of the selection process for this literature review.

3 Results and Discussion

3.1 Occurrence of ibuprofen in Southeast Asian water environment

The distribution of accepted "Grade A" publications for each country in Southeast Asia resulted in this review is shown in Fig. 2. Vietnam and Indonesia had the most publications out of 11 countries in Southeast Asia that quantified ibuprofen in the environment. Meanwhile, no publications are reported from countries like Laos, Myanmar, Cambodia, Brunei, and East Timor.

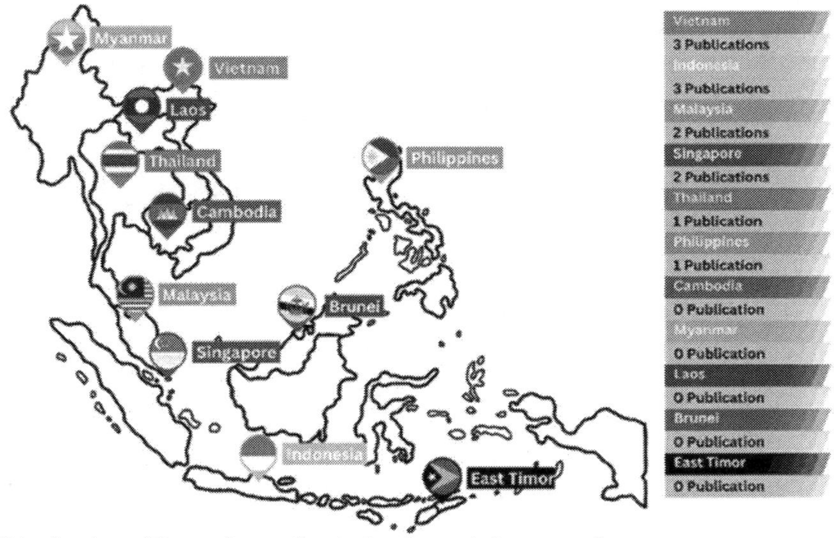

Fig. 2. Distribution of ibuprofen studies in Southeast Asian countries.

Table 1. Ibuprofen occurrence in Southeast Asia water environment.

Country	Concentration (ng/L)	Matrices	Reference
Indonesia	2,315 ± 17	Wastewater	[21]
	< 1.0 – 22	Surface seawater	[22]
	30 – 1,700	Surface seawater	[23]
Malaysia	7,400 – 15,100 (STP1) 2,400 – 8,300 (STP2)	Wastewater Wastewater	[24]
	344 ± 138	Wastewater	[25]
Singapore	< MQL – 2,445 < MQL – 111 < MQL	Wastewater Surface water Groundwater	[26]
	29,075 – 55,975	Wastewater	[27]
Philippines	max of 276 ± 15 to 838 ± 50	Surface water	[28]
Thailand	385 – 1,260	Wastewater	[29]
Vietnam	210 – 1,500 (S1) 340 – 950 (S2) 350 – 1,850 (S3) 260 – 580 (S1) 100 – 1,100 (S2) 210 – 950 (S3)	Wastewater Wastewater Wastewater Lake water Lake water Lake water	[30]
	780 – 1,120 (WWTP A) 160 – 1,700 (WWTP B)	Wastewater Wastewater	[31]
	max of 39,100	Surface water	[32]

Ibuprofen have been detected in different matrices of Southeast Asia water environment at varied concentrations. The summary of ibuprofen occurrence in Southeast Asia reviewed in this study is presented in Table 1. The highest concentration of ibuprofen has been reported by Tran and Gin [27] at the maximum of 55,975 ng/L detected in raw influent of water reclamation plant in Singapore. The ibuprofen concentrations in South East Asian water environment were comparable to ibuprofen recorded in influent found in most of developing countries, such as Kenya (30,000 ng/L) [33] and South Africa (84,600 ng/L) [34].

Meanwhile, different concentrations were reported in another regions, i.e. Europe, North America and other developed Asian countries. Kodom et al. [35] reported ibuprofen with concentration of 13,100 ng/L in the influent of WWTPs in Finland. Ibuprofen was also detected in the concentration range of 381 – 1130 ng/L in Japan's wastewater [36]. A relative significant concentration of ibuprofen (4374 ng/L) was also observed by Carmona et al. [37]. The occurrence of this compound in the influent of WWTPs is primarily influenced by consumption patterns in each country, population density, climatic conditions, water consumption, etc., hence, the significance of different exposure pathways also varies geographically.

Not only in wastewater, ibuprofen was also observed in water bodies, such as surface water and ground water, as well as soil, e.g. irrigation land and sludge in WWTPs. Ibuprofen was observed pollutes surface water in Taiwan with concentrations in the range of 5 – 280 ng/L [38]. Surface water in China is also polluted by ibuprofen with concentrations reaching 1417 ng/L [39]. Ibuprofen with maximum concentration of 385 ng/L was also reported to pollute groundwater in Europe [40]. Furthermore, Ashfaq et al. [41] reported that as much as 321 – 610 μg/kg of ibuprofen was detected in soil in Pakistan. Soil irrigation in Spain was also found to contain 213 ng/L ibuprofen [42]. Ibuprofen was also observed in sediment samples along Cao River in Vietnam with concentration of less than 10 μg/kg [43]. This compound is a moderate hydrophobic substance according to its log K_{ow} value (3.97) [44], thus ibuprofen will have moderate adsorption potential. Substances with this value of log K_{ow} have a low affinity with water and tend to adsorbed on soils or sediments. This explains the reason why ibuprofen was also observed in soils and sediment. However, even though ibuprofen was detected at very low levels in the effluent of WWTPs, the potential risks caused by the presence of ibuprofen in the environment cannot be ruled out.

3.2 Sources of ibuprofen

The presence of ibuprofen is attributed to anthropogenic activities including discharge of pharmaceutical industry, treated effluents from household and hospital WWTPs, sewer leaking, sewer overflow and surface runoff (Fig. 3). Ibuprofen is not thoroughly metabolized and roughly 15% of ingested ibuprofen gets excreted after consumed by human and animals. as unchanged molecule and its metabolites [13]. After the excretion, both ibuprofen and its metabolites discharged into WWTPs, STPs, even river, lake, oceans, soil, groundwater etc. The wastewater treatment system at WWTPs is designed to eliminate organic matters (such as BOD) and suspended solids to generate effluent within the standard quality. However, conventional wastewater treatment systems are unable to eliminate ibuprofen completely and other emerging contaminants because their complex structure and toxicity make these compounds resistant to microorganism activity. As the results, ibuprofen and the transformation products of this pollutant ends up in the effluent of wastewater treatment systems and pollute water bodies, soil and other ecosystems [45].

As the world's third most salable drug, ibuprofen become the most consumed medicine worldwide even being sold without prescription [46]. In Malaysia, 14.2% of the adults consume NSAIDs drugs and 4.2% of them are frequent user [47]. Ibuprofen consumption rate in Sleman, one of the regions in Indonesia, averaged 156.236 DDD (defined daily dose) per 1000 population [48]. It means that 157 people out of 1000 will consume this medicine. Meanwhile, average of 3.6 – 4.6 g/1000 people of ibuprofen was consumed daily in Vietnam [49]. Thus, ibuprofen contained in human excretion-based wastewater further contributing to the concentration of ibuprofen detected in wastewater.

Fig. 3. Ibuprofen sources of occurrence.

3.3 Potential risks of ibuprofen

Although no adverse human health has been reported, exposure to ibuprofen through wastewater, surface water, and groundwater may pose potential risks to some aquatic organisms. This substance will give adverse effect on aquatic organisms due to cytotoxic and genotoxic damage, high oxidative cell stress, and unfavorable effects on growth and reproduction as shown in Fig. 4.

Many researchers have assessed the potential risks of ibuprofen at different dosages on invertebrates, mainly mollusks and crustaceans, and vertebrates, including specific species of fish. Studies on daphnia (crustacean water fleas) and fish have examined the acute (up to 24 hours of exposure) and chronic (repeated exposure) toxic responses to ibuprofen. Long-term exposure will bring about an imbalance in cell oxidation followed by damage to cells, proteins and DNA [50]. Exposure to a sub-chronic environmental ibuprofen concentration (0.8 μg/L) for 14 days decreased the protein carbonyl concentration and lysosomal integrity in the digestive glands of bivalve molluscs (Unio tumidus, typically used as bioindicators of freshwater quality). Additionally, the exposure induced cholinesterase [51]. It also has been observed that ibuprofen decreases fish sperm and egg production while simultaneously increasing the number of eggs Oryzias latipes, a type of medaka fish [52]. Furthermore, ibuprofen exposure has also been linked to membrane damage in the digestive glands and elevated levels of lipid oxidation in shellfish [53]. Ibuprofen was also observed instigating substantial destructive impact on Chrinomus riparious (harlequin fly) metabolism. This substance will prevent insects from synthesizing prostaglandin. Even at trace levels, ibuprofen disrupts physiology, immunity, and reproduction because prostaglandins play critical roles in these processes [54].

Ibuprofen undergoes oxidative and photolytic breakdown into a variety of degradation metabolite chemicals. These metabolite chemicals may have effects that are more harmful than the original compound [55]. For example, ibuprofen conjugated with diacylglycerol discouraged cell division and the nondisjunction of several chromosome pairs [56]. As illustrated in Fig. 5, ibuprofen's metabolism consists of conjugation with glucuronic acid and oxidation, producing two major products: hydroxyibuprofen (OH-IBU) and carboxyibuprofen (CX-IBU). In activated sludge, OH-IBU was formed in an aerobic environment, whereas CX-IBU was produced in an anoxic environment [57]. Ibuprofen metabolites were also detected in photo-based processes, i.e. photolysis in the presence UV light, photo-H_2O_2 oxidation with UV light and photo-Fenton oxidation with visible light radiation [58]. Up to 15 different metabolite types were found during the advanced oxidation processes (AOPs)-mediated breakdown of ibuprofen. The toxic intermediate in the form of 4-isobutylacetophenone was only found in the photolysis process via decarboxylation or hydroxylation reactions. Ferrando-Climent et al. [59] investigated the presence of ibuprofen and its main metabolites in a batch activated sludge system. While CX-IBU disappeared more quickly than hydroxylated metabolites, 1-hydroxylated ibuprofen (1-OH-IBU) and 2-hydroxylated ibuprofen (2-OH-IBU) were shown to remain in the system after degradation. These metabolites were also detected in water bodies. The effect of ibuprofen and its metabolites on wastewater treatment was investigated by Jia et al. [60]. The results showed that in an aerobic sequencing batch reactor (SBR) system ibuprofen underwent biotransformation into 1-OH-IBU and 2-OH-IBU as well as CX-IBU. Increasing ibuprofen concentration in the influent was followed by decreasing of COD removal suggesting inhibitory effect of ibuprofen on aerobic microorganism. However, no significant effect was found on nitrogen removal.

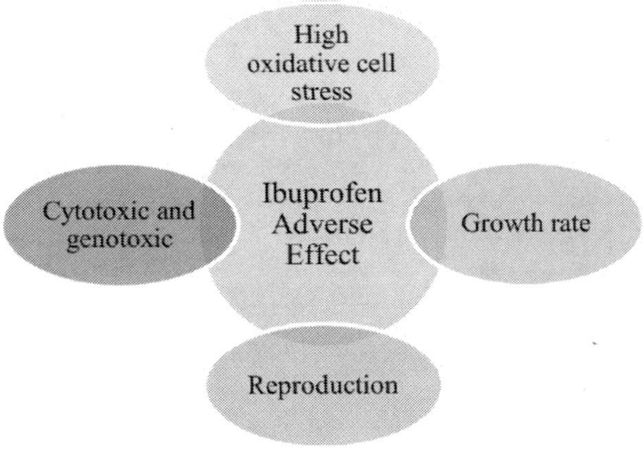

Fig. 4. Ibuprofen adverse effect on ecosystem.

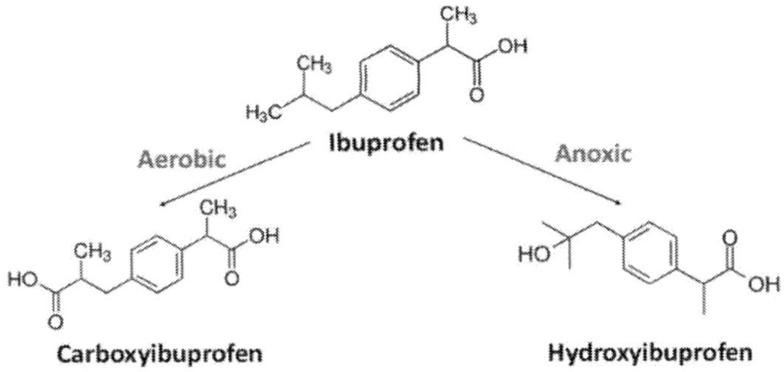

Fig. 5. Major metabolic pathway of ibuprofen.

Even while WWTPs are able to remove over 80% of ibuprofen from raw influent, a considerable quantity still remains in the effluent, which may have a negative impact on the receiving environment [61]. Repeated use of sewage sludge and reclaimed water for irrigation may result in ibuprofen accumulation in agricultural soil. Abiotic stress due to contaminated environment is being held responsible for morphological and physiological changes on plants [62]. Wijaya et al. [63] investigated the effects of ibuprofen on vegetative stage of Vigna unguiculata (cowpea). Significant abatement of chlorophyll (chl) a, chl b, total chlorophyll, and total carotenoid contents in cowpea leaves were inevitable following ibuprofen uptake by the plant. These pigments are compulsory for photosynthesis process.

This result is in accordance with those investigated by Iori et al. [64] who investigated ibuprofen effects on Salix alba (willow). Utilization of treated wastewater effluent from WWTPs that contain ibuprofen was also investigated by Picó et al. [65]. It was reported that 46 metabolites of ibuprofen, including 1-OH and 2-OH-IBU, were confirmed in Vigna unguiculata. High solubility and intermediate hydrophobicity properties are proclaimed to encourage its mobility in wastewater and potential uptake by plants.

4 Conclusion

The present study provides literature overview on the current state of knowledge on the exposure of pharmaceuticals in aquatic environment. Through this review it was found that the research about ibuprofen quantification, one of the most salable pharmaceuticals, in Southeast Asia is hardly any. Among 534 publications about emerging contaminants in this region, only 12 publications that focusing on this topic while it's still barely any in countries such Laos, Myanmar, Cambodia, Brunei, and East Timor. However, available publications displayed the presence of ibuprofen in wastewater and water bodies in Southeast Asia regions. Considering the potential adverse effects due to prolonged exposure to ibuprofen demonstrated in publications from other regions, the risks caused by the presence of this contaminants in the aquatic environment cannot be ruled out. Therefore, in order to guarantee the safety and security of water resources in the future, it will eventually be necessary to develop local ability on identifying emerging contaminant pollution and then completely remove this source of contamination.

Acknowledgement

Yenni Trianda wishes to thank Indonesia Endowment Fund for Education/Lembaga Pengelola Dana Pendidikan (LPDP) from the Ministry of Finance Republic Indonesia for granting the scholarship and supporting this publication.

References

1. J.P.S Sidhu et al., Sci. Total Environ. **463-464**, pp. 488-496 (2013)
2. A. Laffite et al., Front. Microbiol. **7**, p. 1128 (2016)
3. D. Saidulu, B. Gupta, A.K. Gupta and P.S. Ghosal, J. Environ. Chem. Eng. **9(4)**, p. 105282 (2021)
4. N.Z. Arman et al., Water **13(22)**, p. 3258 (2021)
5. O.F.S. Khasawneh and P. Palaniandy, Process Saf. Environ. Prot. **150**, pp. 532-556 (2021)
6. M. Samer, *Biological and Chemical Wastewater Treatment Processes*, in *Wastewater Treatment Engoneering*, Chapter 1 (2015)
7. B.S. Rathi, P.S. Kumar and P.L. Show, J. Hazard. Mater. **409**, p. 124413 (2021)
8. F. Riva, S. Castiglioni, E. Fattore, A. Manenti, E. Davoli and E. Zuccato, Int. J. Hyg. Environ. Health **221(3)**, pp. 451-457 (2018)
9. E. Aloi, B. Rizzuti, R. Guzzi and R. Bartucci, Arch. Biochem. Biophys. **654**, pp. 77-84 (2018)
10. A. Marchlewicz, U. Guzik and D. Wojcieszyńska, Water Air Soil Pollut. **226**, p. 355 (2015)

11. R. Qu, IOP Conf. Ser.: Earth Environ. Sci. **242**, p. 052001 (2019)
12. M. Tanveer, G.T. Guyer and G. Abbas, Water Environ. Res. **91(9)**, pp. 822-829 (2019)
13. S. Chopra and D. Kumar, Heliyon **6(6)**, p. e04087 (2020)
14. R.W. Murdoch and A.G. Hay, Biodegradation **26(2)**, pp. 105-113 (2015)
15. J. Jan-Roblero and J.A. Cruz-Maya, Molecules **28(5)**, p. 2097 (2023)
16. T. Austin, F. Bregoli, D. Höhne, A.J. Hendriks and A.M.J. Ragas, Environ. Res. **209**, p. 112777 (2022)
17. H. Schwartz et al., Can. J. Public Health **112**, pp. 133-153 (2021)
18. S. Liu, X. Zhou, Z. Wang, D. Xie and Z. Li, Mar. Dev. **1**, p. 5 (2023)
19. G. Oberg and A. Leopold, Environ. Int. **131**, p. 104960 (2019)
20. L. Gaston, D.J. Lapworth, M. Stuart and J. Arnscheidt, Environ. Sci. Technol. **53(11)**, pp. 6107-6122 (2019)
21. M.P. Astusi, S. Notodarmojo, C.R. Priadi and L.P. Padhye, Environ. Sci. Pollut. Res. **30**, pp. 21512-21532 (2023)
22. A. Sudaryanto et al., IOP Conf. Ser.: Earth Environ. Sci. **1137**, p. 012050 (2023)
23. L. Dsikowitzky, M. Sträter, Dwiyitno, F. Ariyani, H.E. Irianto and J. Schwarzbauer, Pollut. Bull. **110(2)**, pp. 654-664 (2016)
24. Z.M. Hanafiah et al., Chemosphere **287(2)**, p. 132134 (2022)
25. S.S. Rashid and Y.Q. Liu, Sci. Total Environ. **756**, p. 143849 (2021)
26. N.H. Tran, J. Hu and S.L. Ong, Talanta **113**, pp. 82-92 (2013)
27. N.H. Tran and K.Y.H. Gin, Sci. Total Environ. **599-600**, pp. 1503-1516 (2017)
28. J.D. Villanueva, I. Allan, M. Reid, N. Peyraube and P.L. Coustumer, Int. J. Environ. Sci. **2**, pp. 248-257 (2017)
29. S. Tewari, R. Jindal, Y.L. Kho, S. Eo and K. Choi, Chemosphere **91**, pp. 697-704 (2013)
30. N.H. Tran, T. Urase and T.T. Ta, Clean-Soil Air Water **42(3)**, pp. 267-275 (2013)
31. H.T. Nguyen, P.T. Khai, S.L. Kaserzon, J.W. O'Briend, G. Eaglesham and J.F. Mueller, Sci. Total Environ. **631-632**, pp. 469-475 (2018)
32. M.D. Le, H.A. Duong, M.H. Nguyen, J. Sáiz, H.V. Pham and T.D. Mai, Talanta **160**, pp. 512-520 (2016)
33. K.O. K'oreje, K. Demeestere, P.D. Wispelaere, L. Vergeynst, J. Dewulf and H.V. Langenhove, Sci. Total Environ. **437**, pp. 153-164 (2012)
34. S. Motongo, G. Birungi, B. Moodley and P. Ndungu, Chemosphere **134**, pp. 133-140 (2015)
35. K. Kodom, F. Attiogbe and F.A. Kuranchie, Heliyon **7(11)**, p. e08385 (2021)
36. N. Nakada, T. Tanishima, H. Shinohara, K. Kiri and H. Takada, Water Res. **40(17)**, pp. 3297-3303 (2006)
37. E. Carmona, V. Andreu and Y. Picó, Sci Total Environ. **484**, pp. 53-63 (2014)
38. A.Y.C. Lin, Y.T. Tsai, T.H. Yu, X.H. Wang and C.F. Lin, Desalin. Water Treat. **32(1-3)**, pp. 57-64 (2011)
39. X. Peng, Y. Yu, C. Tang, J. Tan, Q. Huang and Z. Wang, Sci. Total Environ. **397(1-3)**, pp. 158-166 (2008)
40. R. Loos, G. Hanke, G. Umlauf and S.J. Eisenreich, Chemosphere **66(4)**, pp. 690-699 (2007)

41. M. Ashfaq et al., J. Hazard. Mater. **357**, pp. 401-407 (2018)
42. P. Vazquez-Roig, V. Andreu, C. Blasco and Y. Picó, Sci. Total Environ. **440**, pp. 24-32 (2012)
43. T.H. Ngo, D.A. Van, H.L. Tran, N. Nakada, H. Tanaka and T.H. Huynh, Environ. Sci. Pollut. Res. **28**, pp. 12082-12091 (2021)
44. G.C. Nallani, P.M. Paulos, L.A. Constantine, B.J. Venables and D.B. Huggett, Chemosphere **84(10)**, pp. 1371-1377 (2011)
45. V. González-Naranjo, K. Bolters and M. Biel M, Plant Soil Environ. **59(2)**, pp. 68-73 (2013)
46. A. Marchlewicz, D. Domaradzka, U. Guzik and D. Wojcieszyńska, Water air Soil Pollut. **227**, p. 197 (2016)
47. S.M. Awaluddin et al., Malaysian J. Public Health Med. **17(3)**, pp. 58-65 (2017)
48. A.A. Fijasandra, *Evaluation of Analgesic Drug Use in Sleman District Health Center in 2015-2019 using ATC/DDD and DU90% Method*, Undergraduate Thesis Department of Pharmacy Universitas Islam Indonesia, Yogyakarta (2020)
49. H.T. Nguyen, P.K. Thai, S.L. Kaserzon, J.W. O'Brien, G. Eaglesham and J.F. Mueller, Sci. Total Environ. **631-632**, pp. 469-475 (2018)
50. M. Parolini, A. Binelli and A. Provini, Ecotoxicol. Environ. Saf. **76(6)**, pp. 1586-1594 (2011)
51. V. Martyniuk et al., Ecotoxicol. **31(9)**, pp. 1369-1381 (2022)
52. J.L. Flippin, D. Huggett and C.M. Foran, Aquat. Toxicol. **81(1)**, pp. 73-78 (2007)
53. M. Gonzalez-Rey and M.J. Bebianno, Environ. Toxicol. Pharmacol. **33(2)**, pp. 361-371 (2012)
54. A.B. Muñiz-González, Environ. Toxicol. Pharmacol. **81**, p. 103537 (2021)
55. K. Fenner et al., ACS ES T Water **1(7)**, pp. 1541-1554 (2021)
56. M.A. Kayani, J.M. Parry, S. Vickery and P.F. Dodds, Environ. Mol. Mutagen **50(4)**, pp. 277-284 (2009)
57. S.C. Tan, S.H. Jackson, C.G. Swift and A.J. Hutt, J. Chromatogr. B Biomed. Sci. Appl. **701(1)**, pp. 53-63 (1997)
58. S. Adityosulindro et al., Ultrason. Sonochem. **39**, pp. 889-896 (2017)
59. L. Ferrando-Climent et al., Sci. Total Environ. **438**, pp. 404-413 (2012)
60. Y. Jia, S.K. Khanal, L. Yin, L. Sun and H. Lu, Environ. Int. **146**, p. 106265 (2021)
61. T.D. Březinova, J. Vyzamal, M. Koželuh and L. Kule, Ecol. Eng. **120**, pp. 1-5 (2018)
62. N.J. Atkinson and P.E. Urwin, J. Exp. Bot. **63**, pp. 3523-3543 (2012)
63. L. Wijaya et al., Plants **9(11)**, p. 1473 (2020)
64. V. Iori, M. Zacchini and F. Pietrini, J. Hazard. Mater. **262**, pp. 796-804 (2013)
65. Y. Picó, R. Alvarez-Ruiz, L. Wijaya, A. Alfarhan, M. Alyemeni and D. Barceló, Anal. Bioanal. Chem. **410(3)**, pp. 1163-1176. (2018)

Toward Zero Agro-Waste: A Business Model for Sugarcane Leaves Management in Thailand

Kanjanapon Borisoot[1,2], *Pichai* Niltarach[1], *Kaittisak* Pajampa[1,2], *Kittipong* Laloon[3], *Amnart* Suksri[1,4] and *Tanakorn* Wongwuttanasatian [1,5,*]

[1] Centre for Alternative Energy Research and Development, Khon Kaen University, Khon Kaen 40002, Thailand
[2] Energy Engineering Program, Faculty of Engineering, Khon Kaen University, Thailand
[3] Department of Agricultural Engineering, Faculty of Engineering, Khon Kaen University, Khon Kaen 40002, Thailand
[4] Department of Electrical Engineering, Faculty of Engineering, Khon Kaen University, Khon Kaen 40002, Thailand
[5] Department of Mechanical Engineering, Faculty of Engineering, Khon Kaen University, Khon Kaen 40002, Thailand

Abstract. Sugarcane is one of the oldest crops ever cultivated by humans and has a long history. Currently, the prevalent practice of burning sugarcane fields increases environmental concerns because many farmers frequently use this method to facilitate the harvest process and prepare the area for the next planting crop. Thus, the effective use of sugarcane leaves with proper management can overcome this problem. This paper represents sugarcane leaf management by developing a business model focused on collection, densification, and transportation, explicitly targeting the conversion of waste sugarcane leaves into fuel for a biomass power plant. The business model is categorized into two primary groups: the farmer group and the middleman group. Both groups are incentivized to invest in tools and equipment for collecting and densifying biomass, thereby facilitating the effective performance of the model. Furthermore, an economic analysis has been conducted encompassing an investment cost evaluation and the estimation of the payback period for this business model. The results indicate that the farmer group demonstrates a payback period of 1.72 years, while the middleman group reveals a longer payback period of 2.06 years. This strategic approach not only enhances the value of biomass waste but also significantly reduces the unwanted burn on farming fields, effectively mitigating air pollution.

Keywords: Biomass, Sugarcane, Zero-waste, Management, Business model

* Corresponding author: tanwon@kku.ac.th

1 Introduction

Thailand is acknowledged as an agricultural-focused nation with a robust industry dedicated to converting agricultural resources. This industry encompasses processing agricultural commodities into food products, raw materials, and various items and tools. Currently, Thai farmers primarily concentrate on cultivating sugarcane [1]. While agriculture has long existed in Thailand, many farmers still lack the knowledge to effectively handle the agricultural waste left behind after harvesting, which leads to the improper disposal of these remnants. During the harvest season, there is a prevalent practice of incinerating these residuals, which gives rise to environmental problems that directly affect the daily lives of the population [2]. Despite the ongoing government efforts to regulate sugarcane harvests by burning methods for sale in factories [3], the issue is only partially resolved. Several factors continue to contribute to the burning of sugarcane plantations in Thailand. These factors include a labor shortage for harvesting, a lack of knowledge about biomass management from sugarcane leaves, and farmers who fail to recognize the value of utilizing sugarcane leaf residuals to create products. It is important that farmers have the appropriate knowledge and skills to handle challenging post-harvest biomass. These leftover residuals can be utilized to sustainably increase the value of biomass waste.

In accordance with the Alternative Energy Development Plan 2018–2037 (AEDP2018), the Thai government has instituted a policy aimed at promoting the utilization of renewable energy sources for electricity generation, with a specific focus on the use of biomass fuels [4]. Saleem [5] highlighted that agricultural waste presented a viable and sustainable fuel source for energy generation, with the energy content varying depending on the specific agricultural residuals. This biomass comprised agricultural waste left after harvest, such as white straw, rice husks, tea leaf remnants, and sugarcane leaves discarded post-harvest. Initial investigations revealed that sugarcane leaves exhibited a heat value of up to 17.41 MJ/kg. Meanwhile, water hyacinth, rice straw, rice husk, and tea leaf residuals possessed calorific values of 14.81 MJ/kg, 15.09 MJ/kg, 16.47 MJ/kg, and 17.10 MJ/kg, respectively [6]. Thailand, being the second-largest sugar exporter globally after Brazil [7], experiences a significant surplus of sugarcane leaves post-harvest. As reported by Dangprok et al. [8], the annual average quantity of sugarcane leaves remaining after harvesting in Thailand was approximately 17,016 ktons. Regrettably, this considerable volume of agricultural waste currently goes unused.

Numerous studies have demonstrated the utility of sugarcane leaves in various applications. Firstly, sugarcane leaves were extensively employed for biofuel production. For instance, Go et al. [9] conducted a study on the utilization of sugarcane leaves (SCL) as a biofuel, encompassing bioethanol and biomethane production in the Philippines. The findings indicated that the use of SCL could potentially have reduced coal consumption in electricity generation by up to 75% for the Visayas region. The study also highlighted that the potential of SCL was substantial enough to meet the demand for fuel ethanol and biomethane. Martins et al. [10] researched the natural senescence process of sugarcane leaves, which included the remobilization of nutrients and the effect on cell walls of producing second-generation bioethanol (2G-bioethanol) from sugarcane leaves. Sugarcane leaves were also employed for electricity generation. Silalertruksa et al. [11] introduced the bio-circular-green economy (BCG) for sugarcane to maximize waste usage after harvest. The study assessed the environmental sustainability of five models: sugarcane leaves for electricity generation, biochar, and soil conditioner, and vinasse for bio-fertilizer and electricity generation, with sugarcane leaves used for electricity generation exhibiting the least environmental impact. Fioranelli et al. [12] also looked into and used biomass integrated gasifier-gas turbine combined cycle (BIGCC) technology in Brazil's sugarcane industry. This work used bagasse and sugarcane leaves to make electricity; the technology was successfully integrated into sugarcane industrial processes with a small power capacity of 15 MW.

In the context of previous research on biomass waste management, utilizing sugarcane leaves for electricity generation has become a compelling topic in Thailand, primarily due to the substantial annual quantity of sugarcane leaves left after harvesting. The utilization of sugarcane leaves for electricity generation signifies a proactive measure to diminish dependence on finite fossil fuels. However, in underdeveloped regions, the limited access to technology, machinery, and tools poses significant obstacles in the collection and transportation processes, hindering the effective use of agricultural waste as a bio-energy source. For Thai farmers, there is a lack of knowledge concerning managing waste sugarcane leaves, including collection, densification, and transportation to fuel biomass power plants. Consequently, a significant portion of these agricultural residues are left abandoned and subjected to burning in open fields, leading to environmental concerns, particularly during harvesting seasons.

This study introduces a method for managing agricultural waste through collecting and densifying processes (referred to as processing) and transportation processes. These procedures are demonstrated within a sample group, aligning with the zero-biomass waste management model. The primary goal of this model is to foster knowledge development among Thai farmers, empowering them with the skills to manage sugarcane leaves left over from harvesting effectively. The model also seeks to promote the utilization of existing technology, encouraging farmers to leverage agricultural waste for sustainable practices. The contributions of this study can be summarized as follows:

- **Business Model:** This paper presents the business model for the utilization of benefits from sugarcane leaves to facilitate access to tools and equipment for the processing of biomass. This approach helps address the issue of agricultural waste incineration after harvest and provides income-generating opportunities for farmers.
- **Economic Analysis:** To demonstrate the best management solution, the study conducts a comprehensive analysis of the economics, expenses, and average returns for two selected sample groups: the farmer and middleman groups. This analysis enables a comparison of costs and profits, guiding sellers toward the most favorable model for achieving economic sustainability.

2 Methodology

The fundamental principle of the zero-biomass waste management model is to empower farmers to appropriately dispose of biomass after harvesting and ensure the availability of biomass for electricity generation in biomass power plants. This paper establishes guidelines for the effective management of biomass waste, with a specific focus on the residues derived from sugarcane leaves after harvesting. The zero-biomass waste management model comprises both sellers and buyers. The sellers are persons who own or process the biomass, which are considered into 2 groups: a farmer who owns the biomass designated hereafter as A, and a middleman who buys the biomass from the farmer and processes it for sale hereafter called B. On the other side, buyers of biomass who have biomass demand for use as fuel include biomass power plants in sugar factories.

2.1 Business model for utilization of sugarcane leaves

The business model for the utilization of sugarcane leaves is strategically designed to capitalize on knowledge regarding biomass management techniques and technology, fostering effective learning among farmer groups. To implement and assess the model, two specific sample groups are chosen, A and B, as mentioned above. The A-group comprises sugarcane farmers, collectively cultivating an area of 197.63 acres (referred to as farmers). Concurrently, the B-group is called middlemen, who acquire biomass from farmers, overseeing the same area of 197.63 acres.

To facilitate the implementation of the model, both groups are provided with the essential tools and equipment for biomass collection and densification. This work aims to establish a framework for efficient biomass utilization including collecting, densifying, and transportation. Fig. **1** visually represents the zero-biomass waste management model, outlining the interconnected processes inherent in the model. In addition to enhancing the value of agricultural waste, the model takes a proactive approach to address the urgent issue of incineration.

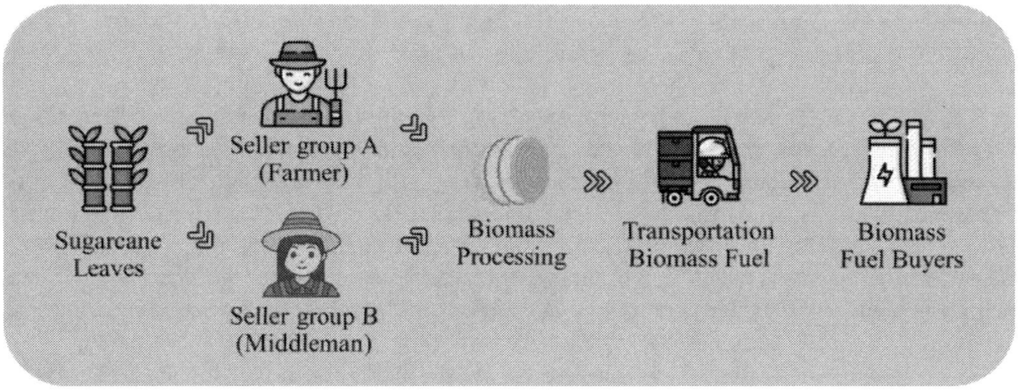

Fig. 1. Zero-biomass waste management model

Sugarcane, recognized as one of the most economically significant crops in the world, holds value across diverse applications, serving as a primary source of sustenance for both humans and animals. Beyond sugar production, various components of sugarcane, including its leaves, present valuable by-products. Northeastern Thailand is notable for extensive sugarcane cultivation across numerous areas, with Udon Thani Province, Khon Kaen Province, Nakhon Ratchasima Province, Chaiyaphum Province, and Kalasin Province ranking highest in sugarcane cultivation within this region [13]. While the widespread cultivation of sugarcane contributes to increased agricultural production, it also gives rise to the challenge of sugarcane field burning, particularly during the annual harvest. The practice of burning sugarcane is primarily attributed to the scarcity of labor for harvesting fresh sugarcane, a task complicated by the sharp and thorny nature of sugarcane leaves, thus extending the harvesting process. Moreover, post-harvest, the residual sugarcane leaves are burned to prevent fires that could impact the germination of the next generation of sugarcane plants [3]. In light of these challenges, this paper proposes the utilization of prototype areas in Udon Thani and Kalasin provinces, as depicted in Fig. 2, to serve as a model for the study of zero-waste biomass management.

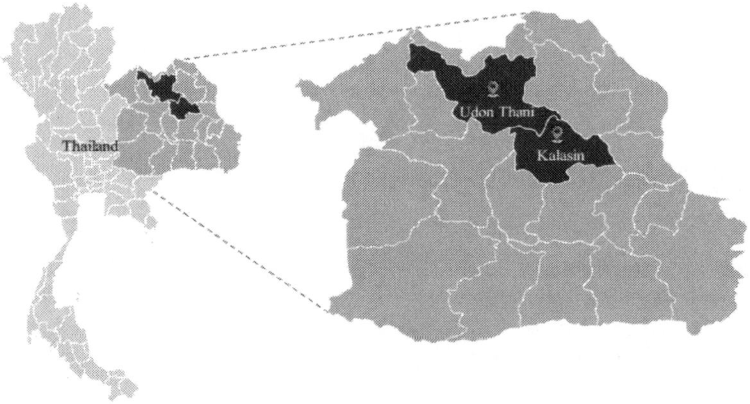

Fig. 2. Location of the selected groups

2.2 Economic computations of typical returns and expenses

The payback period is a commonly employed metric for evaluating economic cost-effectiveness, providing insights into the duration required for the net operating benefits or net profit (after deducting various expenses) to offset the initial investment cost. Denoted as the payback period ($Y_{Payback}$), this metric serves as a crucial indicator in financial assessments. It can be calculated by dividing the initial investment (*Initial investment*), representing costs like the purchase of agricultural equipment, by the annual net return (*Annual net profit*), as demonstrated in Equations (1) and (2). A shorter payback period suggests that the investment is financially advantageous, as it signifies a quicker recovery of the initial investment through the generated returns. Understanding and optimizing the payback period is integral to making informed financial decisions and assessing the viability of investments.

$$Y_{Payback} = \frac{\text{Initial investment}}{\textit{Annual net profile}} \tag{1}$$

$$\textit{Annual net profile} = W_{Leaves,year} \times P_{net} \tag{2}$$

where $W_{Leaves,year}$ is sugarcane leaves weight (tons/year), and P_{net} is net profit (\$/ton).

The annual weight of sugarcane leaves ($W_{Leaves,year}$) can be determined using Equation (3). In this study, the total number of areas (A_{farm}) is assumed to be 197.63 acres per year. Net profit is calculated based on Equations (4) and (5). The total operating cost ($C_{Operation}$) includes various components such as fuel expenses for the sugarcane leaf sweeper truck ($C_{Fuel,sweep}$), fuel expenses for the sugarcane leaf roll truck ($C_{Fuel,roll}$), expenses of rope for rolling sugarcane leaves (C_{Rope}), labor expenses (C_{Labor}), expenses for procuring sugarcane leaves from farmers ($C_{Biomass,buy}$), transportation expenses ($C_{Transportation}$), and maintenance expenses ($C_{Maintenance}$). All costs are in \$/ton. It is noted that the $C_{Biomass,buy}$ is only for the B group, which is an extra cost for the middleman model.

$$W_{Leaves,year} = A_{farm} \times W_{lumps} \times N_{lumps} \tag{3}$$

$$P_{net} = P_{sell} - C_{Operation} \tag{4}$$

$$C_{Operation} = C_{Fuel,sweep} + C_{Fuel,roll} + C_{Rope} + C_{Labor} + C_{Biomass,buy} + C_{Transportation} + C_{Maintenance} \tag{5}$$

where W_{lumps} is the Weight of sugarcane leaves lumps (tons/lump), N_{lumps} is the average number of sugarcane leaves lumps (lump/acre), and P_{sell} is biomass selling price ($/ton).

3 Results and Discussions

The business model for biomass processing (collecting and densifying as rolls) is divided into two distinct groups covering the selected area in Northeastern Thailand:

- Farmer Group (Group A):

Sub-Group A1: This sub-group operates on 197.63 acres in Udon Thani Province, designated as area A1, which is 30 kilometers from the biomass power plant.

Sub-Group A2: Another sub-group, also part of Group A, operates on an additional 197.63 acres in Udon Thani Province, designated as area A2, which is 50 kilometers from the biomass power plant.

- Middleman Group (Group B):

Sub-Group B1: This sub-group engages in the purchase of biomass from farmers for further processing. It operates within a 197.63 acres area in Kalasin Province, 30 kilometers from the biomass power plant.

The selected groups were well-appointed with information and guidance pertaining to the preparation of sugarcane leaves, their collection and storage, as well as machinery operation and maintenance. In order to facilitate the collection of the sugarcane leaves, a sugarcane leaf sweeper truck and a sugarcane leaf roll truck were provided to both Group A and Group B. Based on the results obtained from processing sugarcane leaves post-harvest, it was observed that Group A1 achieved an average yield of 20.59 lumps/acre from processing sugarcane leaves. The processing time averaged 120.17 minutes/acre, with the sugarcane leaf bales having an average weight of 0.168 tons/lump. Group A2, on the other hand, obtained an average yield of 19.08 lumps/acre from sugarcane leaf processing. The processing time averaged 120.68 minutes/acre, with sugarcane leaf lumps weighing an average of 0.170 tons/lump. Group B1, representing the middleman group, achieved an average yield of 19.35 lumps/acre from sugarcane leaf processing. The average processing time was 105.43 minutes/acre, and the sugarcane leaf bales had an average weight of 0.164 tons/lump. Fig. **3** illustrates the process of collection and densification. For these above, it was found that Group A did not have experience in the operation. Therefore, it was necessary to have a research team to assist and resolve on-site issues. On the other hand, the middleman group (Group B) has previous experience in sugarcane leaf rolling, enabling them to process faster. Detailed information is provided in Table 1.

(a) (b)

Fig. 3 (a) Sweeping process; (b) Densification process

Table 1. Details of sugarcane leaves densification process

Details	A		B	Unit
	A1	**A2**	**B1**	
Average number of sugarcane leaves lumps	20.59	19.08	19.35	lump/acre
Average working time	120.17	120.68	105.43	minutes/acre
Fuel consumption rate	5.19	5.06	4.86	liters/acre
Average time to sweep sugarcane leaves	27.63	27.83	19.28	minutes/acre
Average time to roll sugarcane leaves	5.80	6.31	4.44	minutes/lump
Average lump size	D120 x 120	D120 x 120	D120 x 120	centimeter
Weight of sugarcane leaves lumps	0.168	0.170	0.164	tons/lump
Average yield of sugarcane leaves	3.47	3.24	3.18	tons/acre
Annual yield of sugarcane leaves	685	640	628	tons/year

Table 2. Payback period of 2 proposed business sub-models

Details	A		B	Unit
	A1	**A2**	**B1**	
Machine cost				
Investment in agricultural machinery	17,202.53	17,202.53	17,202.53	$*
Operation costs				
Fuel costs for sugarcane leaves sweeper truck	0.25	0.26	0.20	$/ton
Fuel costs for sugarcane leaves roll truck	0.99	0.99	1.07	$/ton
Cost of rope for sugarcane leaves roll	2.55	2.53	2.62	$/ton
Labor cost	1.36	1.35	1.40	$/ton
Cost of buying sugarcane leaves from farmers	-	-	1.14	$/ton
Transportation cost	8.60	9.46	8.60	$/ton
Maintenance cost	0.31	0.34	0.34	$/ton
Total operation costs	14.07	14.93	15.38	$/ton
Biomass selling price	**28.67**	**28.67**	**28.67**	**$/ton**
Net profit	14.60	13.74	13.30	$/ton
Payback period	1.72	1.96	2.06	year

* Remark: $1 = 34.89 Baht

The economic evaluation of sugarcane leaf business models was carried out. The investment cost of all three models was equal since all the equipment required was the same. A purchase price for processed sugarcane leaves of 28.67 $/ton has been based on the rates set by the Office of the Cane and Sugar Board under the Ministry of Industry of Thailand [14], which will be used as fuel in biomass power plants. This pricing strategy aims to reduce the reliance on fossil fuels in electricity production and contribute to the establishment of national energy security. Based on the operational outcomes, it was noted that the farmer

Group A1 generated a net profit of 14.60 $/ton and could pay back their initial investment within an estimated time frame of 1.72 years. The sugarcane leaf processing revenue of Group A2 amounted to 13.74 $/ton, with an approximate payback period of 1.96 years. The payback period is expected to be faster if these two groups of farmers are able to expand their operations to encompass more than 197.63 acres, making more profit per year. For the middleman group (Group B1), there was an extra cost of sugarcane leaf bought from farmers for 1.14 $/ton. Therefore, this group yields less profit of 13.30 $/ton, resulting in a longer payback period of 2.05 years. The data indicates that farmer groups have a shorter payback period compared to the middleman group. This difference can be primarily attributed to the expenses associated with acquiring sugarcane leaves from farmers. Consequently, investing in farmer groups appears to be a more attractive option due to their comparatively quicker return on investment. Exhaustive details are available in Table 2.

4 Conclusion

The collection and densification of sugarcane leaves, which are established as business models, are proposed in two different concepts called the farmer and the middleman groups. The middleman group showed faster sweeping, collecting, and densifying of the sugarcane lump than the farmer groups due to its expertise and more experience. In both A and B groups, with the same targeted area of 197.63 acres, the yields of sugarcane leaves were approximately 628 to 685 tons/year. Consequently, the calculated payback period was to be within the range of 1.72 to 2.06 years. Thus, Group A1 exhibited a notably short payback period because there were no expenses related to purchasing sugarcane leaves, a cost incurred by the middleman group. Additionally, the shorter biomass transportation distance in Group A1 compared to Group A2 facilitated a quicker payback. In summation, the farmer model, where they owned sugarcane leaves, is recommended due to a potential shorter payback period. Moreover, if they gain more experience and expand to a larger area, the return on investment will be shorter.

Finally, the recommended course of action to further develop this work was to focus on creating a biomass trading platform with the aim of optimizing the efficiency of trading processes due to transportation costs being an important factor influencing the payback period. Presently, the trading options available to buyers and sellers are somewhat constrained, lacking considerations for transport distance, purchase price, biomass quantity, and biomass type. Considering the aforementioned elements in biomass trading, it laid the groundwork for a more efficient and effective biomass trading process.

The research team would like to thank the Center of Alternative Energy Research and Development, Khon Kaen University, for supporting our work.

References

1. Agricultural Development Policy and Planning Division, *Agricultural Economic Conditions in the First Quarter of 2023*, vol. 6. Bangkok: Office of Agricultural Economics, 2023.
2. G. Geldenhuys, J. Orasche, G. Jakobi, R. Zimmermann, and P. B. C. Forbes, "Characterization of Gaseous and Particulate Phase Polycyclic Aromatic Hydrocarbons Emitted During Preharvest Burning of Sugar Cane in Different Regions of Kwa-Zulu Natal, South Africa," *Environ. Toxicol. Chem.*, vol. 42, no. 4, pp. 778–792, 2023.

3. S. Rodmanee and C. Sungsing, "Management of Pre-harvest Sugarcane Burning," *Acad. J. Manag. Technol.*, vol. 3, no. December, pp. 60–74, 2022.
4. Department of Alternative Energy Development and Efficiency, *Alternative Energy Development Plan (AEDP2018)*. 2020.
5. M. Saleem, "Possibility of utilizing agriculture biomass as a renewable and sustainable future energy source," *Heliyon*, vol. 8, no. 2, p. e08905, 2022.
6. D. R. Nhuchhen and P. Abdul Salam, "Estimation of higher heating value of biomass from proximate analysis: A new approach," *Fuel*, vol. 99, pp. 55–63, 2012.
7. Office of Agricultural Economics, *Information on Agricultural Products by Commodity 2022*. Ministry of Agriculture and Cooperatives: Bangkok, Thailand, 2022.
8. B. Dangprok, K. Y. Tippayawong, and N. Tippayawong, "Development of a cost optimization model for power generation from agricultural residual biomass in Thailand," *Energy Reports*, vol. 9, pp. 55–62, 2023.
9. A. W. Go and A. T. Conag, "Utilizing sugarcane leaves/straws as source of bioenergy in the Philippines: A case in the Visayas Region," *Renew. Energy*, vol. 132, pp. 1230–1237, 2019.
10. M. T. B. Martins *et al.*, "Characterization of sugarcane (Saccharum spp.) leaf senescence: Implications for biofuel production," *Biotechnol. Biofuels*, vol. 9, no. 1, pp. 1–17, 2016.
11. T. Silalertruksa, C. Wirodcharuskul, and S. H. Gheewala, "Environmental Sustainability of Waste Circulation Models for Sugarcane Biorefinery System in Thailand," *Energies*, vol. 15, no. 24, p. 9515, Dec. 2022.
12. A. Fioranelli and W. A. Bizzo, "Generation of surplus electricity in sugarcane mills from sugarcane bagasse and straw: Challenges, failures and opportunities," *Renew. Sustain. Energy Rev.*, vol. 186, p. 113647, Oct. 2023.
13. Information and Communication Technology Group Strategy and Planning Division, *Situation report on sugarcane cultivation*. Bangkok: Office of The Cane and Sugar Board, 2023.
14. Office of The Cane and Sugar Board, *Sustainable accurate sugarcane farming guide*. 2020.

Optimization of the cocoa shell, inoculum, and water mixture for biogas production with 40-70% methane concentrations

C Matute Canales*, *J* Díaz Chinchilla and *H* Villatoro Flores

Faculty of Engineering, Universidad Tecnológica Centroamericana, San Pedro Sula, Honduras

Abstract. Biomass has been a valuable source of energy over the years and its application includes many areas, whether for cooking, heating, or electricity generation. In Honduras, especially in the rural sector, biomass, mainly firewood, became the main source of energy, which has led to a high consumption of firewood and, consequently, to an increase in the rate of deforestation. In this study, we proposed to use a more specific biomass, cocoa husk, as feedstock for biogas generation. It is important to note that biogas does not only consist of methane; it also contains carbon dioxide (CO_2), nitrogen (N_2), hydrogen sulfide (H_2S) and oxygen (O_2). However, to be considered of quality, the concentration of methane in the biogas must range between 40% and 70%. The main objective of this research has been to optimize the mixture of raw materials, including cocoa husk, water and inoculum, with the purpose of obtaining a methane (CH_4) concentration in the biogas produced, located in the range of 40%-70%. To achieve biogas production, several tests were carried out in four biodigesters, each with different proportions of water, inoculum, cocoa husk and grinding level. This made it possible to obtain varying amounts of biogas and to measure the methane gas concentration in each test. During the operation of the biodigesters, the temperature and pH of the mixture were constantly monitored. This was possible because methanogenic bacteria, responsible for methane generation, thrive at temperatures between 30 and 40 degrees Celsius and in a pH range of 6 to 8. The results obtained have the potential to demonstrate that cocoa shells can be a viable feedstock for biogas generation in Honduras. Furthermore, these findings could serve as a basis for future research in other cities around the world, thus contributing to the search for energy generation solutions.

1 Introduction

The search for renewable and sustainable energy sources is a global priority due to the growing demand for energy, the depletion of fossil resources and concerns about climate change.

Approximately 80% of the world's energy consumption comes from fossil fuels, whose reserves are depleting [1]. In 2016, around 1.6 billion people, mostly in rural areas, lacked

* Corresponding author: christopher_matute@unitec.edu

 https://doi.org/10.1051/e3sconf/202453003002

access to electricity, while 2.5 billion still relied on traditional fuels such as wood and dried dung for cooking and heating, with negative consequences for the environment and health, as well as limiting economic opportunities to combat poverty [2]. Even though greenhouse gas emissions typically come from burning fossil fuels, in Honduras 47% of the population, equivalent to 1,127,000 households, relies on firewood and other organic waste for cooking [3]. According to the 2019 Multipurpose Household Survey, 87% of the rural population and 26% of the urban population used firewood for cooking, often in conjunction with electricity or liquefied petroleum gas (LPG). In 2011, 66% of urban households purchased firewood, in contrast to 49% in rural areas, since in this area, the majority collect it directly [4].

Studies have been conducted on the production of biogas from cocoa shells. Acosta [5] evaluated its viability to generate renewable energy through anaerobic digestion, achieving 60% methane in two reactions, one dry anaerobic and the other with distilled water. Carreño Vesga [5] analysed the pre- preparation of cocoa lignocellulosic biomass using a specific methanogenic activity (SMA) test and found that a cow inoculum ratio of 1:3 resulted in higher methane production due to volatile solids content. These studies highlight the potential of cocoa as a renewable energy source[6]. In this research, biogas production from organic wastes is explored, focusing on the use of cocoa shells as feedstock. Four biodigesters will be used to perform anaerobic reactions with various mixtures of cocoa husk, inoculum, and water. Cocoa, which comes from Theobroma cacao trees, is economically important in various industries such as food, pharmaceuticals, and cosmetics. However, cocoa shells, which are rich in organic matter, are often discarded without efficient utilization[7]. This thesis focuses on analysing the critical factors affecting biogas production from cocoa shells, inoculum, and water. This includes determining the optimum ratio between these components and the conditions for anaerobic digestion[8]. It also seeks to achieve methane concentrations in the biogas that range between 40% and 70%, a key indicator of its quality and usefulness as a fuel. This study has important implications for waste management, as it can offer a solution for the management of cocoa shells while generating a renewable energy source. The thesis aims to promote future research on the generation of biogas from biomass as a main source of energy and to serve as a model for studies in various cities around the world.

2 Context

Today, renewable energy has become more prominent as a source of energy for many countries. Asshown in Fig. 1.

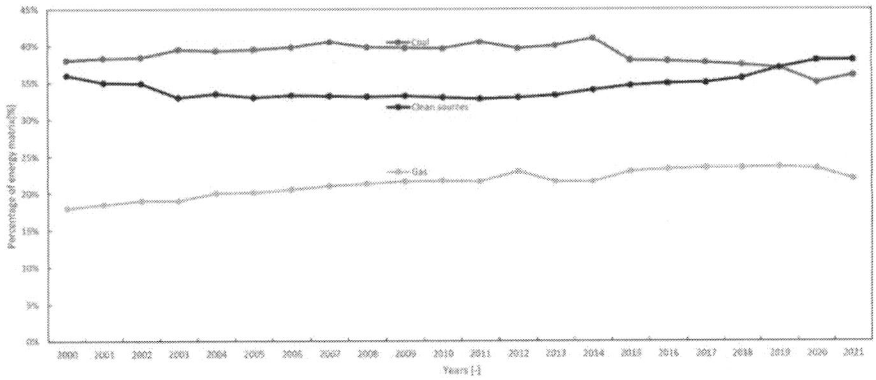

Fig.1. Comparison of world energy matrix evolution *Clean sources include hydro, nuclear, wind, solar and bioenergy.

Biomass has become an attractive alternative for energy generation, as it produces biogas through a process called anaerobic decomposition [9]. This process is useful for dealing with waste that is biodegradable, as this process oxidizes the glucose in the organic compounds generating methane. Honduras has many resources such as sugarcane bagasse, King Grass, African palm, sawdust, and cocoa acorns to produce biomass, highlighting the generation of biogas from cocoa shells [10]. The country has approximately 7,000 hectares of cocoa, which are distributed in nine departments: Cortés, Santa Bárbara, Copán, Copán, Yoro, Atlántida, Colón, Olancho, El Paraíso and Gracias a Dios [11].

That said, a study was conducted in Ecuador in which it was shown that, thanks to the cocoa waste available in the area, 8,341 MWh of electricity could be produced, satisfying 88% of the current electricity demand in Balao. This study highlighted the potential of cocoa waste as a renewable energysource in rural areas[12].

Although there are few studies on the anaerobic reaction using cocoa shells, a study funded by the National Council for Technological and Scientific Development of Brazil analyzed the potential of cocoa waste for energy conversion through anaerobic digestion and the evaluation of strategies for the optimization of feedstock and substrate mixtures, thus improving methane production yields. This study concluded that cocoa waste biomass has a high potential for energy conversion through anaerobic digestion, being a sustainable process to treat these wastes and an alternative to the use of fossil fuels [13].

3 Methodology

3.1 Approach

A quantitative approach methodology has been adopted, which means that the focus has been on the collection and analysis of numerical data to address the hypotheses put forward. By using this approach, a more precise and verifiable understanding of the phenomena studied is sought, which provides a solid basis to support the conclusions and recommendations of the research.

3.2 Research variables

Dependent Variable

Concentration of methane (CH_4) in the biogas produced because it is the variable to be optimized and represents the proportion of methane in the biogas resulting from the anaerobic digestion process.

Independent Variable

Cocoa husk: Refers to the different combinations and proportions of organic materials used in the biogasproduction process.

Water: The amount of water added to the system can influence the efficiency of anaerobic digestion andthus affect methane production.

Inoculum: Refers to the amount and type of microorganisms added to the anaerobic reactor to initiatethe process of organic matter decomposition.

3.3 Technique and instruments

Within the framework of this project, pH strips and a digital thermometer were used as control variables to monitor the pH and temperature of the biodigester. By using the pH strips, it was possible to adjust and maintain the pH within the ideal parameters, which are above six and below eight during the entire test period, since a pH within this range favors

 https://doi.org/10.1051/e3sconf/202453003002

microbial activity and methane production, while significant deviations from the optimal range can negatively affect biogas production. Temperature directly influences microbial activity and thus biogas production; an ideal temperature is considered to be above 30°C-40°C [14]. A laboratory oven was used to analyze the percentage of moisture in the cocoa shells. Also, in order to perform gas measurements, as shown in fig.2. The gas analyzer MRU Optima 7 was used. These tests consisted of the analysis of two types of two samples: the first sample was made with fermented cocoa shells and the second with unfermented cocoa shells, since there was a large amount of cocoa, the average percentage of moisture in the shell was determined for both samples,which was 75.5%.

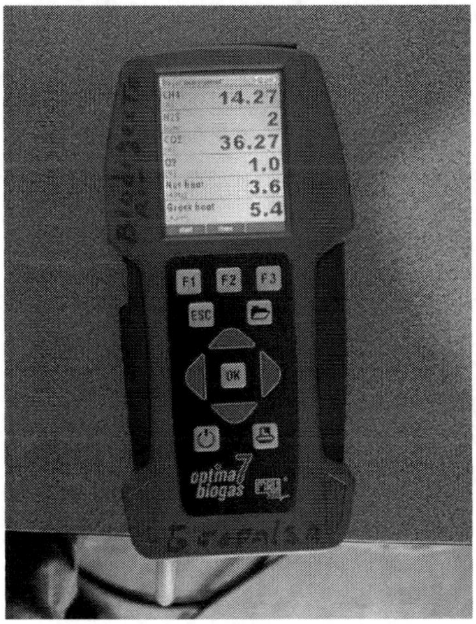

Fig.2. Gas analyzer MRU Optima 7.

3.3.1 Moisture content

Since there was a large quantity of cocoa shells (150 kg), it was decided to extract two cocoa acorns from each of the five bags. The acorns were then crushed, placed in the oven and left to rest for 24 hours. The same process was carried out for the unfermented cocoa shells. The inclusion of both types of samples in the tests made it possible to evaluate how the fermentation process influences the moisture content of the cocoa shell. To calculate the percentage moisture content of a fresh plant material, the following formula 1 can be used:

$$Humidity\ percentage = \left(\frac{initial\ weight - fianl\ weight}{initial\ weight}\right) \times 100\%$$

Equation 1- Moisture percentage

3.4 Materials

The biodigester used was based on the design of [15]. To avoid biogas leakage, certain specific modifications were carried out. Also, an adapted tire was implemented to perform tests on the chemicalcomposition of the biogas generated from cocoa husk.

3.5 Population and sample

In Honduras, cocoa is found on approximately 7,000 hectares distributed in nine departments due to factors that favor its harvest, such as suitable climatic conditions, soils suitable for cultivation and its cultural importance. The cocoa used was extracted from two farms located in different areas of Honduras. Both farms have mixed cocoa, which means that the crop includes different varieties of cocoa. The first farm is in Trascerros, Santa Barbara, Honduras, while the second is located in La Másica, Atlántida, Honduras as shown in Fig.3.

Fig.3. Ground cocoa and cocoa sample from Trascerros Santa Barbara.

3.6 Study methodology

A detailed record of the research is provided below, covering each stage, from the initial configuration of the biodigesters, the collection of the feedstock, the level of grinding of the feedstock, the mixture that was fed into each biodigester, the parameters that were measured, which were pH and temperature, and finally the analysis of the biodigesters as shown in Fig 4.

Fig.4. Study methodology record.

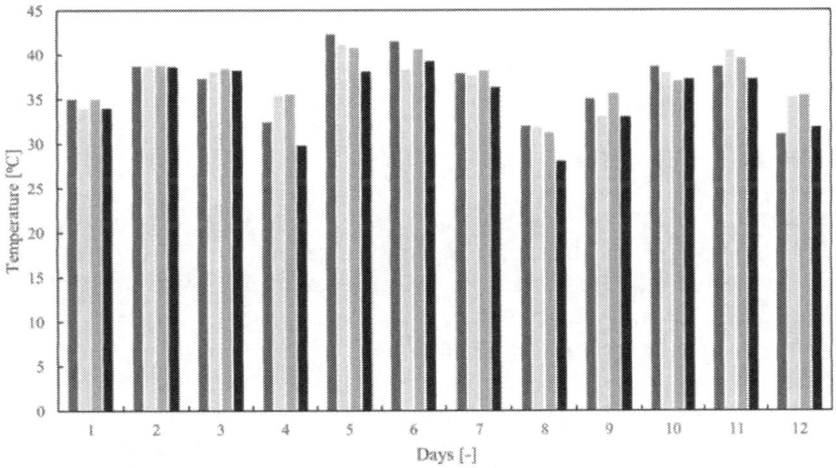

Fig.5. Temperature of biodigesters

4 Results and analysis

Monitoring was carried out every two days after the start-up of the biodigesters, in order to obtain the ideal temperature data inside the digester. Each bar represents the biodigesters in which measurements were taken over a 12-day period. The ideal temperature data to be maintained by the biodigesters is between 30 and 40 degrees Celsius. As shown in fig.5. The temperature graph below displays the measured values for each biodigester, with each color corresponding to a biodigester, ranging from biodigester one to four.

4.1 PH Level of the Ratio of Each Biodigester

The pH level within each biodigester was calculated for the purpose of determining if the bacteria present in the mixture of water, cocoa husk and inoculum are in the proper conditions to carry out the anaerobic reaction. The ideal range of pH levels should be between six and eight, with six being the lowest possible value and eight being the highest. As shown in fig.6. The pH graph below displays the measured values for each biodigester, with each color representing the days on which the acidity level of each biodigester was measured.

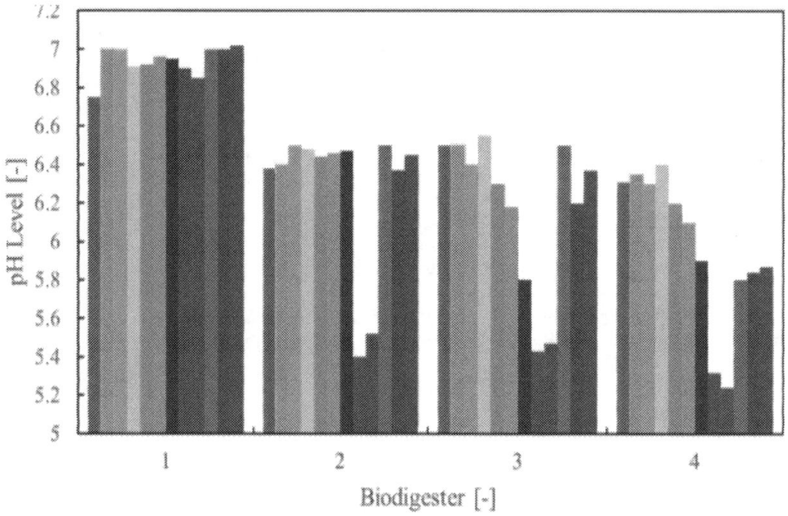

Fig.6. pH level in the biodigesters.

4.2 Analysis of the Biodigesters

As can be seen in the diagram, various types of mixtures were introduced into each of the biodigesters. Each number in the diagram represents a gallon, as it was not hermetically sealed properly due to the mixture introduced, and it is associated with its corresponding grind level. In the case of biodigester number one, a unique situation was encountered because it experienced failures due to improper sealing.

In biodigester two, the results in terms of methane (CH_4) generation were considerably lower than ideal, reaching only 12.87% methane gas. However, the level of hydrogen sulfide (H_2S) particles was reduced to one, with a percentage of carbon dioxide (CO_2) of 19.62% and oxygen (O_2) of only about three points two percent. It is important to note that this ratio is based on a 5:10:1 ratio, where each number represents one gallon. In addition, the amount of feedstock, in this case cocoa husk, was lower compared to the other biodigesters. Regarding the crushing of the feedstock, it can be stated that it was moderately crushed (300 mortar strokes) and that the husk was in a mature state.

In biodigester three, the results for methane (CH_4) generation were also lower than ideal but the highest percentage of methane with 14.27% methane gas. The level of hydrogen sulfide (H_2S) particles was two, and the percentage of carbon dioxide (CO_2) was 36.27% and oxygen (O_2) was only one percent. It is important to note that this ratio is based on a ratio of 7 ½ ,12 ½ ,2 where each number represents one gallon. Regarding the crushing of the raw material, it can be stated that it was moderately crushed (300 mortar blows) and that the peel was in a mature state.

As can be seen in table 1. The ratio of cocoa hulls, water and inoculum used in the biodigester 4 was 20:17 ½ :2, respectively, where each number represents one gallon. In other words, 10 gallons of cocoa hulls, 17 ½ gallons of water and two gallons of inoculum were used. In terms of the crushing of the raw material, it can be stated that it was adequately crushed (500 mortar strokes) and that the husk was in a mature state. As shown in fig.7 and 8. All biodigester diagrams are identical.

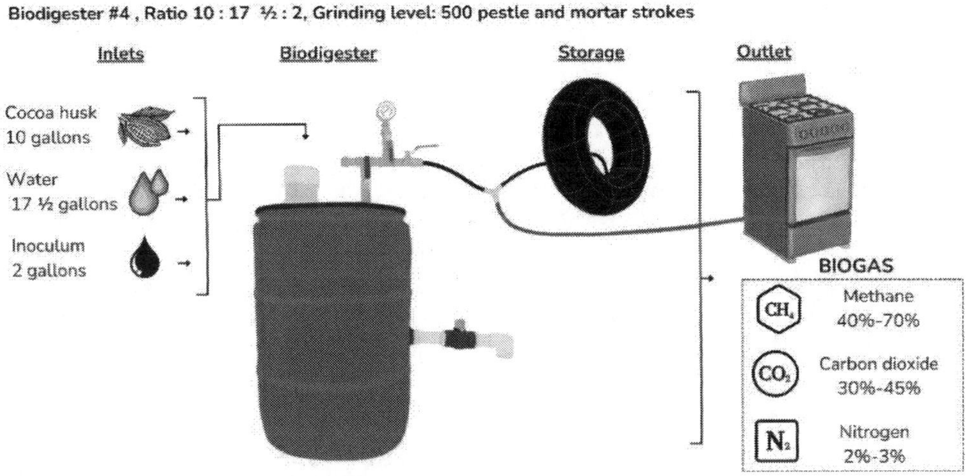

Fig.7. Concept Diagram of biodigester 4.

Fig.8. Biodigester 4 based on [15]

Table 1. Results of biodigester 1-4

N°	COMPONENT	EXPRESSION	BIO 2	BIO 3	BIO 4
1	Methane (CH4)	[%]	12.87	14.27	20.64
2	Hydrogen Sulfide (H2S)	[ppm]	1	2	404
3	Carbon Dioxide (CO2)	[%]	19.62	36.27	74.42
4	Oxygen (O2)	[%]	3.32	1.0	1.91
5	Net heat	[BTU/lb]	3.5	3.6	528
6	Gross heat	[BTU/ft^3]	4.8	5.4	60.7

5 Conclusions

Four biodigester experiments were carried out to compare the amount of methane that can be generated by using cocoa hulls as feedstock. In each of the biodigesters, the composition of the mixture, consisting of cocoa hulls, water and inoculant, was varied in order to determine which ratio produces the best biogas concentration. Monitoring was carried out in the biodigesters in order to measure the temperature and pH level. These measurements were carried out with the objective of maintaining the anaerobic digestion in optimal conditions, which allows the methanogenic bacteria to have the proper environment for the generation of methane in the biogas. The following results were obtained:

1. Three biodigesters were successfully completed. Biodigester number one did not show results, as it had more leaks. Each biodigester was set up and operated according to a pre-established plan, and subsequently underwent a detailed analysis using a gas analyzer.
2. Analyses were carried out on several cocoa shell samples, each weighing 200 grams, in order to determine the moisture content present in the shells. Using an oven at a temperature of 75 degrees Celsius for a period of 24 hours, a moisture content of 76% was obtained in the cocoa shell samples.
3. From the moisture percentage, which was found to be 76%, it was determined that for every 5 gallons of cocoa shells, 1.05 gallons of water should be added at the time of grinding.
4. Monitoring techniques, such as the measurement of acidity in the biodigesters and temperature using a digital thermometer, were implemented to control these crucial parameters for methanogenic activity. When the temperature was below the optimum level, a steam engine was used, and in those biodigesters whose acidity levels were not as desired, lime was added.
5. Methane measurement was carried out in each of the biodigesters using a gas analyzer.
6. The biodigester with the best biogas conditions was biodigester number four with a 10:17.5:2 ratio, which gave us a total of 20.64 methane, since it was the one with the rawest material of the three biodigesters; however, it did not reach the pre-established methane quantity, which was 40-70% methane.

The main limitation of this research was not having an instrument that could grind the cocoa faster and achieve a homogeneous mixture. As could be observed in biodigester number four, which had the best grinding level (500 pestle strokes), it was the one that obtained the most methane from the biogas. Since the higher the grinding level, the faster the methanogenic bacteria decompose the feedstock, the methane gas production could be increased.

Words cannot express my gratitude to my professor and advisor, PhD. Héctor Villatoro, for his invaluable patience and feedback.

References

1. A. Shane, S. H. Gheewala, and S. Phiri, "Rural domestic biogas supply model for Zambia," *Renew. Sustain. Energy Rev.*, vol. 78, pp. 683–697, Oct. 2017, doi: 10.1016/j.rser.2017.05.008.
2. M. Garfí, J. Martí-Herrero, A. Garwood, and I. Ferrer, "Household anaerobic digesters for biogas production in Latin America: A review," *Renew. Sustain. Energy Rev.*, vol. 60, pp. 599–614, Jul.2016, doi: 10.1016/j.rser.2016.01.071.
3. R. J. Tauro, J. L. Caballero, M. Á. Salinas, A. Ghilardi, and J. M. Arroyo, "Evaluación delpotencial energético de los recursos biomásicos en Honduras". 24, 2023. [Online]. Available: https://iris.paho.org/bitstream/handle/10665.2/55998/OPSHND210002_spa.pdf
4. J. P. Carreño-Vesga, "Pretratamiento de Biomasa Lignocelulósica de los Desechos Agrícolas del Cacao Para la Producción de Biogás en una Finca Cacaotera," Mar. 2023, Accessed: Jul. 04, 2023.[Online]. Available: https://repositorio.udes.edu.co/entities/publication/5b18596e-5aed-4f27- 9b57-83542410641c
5. A. I. Akinjokun, L. F. Petrik, A. O. Ogunfowokan, J. Ajao, and T. V. Ojumu, "Isolation andcharacterization of nanocrystalline cellulose from cocoa pod husk (CPH) biomass wastes," *Heliyon*, vol. 7, no. 4, p. e06680, Apr. 2021, doi: 10.1016/j.heliyon.2021.e06680.
6. "Anaerobic digestion using cocoa residues as substrate: Systematic review and meta-analysis," *Energy Sustain. Dev.*, vol. 72, pp. 265–277, Feb. 2023, doi: 10.1016/j.esd.2022.12.007.
7. A. Abraham *et al.*, "Pretreatment strategies for enhanced biogas production from lignocellulosic biomass," *Bioresour. Technol.*, vol. 301, p. 122725, Apr. 2020, doi: 10.1016/j.biortech.2019.122725.
8. O. D. Sierra, M. Engineer, J. S. Ramirez, E. Engineer, and A. R. Duke, "Biogas resource evaluation: analysis of temperature and pH factors to ensure efficient biogas production usingpineapple waste", Accessed: Sep. 28, 2023. [Online]. Available: https://www.laccei.org/LACCEI2022-BocaRaton/full_papers/FP82.pdf
9. L. Vásquez and H. Álvarez, "Estado Actual de la Generación de Energía Eléctrica a través de Biomasa en Honduras. Análisis de 2007 al 2017," *Rev. Esc. Física*, vol. 7, no. 1, Art. no. 1, Jun.2019, doi: 10.5377/ref.v7i1.8261.
10. M. C. I. Amparo, "Diagnóstico sobre la situación actual del Cacao (Theobroma cacao L.) yperspectivas sobre la producción de Cacao fino de aroma En, Honduras".
11. N. Acosta, J. De Vrieze, V. Sandoval, D. Sinche, I. Wierinck, and K. Rabaey, "Cocoa residues asviable biomass for renewable energy production through anaerobic digestion," *Bioresour. Technol.*, vol. 265, pp. 568–572, Oct. 2018, doi: 10.1016/j.biortech.2018.05.100.
12. J. da Cruz Ferraz Dutra *et al.*, "Anaerobic digestion using cocoa residues as substrate: Systematicreview and meta-analysis," *Energy Sustain. Dev.*, vol. 72, pp. 265–277, Feb. 2023, doi: 10.1016/j.esd.2022.12.007.

13. Y. I. Lijarza Galvez, “Produccion de biogas a partir de la cascara del fruto del cacao (Theobromacacao L.) a nivel de laboratorio,” *Univ. Nac. Agrar. Selva*, 2022, Accessed: Jun. 30, 2023. [Online]. Available: http://repositorio.unas.edu.pe/handle/20.500.14292/2352
14. R. Jyothilakshmi and S. V. Prakash, “Design, Fabrication and Experimentation of a Small Scale Anaerobic Biodigester for Domestic Biodegradable Solid Waste with Energy Recovery and SizingCalculations,” *Procedia Environ. Sci.*, vol. 35, pp. 749–755, Jan. 2016, doi: 10.1016/j.proenv.2016.07.085.

Waste-to-Energy Online Marketplace:Leveraging AI Recommendation Matchmaking for Enhanced Biomass Sourcing in Bioenergy Production

Kanjanapon Borisoot[1,2], *Wanida* Kanarkard[1,3,*], *Tanakorn* Wongwuttanasatian[1,4], *Pichai* Niltarach[1], *Amnart* Suksri[1,5], *Kitt* Tientanopajai[3] and *Denpong* Soodphakdee[4]

[1]Centre for Alternative Energy Research and Development, Khon Kaen University, Thailand
2 Energy Engineering Program, Faculty of Engineering, Khon Kaen University, Thailand
3Department of Computer Engineering, Khon Kaen University, Thailand
4Department of Mechanical Engineering, Khon Kaen University, Thailand
5Department of Electrical Engineering, Khon Kaen University, Thailand

Abstract. This paper introduces a novel platform of Waste-To-Energy Online Marketplace. The platform maintains a comprehensive catalogue of available biomass resources, detailing types, quantities, and geographical locations. This allows bioenergy facilities to identify and select suitable biomass feedstock based on their specific energy production requirements. Through an intuitive online marketplace, stakeholders can negotiate agreements, ensuring a streamlined and mutually beneficial exchange of biomass feedstock for bioenergy production. The online matchmaker by A.I. recommendation engine platform opens new avenues for biomass suppliers and bioenergy facilities to connect beyond traditional geographical and logistical constraints, fostering a more expansive and interconnected market. Efficient matching ensures that biomass resources are utilised optimally, reducing waste, and maximising bioenergy production. The proposed model seeks to enhance the efficiency of converting sugarcane biomass into bioenergy, leveraging digital and A.I. technologies to match biomass producers with bioenergy facilities, optimising the efficient conversion of biomass resources into renewable energy and fostering a reduction in GHG emissions associated with traditional waste disposal methods. This innovative approach has the potential to revolutionise the biomass supply chain, facilitates competitive pricing and cost-effective transactions, benefitting both biomass suppliers and bioenergy producers, promoting sustainability, efficiency, and collaboration in the journey towards a greener and more resilient energy future.

Keywords: Biomass, Bioenergy, Zero waste, Artificial Intelligence, Online Marketplace

---

* Corresponding author: wanida@kku.ac.th

1 Introduction

Thailand is commonly acknowledged as an agricultural-focused nation with a robust industry dedicated to the conversion of agricultural resources. This industry encompasses processing agricultural commodities into food products, raw materials, and a wide range of items and tools. Currently, Thai farmers primarily concentrate on cultivating sugarcane [1], resulting in a significant volume of biomass waste post-harvest. Despite agriculture serving as the historical backbone of Thailand, a considerable number of farmers lack the knowledge necessary for effective handling of agricultural waste generated after the harvest, leading to improper disposal practices. During the harvest season, there is a prevalent practice of incinerating these residuals, which gives rise to environmental problems that directly affect the daily lives of the population [2]. Despite efforts by the government to establish guidelines for the regulated transport of harvested and burned sugarcane to factories [3], the issue remains only partially resolved. Several factors contribute to the persistent burning of sugarcane plantations in Thailand, including insufficient support for harvesting labourers, the need for sugarcane farmers to acquire knowledge about managing biomass residuals from sugarcane leaves, and, most importantly, farmers who fail to recognise the value of utilising sugarcane leaf residuals to create alternative products. Addressing this challenge necessitates providing farmers with the requisite knowledge and skills for effectively managing challenging post-harvest biomass residuals. Moreover, fostering collaboration among diverse farmer groups across geographical locations is crucial for obtaining agricultural waste. Through these measures, the remaining residuals can be processed to enhance the sustainable value of biomass waste. The Alternative Energy Development Plan 2018–2037 (AEDP2018), the Thai government has implemented a policy to promote the use of renewable energy sources to make electricity, focusing on using biomass fuels [4]. Saleem [5] has pointed out that agricultural waste can serve as a viable fuel source for the sustainable generation of energy, with the energy content varying depending on the specific agricultural residuals. The biomass is agricultural waste left after the harvest, such as white straw, rice husks, tea leaf remnants, and sugarcane leaves discarded post-harvest. According to initial investigations, sugarcane leaves exhibit a heat value of up to 17.41 MJ/kg. Meanwhile, water hyacinth, rice straw, rice husk, and tea leaf residuals display calorific values of 14.81 MJ/kg, 15.09 MJ/kg, 16.47 MJ/kg, and 17.10 MJ/kg, respectively [6]. Thailand is the second-largest sugar exporter in the world, following [7], resulting in a significant surplus of sugarcane leaves post-harvest. According to findings by Dangprok et al. [8], the annual average quantity of residual sugarcane leaves left post-harvest in Thailand is recorded at 17,016 Kilotons. This considerable volume of agricultural waste remains underutilised. Currently available technologies and tools can be strategically employed to address this situation. Cano et al. [9] have noted that the upswing in electronic commerce has led to the emergence of electronic markets, serving as intermediaries in the transactional processes of buying and selling. These platforms bring together multiple vendors to offer a diverse range of products and services to consumers. Consequently, this phenomenon has the potential to significantly contribute to the enhanced sustainability of biomass supply. Flak [10] has compiled Information and Communications Technology (ICT), Internet of Things (IoT), and Industry 4.0 technologies used in biomass supply, including numerous exciting technologies such as precision agricultural tools, storage and logistics tools, and trading platforms. A problem for farmers is the need for a market for agricultural waste, which prevents the correct disposal of agrarian refuse. Farmers would benefit significantly from a platform for managing fundamental data, proposals, and sales. According to previous research, there have been efforts to establish an online marketplace to purchase and sell products, bringing buyers and sellers together to form a trading agreement. Qiang et al. [11] constructed a one-stop platform to promote the development of biomass fuel in China using the WeChat application for telephone consultations and access to information on websites, as well as an analysis of the number of fuel types, provincial contribution, and fuel price distribution. Ninda et. al. [12] discusses the use of Artificial Intelligence in a startup matchmaking platform, which aims to bring together startups and partners through a matchmaking process. Lee et. al, [13] deployed the Artificial Intelligence-based job and education institution matching system that matches job seekers to job offers or provides job education information. Beguedou et al. [14] evaluate a mobile app-based digital platform that promotes a circular economy among farmers, suppliers, and industries to reduce trade barriers for alternative fuels and ensure fair and competitive farming business prices. Moreover, in alignment with the Thailand 4.0 policy, the Ministry of Digital Economy and Society launched the National Digital Economy and Society Development Plan and Policy for a 20-year period (2018-2037). This initiative aims to advance the nation towards a digitally empowered economy. Lilavanichakul [15] briefly summarises important policy support, one of which is the Department of Agriculture Extension's "Young Smart Farmers" program. The primary objective of this program is to alleviate marketing risks along with personal and labour risks in the agriculture sector. This is achieved through the enhancement of digital skills and agricultural capabilities, facilitating the integration of digital technology into smart farming systems and commercial operations. Consequently, this initiative has given rise to a regional online marketplace facilitating the buying and selling of various agricultural products. Illustratively, the DGT Farm platform, operated by the National Bureau of Agricultural Commodities and Food Standards, serves as a central online market where a diverse array of products, especially processed agricultural goods, can be acquired. Additionally, other platforms, such as the Thailand Post Mart platform [16] and the Hub Thailand platform [17] cater to local products and those involved in the One Tambon, One Product (OTOP) Project of Thailand, thereby supporting local businesses.

The challenge of utilising agricultural waste is widely recognised, primarily due to the dispersed distribution of such material at various locations within agricultural fields. To support the commercialisation of biomass, an online marketplace platform has been established, accessible through web-based and via a LINE mobile application. The Waste-to-Energy online marketplace platform offers advantageous features, including pricing and streamlining the matchmaking process between sellers and buyers. This aligns with the concept of zero biomass waste management, aimed at promoting the use of biomass. It also encourages sugarcane farmer groups to leverage agricultural waste materials for maximum benefit and sustainability. The main contributions of this study can be summarised as follows: This paper introduces an inventive and user-friendly technology designed for farmers, facilitating their access to digital tools and equipment for efficient storage and processing of biomass. This approach effectively mitigates the problem of agricultural waste incineration post-harvest, concurrently offering income-generating prospects for farmers. The online marketplace platform for biomass trading, operating under the Waste-to-Energy paradigm, is accessible through both a website and a LINE mobile application. this platform facilitates connections between individuals interested in purchasing biomass and those looking to sell biomass, thereby streamlining the trading processes for enhanced convenience. The rest of this paper is organised as follows. Section 2 introduces the Methodology. In Section 3, The Result and Discussion of this paper are presented, and the conclusion is presented in Section 4.

2 Methodology

The core principle of the zero-biomass waste management framework is to empower farmers in the proper disposal of biomass post-harvest, while concurrently providing biomass power plants with the requisite material for electricity generation. This study establishes guidelines for the efficient management of biomass waste, with a specific focus on the residues derived from sugarcane leaves post-harvest. The zero-biomass waste management concept entails the inclusion of biomass processing farmer groups as sellers of biomass, while buyers consist of sugar factories equipped with biomass power plants. Parameters governing transactions on this platform are intelligently determined by Artificial Intelligence recommendation engine based on considerations such as transportation distance, biomass fuel pricing, as well as the quantity and type of biomass fuel involved. Fig. 1 provides the zero-biomass waste management framework.

2.1 The framework of online marketplace for biomass

Upon gathering information about the issues and needs of biomass buyers and sellers, it became apparent that initially, biomass power plants (the buyers) and farmers (the sellers) contacted each other and established agreements to trade directly. However, this approach was limited, as it did not offer adequate options such as transportation distance, purchase price, biomass quantity, and biomass type. The online biomass marketplace platform presented in this paper offers the following primary service functions: selling biomass, purchasing biomass, searching for specific biomass, and suggesting potential biomass trading partners. Moreover, upon the completion of the buying and selling process, users have the opportunity to rate their partners, and the platform can also calculate the heat and energy generated from biomass trading, facilitating the assessment of reduced CO_2 emissions, as shown in Fig. 2. This platform was developed as a LINE mobile application and provided services on the website, making it accessible for use by biomass power plants in the northeastern region.

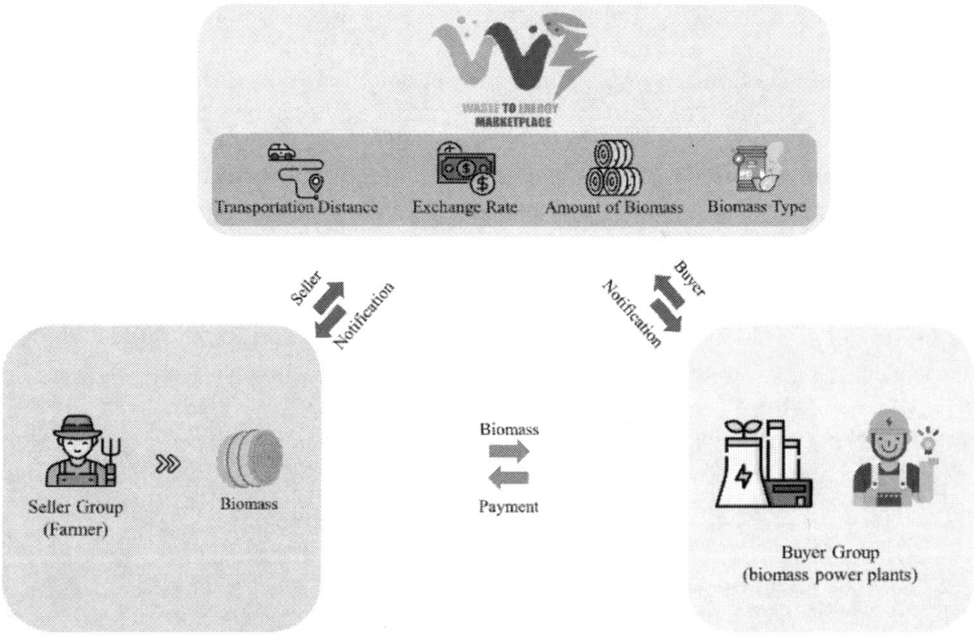

Fig. 1. Zero-biomass waste management concept.

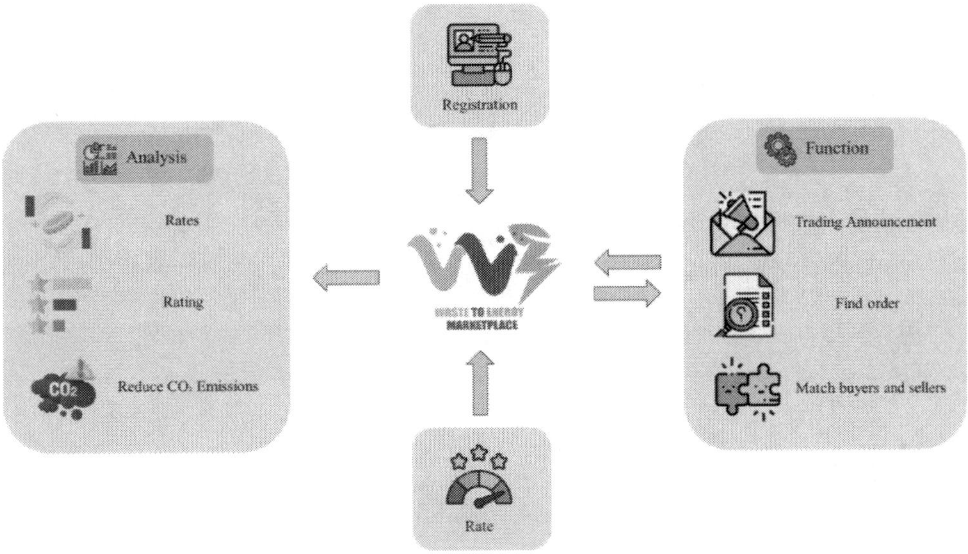

Fig. 2. Workflows of the Waste to Energy Marketplace platform.

2.2 Waste to Energy Marketplace platform

There are vital steps to follow in the Waste to Energy Marketplace within the biomass market platform, including registration and participating in the trading market. Trading agreements involve various details, as depicted in Fig. 3.

Step 1: Begin the registration process by logging in through the website (https://w2e.space) or the LINE mobile application (ID: @biomassmarket).

Step 2: Register using an email or a LINE account during the registration process.

Step 3: Once the registration identity has been verified, the user can log in using the username and password that the user has registered. After that, the user must fill in the profile details. The user must specify whether they are a buyer or seller biomass. Then fill in more information: name, address, telephone number, picture of the biomass of the buyer or seller, and specify the biomass you want to trade.

Step 4: When wanting to trade biomass, the user can provide specific details such as the location, the type of biomass, transportation, the quantity to be traded, and the price per unit. Also, the user can upload a photo of the biomass sample the user intends to buy or sell for searching and matching various buying and selling announcements.

Step 5: The match making system by A.I. recommendation engine provides recommendations and sorts potential matches to meet the needs of both biomass buyers and sellers. This intelligent matchmaking process aims to optimise the pairing of sellers, typically represented by biomass processing farmer groups, with buyers, which are often sugar factories equipped with biomass power plants. The Artificial Intelligence algorithms analyse various parameters, including transportation distance, to ensure efficient logistics and minimise environmental impact. Biomass fuel pricing is considered to establish fair and competitive transactions, fostering a sustainable biomass-to-bioenergy market. Additionally, the quantity and type of biomass fuel are factored into the matching process to align the specific energy needs of the buyers with the available biomass resources.

Step 6: The buyer and seller agree to the trade, which is considered a confirmed buy or sale. Both parties can rate each other. The evaluations given by trading partners who have executed transactions effectively function as a criterion for users to consider when choosing subsequent trading partners, as well as a mechanism for buyers and sellers to improve their transactions in the case of receiving low ratings.

https://doi.org/10.1051/e3sconf/202453003003

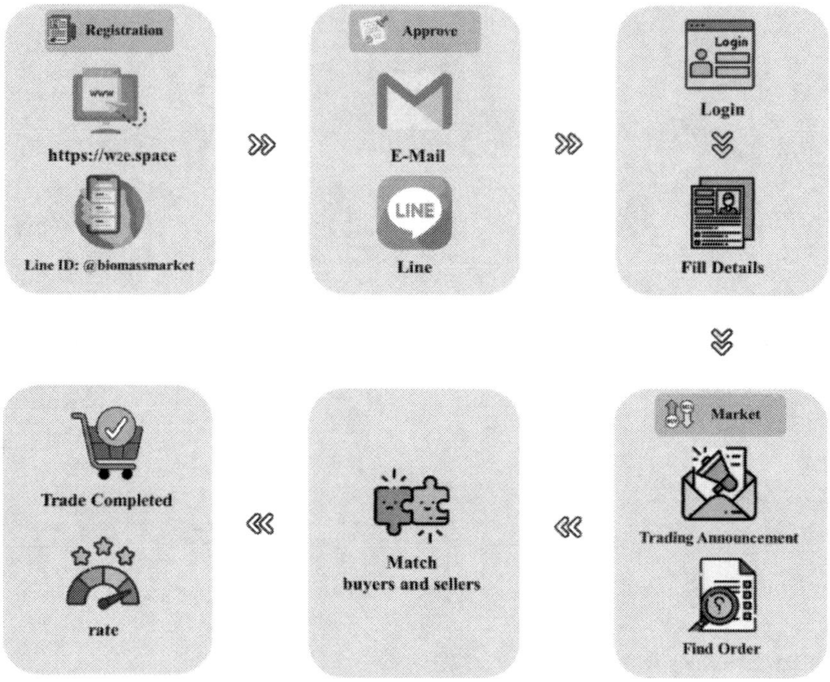

Fig. 3. Workflows involved in utilising the Waste to Energy Marketplace.

3 Results and Discussions

To address the issue of limited access to sources for buying and selling biomass, an innovative online biomass marketplace platform has been developed. This platform is accessible through a dedicated website and a LINE mobile application to ensure convenience and widespread use. The LINE mobile application, a highly popular messaging, and social media platform in Thailand, saw significant adoption. As of 2019, according to the Hootsuite Analytics report [18], 84% of internet users in Thailand have active LINE accounts, making it the 3rd most popular social media platform in the country. This paper has leveraged both the website and LINE for its operations. The user interfaces for the online biomass market platform are presented in Fig. 4(a) for the website version and Fig. 4(b) for the LINE mobile application version.

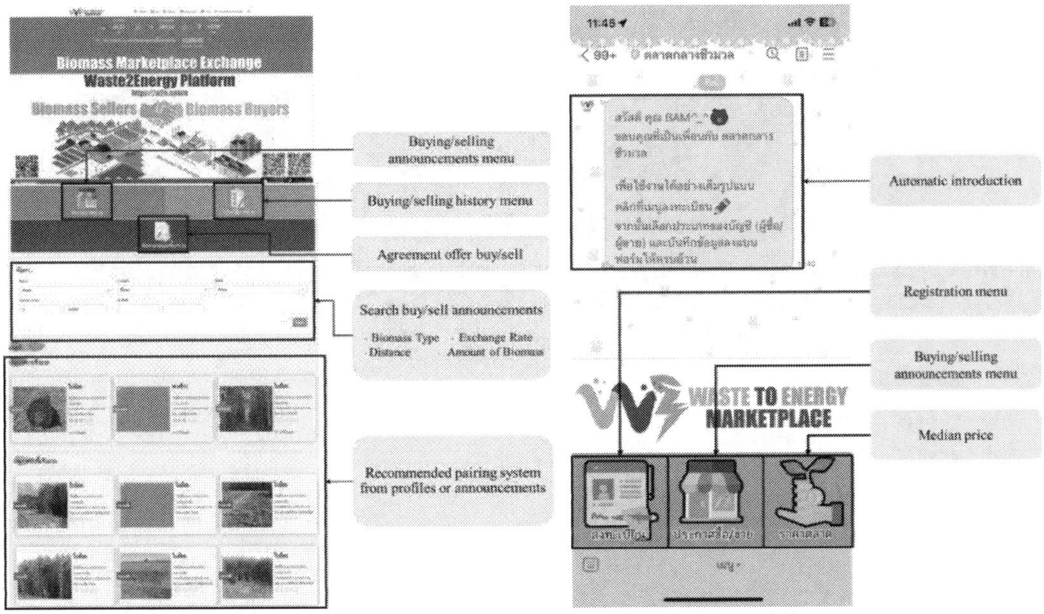

(a) (b)

Fig. 4. User-interfaces for the biomass marketplace platform (a) the website version (b) the LINE mobile application version.

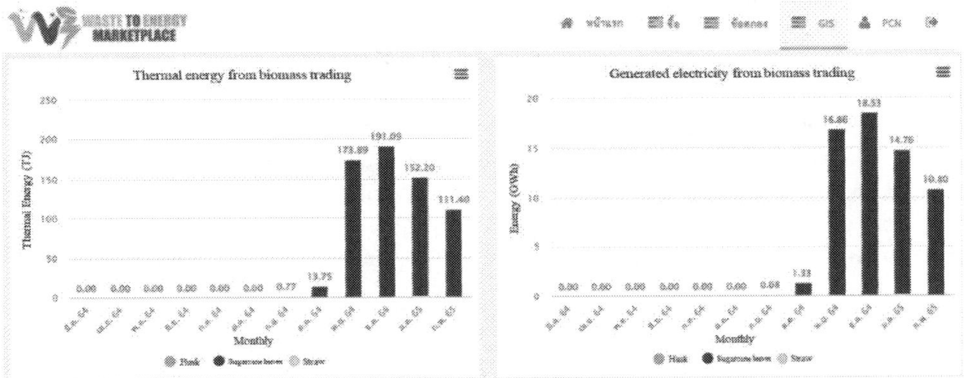

(a) (b)

Fig. 5. Analytical report: the computation of heat and energy generated through biomass trading.

Based on the practical utilisation of the online biomass market platform during the sugarcane harvest season, the online biomass market platform enabled the calculation of the heat and energy generated from biomass trading. These calculations are presented on the website of the platform, as depicted in **Error! Reference source not found.**(a) and **Error! Reference source not found.**(b), respectively. The online biomass marketplace platform is designed to facilitate trading not only in sugarcane leaves but also in other types of fuel, such as rice husks and rice straws, which are common agricultural products in Thailand. Furthermore, the website displays the reduction in carbon dioxide (CO2) emissions resulting from electricity generation using these alternative fuels, as illustrated in **Error! Reference source not found.**.

Fig. 6. The reduction in CO_2 emissions resulting from the utilisation of biomass fuel for electricity generation.

4 Conclusion

This paper aims to establish comprehensive guidelines for the management of residual biomass generated from sugarcane leaves post-harvest under the framework of a zero-biomass management concept. The primary goal is to propose the innovative online biomass marketplace platform to foster the development of channels for biomass fuel trading, leveraging the online marketplace system available in website and LINE mobile application formats. The establishment of an online biomass trading platform serves as a crucial channel for the procurement and sale of biomass fuel. Within this platform, both buyers and sellers benefit from potential matches through A.I. recommendations, which consider criteria such as transportation distance, pricing, biomass quantity, and biomass type. By leveraging Artificial Intelligence in this matchmaking approach, the biomass-to-bioenergy supply chain can be optimised, promoting effective utilisation of biomass resources, minimizing waste, and contributing to a more sustainable and economically viable bioenergy industry. This streamlined process empowers stakeholders to make well-informed decisions regarding biomass trading. Noteworthy benefits of this platform encompass decreased transportation costs and heightened transparency in trading prices. Biomass power plants can, therefore, rely on a steady fuel supply for electricity generation, aligning supply with demand more effectively. For farmers, the platform provides an avenue to utilise existing biomass residuals, leading to additional income generation. Consequently, it establishes a sustainable solution promoting zero biomass waste management. Moreover, the online biomass trading platform possesses the capability to calculate the heat and energy generated post-biomass trading, facilitating the assessment of reduced CO2 emissions.

The research team would like to thank the Centre of Alternative Energy Research and Development, Smart City Operation Centre, Khon Kaen University, for supporting our work.

References

1. Agricultural Development Policy and Planning Division, *Agricultural Economic Conditions in the First Quarter of 2023*, vol. 6. Bangkok: Office of Agricultural Economics, 2023.
2. G. Geldenhuys, J. Orasche, G. Jakobi, R. Zimmermann, and P. B. C. Forbes, “Characterization of Gaseous and Particulate Phase Polycyclic Aromatic Hydrocarbons

Emitted During Preharvest Burning of Sugar Cane in Different Regions of Kwa-Zulu Natal, South Africa," *Environ. Toxicol. Chem.*, vol. 42, no. 4, pp. 778–792, 2023, doi: 10.1002/etc.5579.

3. S. Rodmanee and C. Sungsing, "Management of Pre-harvest Sugarcane Burning," *Acad. J. Manag. Technol.*, vol. 3, no. December, pp. 60–74, 2022.
4. Department of Alternative Energy Development and Efficiency, *Alternative Energy Development Plan (AEDP2018)*. 2020.
5. M. Saleem, "Possibility of utilizing agriculture biomass as a renewable and sustainable future energy source," *Heliyon*, vol. 8, no. 2, p. e08905, 2022, doi: 10.1016/j.heliyon.2022.e08905.
6. D. R. Nhuchhen and P. Abdul Salam, "Estimation of higher heating value of biomass from proximate analysis: A new approach," *Fuel*, vol. 99, pp. 55–63, 2012, doi: 10.1016/j.fuel.2012.04.015.
7. Office of Agricultural Economics, *Information on Agricultural Products by Commodity 2022*. Ministry of Agriculture and Cooperatives: Bangkok, Thailand, 2022.
8. B. Dangprok, K. Y. Tippayawong, and N. Tippayawong, "Development of a cost optimization model for power generation from agricultural residual biomass in Thailand," *Energy Reports*, vol. 9, pp. 55–62, 2023, doi: 10.1016/j.egyr.2022.11.157.
9. J. A. Cano, A. Londoño-Pineda, M. F. Castro, H. B. Paz, C. Rodas, and T. Arias, "A Bibliometric Analysis and Systematic Review on E-Marketplaces, Open Innovation, and Sustainability," *Sustain.*, vol. 14, no. 9, 2022, doi: 10.3390/su14095456.
10. J. Flak, "Technologies for sustainable biomass supply-overview of market offering," *Agronomy*, vol. 10, no. 6, 2020, doi: 10.3390/agronomy10060798.
11. J. Qiang, Y. Zhou, Y. Yu, Y. Deng, Y. Fu, and W. Wu, "Design of a One-Stop Platform for on-Demand Supply of Biomass Pellet Fuel in China," *E3S Web Conf.*, vol. 406, p. 01010, Jul. 2023, doi: 10.1051/e3sconf/202340601010.
12. N. Lutfiani, S. Wijono, U. Rahardja, A. Iriani, and E. A. Nabila, "Artificial Intelligence Based on Recommendation System for Startup Matchmaking Platform," in *2022 IEEE Creative Communication and Innovative Technology (ICCIT)*, IEEE, Nov. 2022, pp. 1–5. doi: 10.1109/ICCIT55355.2022.10118708.
13. D. Lee, M. Kim, and I. Na, "Artificial intelligence based career matching," *J. Intell. Fuzzy Syst.*, vol. 35, no. 6, pp. 6061–6070, 2018, doi: 10.3233/JIFS-169846.
14. E. Beguedou, S. Narra, E. Afrakoma Armoo, K. Agboka, and M. Kongnine Damgou, "E-Technology Enabled Sourcing of Alternative Fuels to Create a Fair-Trade Circular Economy for Sustainable Energy in Togo," *Energies*, vol. 16, no. 9, 2023, doi: 10.3390/en16093679.
15. A. Lilavanichakul, "Development of Agricultural E-commerce in Thailand," *FFTC J. Agric. Policy*, vol. 1, no. 1, pp. 7–17, 2021, doi: 10.56669/otdt7194.
16. Thailand Post Company Limited, "Thailandpostmart platform." Accessed: Sep. 12, 2023. [Online]. Available: https://www.thailandpostmart.com/
17. General Electronic Commerce Services, "The Hub Thailand platform." Accessed: Oct. 20, 2023. [Online]. Available: https://thehubthailand.biz/home/home.html
18. Steve Callerame, "The LINE Phenomenon in Thailand – By the Numbers 2019." Accessed: Oct. 18, 2023. [Online]. Available: https://lexiconthai.com/blog/the-line-phenomenon-in-thailand-by-the-numbers/

Bioconversion of carbon dioxide and zero valent iron to methane by anaerobic sludge: kinetics and archaeal consortium

Vichai Domrongpokkaphan[1,2], *Chantaraporn* Phalakornkule[3,4], and *Maneerat* Khemkhao[2,5*]

[1]Department of Agro-Industrial, Food and Environmental Technology, King Mongkut's University of Technology North Bangkok, Bangkok 10800, Thailand
[2]Microbial Informatics and Industrial Product of Microbe Research Center, King Mongkut's University of Technology North Bangkok, Bangkok 10800, Thailand
[3]Department of Chemical Engineering, King Mongkut's University of Technology North Bangkok, Bangkok 10800, Thailand
[4]Research Center for Circular Products and Energy, King Mongkut's University of Technology North Bangkok, Bangkok 10800, Thailand
[5]Rattanakosin College for Sustainable Energy and Environment, Rajamangala University of Technology Rattanakosin, Nakhon Pathom 73170, Thailand

Abstract. Using carbon dioxide (CO_2) as a carbon source for renewable energy production has potential applications for CO_2 sequestration and greenhouse gas (GHG) emission reduction. In biological conversion, CO_2 can be transformed into methane (CH_4) by hydrogenotrophic methanogens with hydrogen (H_2) as an energy source. In this study, zero-valent iron (ZVI) of 16, 32, 64, and 96 g/L was used as the H_2 energy source for a bioconversion of CO_2 to CH_4. When the ZVI dosage was increased, a decrease in CO_2 in the headspace occurred simultaneously with the increase in CH_4. The presence of CH_4 in both CO_2/H_2 and CO_2/ZVI indicates that hydrogenotrophic methanogens can utilize both ZVI and H_2 as electron donors and convert CO_2 to CH_4. The highest methane yield of 1.728 mmol CH_4/mmol CO_2 was observed for the CO_2/ZVI 96 g/L. The modified Gompertz equation fitted the cumulative CH_4 production curves of CO_2/H_2 and CO_2/ZVI very well, where R^2 was 0.9915 and 0.9903-0.9968, respectively. 16S rDNA high-throughput sequencing results revealed that ZVI addition facilitated the increase of the family Methanobacteriaceae, which became the most abundant among other archaea. It points out that this family favors ZVI and utilizes electrons more effectively from ZVI than H_2.

1 Introduction

Concerning environmental issues and climate change, alternative energy sources are increasingly interested in biofuels. That is, bio-alcohols and biodiesel are considered for automotion, while biomass and biogas are considered for electrical power generation [1].

* Corresponding author: maneerat.khe@rmutr.ac.th

Biogas from anaerobic digestion of sludge in a wastewater treatment plant of organic wastes or residual crops is a renewable energy source. It is composed of various gases such as combustible methane (CH_4) of 50-75%, non-combustible carbon dioxide (CO_2) of 25-50%, and traces of nitrogen (N_2) <1% and hydrogen sulfide (H_2S) ranging from 10 to 2,000 ppm. H_2S can cause corrosion, damaging energy co-generation equipment or other installations [2-3]. Biogas composition depends on the type of feedstocks and operating conditions [4].

CH_4 is an odorless, colorless, or combustible gas that is lighter than air. Its calorific value is 37,781.6 kJ/Nm^3, and its energy generation capacity is 5 kWh/Nm^3 [2]. In order to be used as fuel for vehicles or for internal combustion engines, CH4 concentration should be greater than 90% [4] because biogas calorific power is proportional to CH_4 content.

If the percentage of non-combustible CO_2 in biogas is high, approximately 40%, biogas's caloric or heating value will be low. This leads to a reduction of engine power output, resulting in a limit on the use of biogas in electrical power plants or combustion engines [2-3]. Not only does CO_2 decrease the energy content of biogas, but the condensate water accumulates in the pipeline [2]. Therefore, biogas should be purified, and its pollutants should be eliminated to obtain a high percentage of CH_4 and reach a standard natural gas or vehicle fuel.

Biogas upgrading needs to increase the heating value and standardize the biogas quality. There are several methods of CO_2 removal, such as physical absorption, chemical absorption, biological absorption, pressure swing adsorption with activated carbon or molecular sieves, use of organic solvents, cryogenic separation, and membrane purification [2, 5]. However, CO_2 from these processes contributes to the greenhouse effect.

Recently, CO_2 has been considered substrate or chemical feedstock but is not a waste that must be treated [5]. In order to reduce CO_2 emission into the atmosphere, there are several strategies to utilize CO_2, including the production of urea (~70 Mt CO_2/y), inorganic carbonates and pigments (~30 Mt CO_2/y), salicylic acid (~20 Mt CO_2/y), methanol (~6 Mt CO_2/y) and propylene carbonate (a few kt CO_2/y) [6].

The utilizing CO_2 as a carbon source is attractive. Biological transformation of CO_2 into value-added products such as biomethane. In biological processes, CH_4 is a handy energy source produced by hydrogenotrophic methanogens using CO_2 or CO as carbon and H_2 as energy sources [7-8]. In addition, several studies have focused on a methanation reaction because of its important implications for energy and the environment [9]. Therefore, CO_2 methanation driven by microbial metabolism is a highly attractive option for renewable sources. CO_2 methanation requires an H_2/CO_2 ratio of 4 to yield a complete CH_4 of 100% [10]. However, a required H_2, supplemented with CO_2 methanation, is expensive, and H_2 storage is problematic because of safety issues and requires new infrastructure [11]. Therefore, H_2 should be produced from renewable resources. For example, H_2 can be generated from water electrolysis, and the energy required for electrolysis should also be renewable [10].

Another option for an H_2 source is the addition of zero-valent iron (ZVI), which can produce H_2 and release Fe^{2+} from Fe^0 [12]. In a previous study, adding ZVI enhanced CH_4 production in an anaerobic digestion of waste sludge. That is, CH_4 production was raised to 21% at 10 g/L of iron scrap added into the anaerobic sludge digester [13]. In the study by Domrongpokkaphan et al. [14], ZVI 16 g/L promoted the production of CH_4 from decanted palm oil mill effluent (POME) with an initial pH of 4.2 and increased the average CH_4 content in biogas to 76% in fed-batch mode. For the continuous mode in the completely stirred tank reactor (CSTR), the addition of ZVI increased the CH_4 production to 86%, the CH_4 content in biogas to 75%, and the total chemical oxygen demand (tCOD) removal efficiency to 88–95%. An addition of ZVI could efficiently recover the methanogenic activity suppressed by heat pre-treatment and alkali conditions. In addition, ZVI provided a short lag time for CH_4 production and increased the CH_4 yield by 91.5% [12].

In this study, ZVI powder was employed for CO_2 methanation. The CH_4 production from CO_2 was predicted using the modified Gompertz kinetic model and the archaeal changes of methanogens was characterized. Scanning electron microscopy was used to assess the micro-distribution of microorganisms on ZVI, while next-generation sequencing was used to reveal the archaeal community.

2 Materials and methods

2.1 Anaerobic sludge source

The anaerobic sludge used in this study was taken from a water quality control plant (Bangkok, Thailand). The sludge was washed and sieved to remove fine particles and excess water. The content of volatile suspended solids (VSS) was also analyzed. The anaerobic sludge was stored at 4 ℃ before further testing.

2.2 Anaerobic basal mineral medium

The anaerobic medium for methanogenic bioassay contained (per liter) 280 mg of NH_4Cl, 5,000 mg of $NaHCO_3$, 250 mg of K_2HPO_4, 10 mg of $CaCl_2.2H_2O$, 83 mg of $MgCl_2.6H_2O$, 100 mg of yeast nitrogen base and 1 mL of trace elements [15-16]. The pH of the anaerobic basal medium was adjusted to 7.2. The trace element solution contained (per liter) 50 mg of H_3BO_3, 2,000 mg of $FeCl_2.4H_2O$, 50 mg of $ZnCl_2$, 50 mg of $MnCl_2.4H_2O$, 50 mg of $(NH_4)6Mo_7O_{24}.4H_2O$, 90 mg of $AlCl_3.6H_2O$, 2,000 mg of $CoCl_2.6H_2O$, 50 mg of $NiCl_2.6H_2O$, 30 mg of $CuCl_2.2H_2O$, 100 mg of $NaSeO_3.5H_2O$, 1,000 mg of EDTA, 200 mg of resazurin as a redox indicator and 1 mL of 36%HCl [15, 17]. The anaerobic medium was boiled in a microwave for 5-10 min to release O_2 and then flushed with nitrogen (N_2)/CO_2 (80:20, v/v) to make it anoxic.

2.3 Batch experiments on the influence of ZVI on CO_2 reduction to CH_4

In an assay with ZVI on methane production from CO_2, a 120 mL serum bottle with a working volume of 45 mL was used. The serum bottle contains ZVI powder with a final concentration of 16, 32, 64, and 96 g/L, 30 mL of anaerobic inoculum (1.5 g VSS/L), and 15 mL of the anaerobic medium. After that, a serum bottle was flushed with the gas mixture N_2/CO_2 (80:20, v/v) for 1 min until the pink color disappeared. The serum bottles were sealed with a butyl rubber stopper and aluminum cap.

After the disappearance of the pink color, the serum bottles containing various ZVI dosages were supplied by pressurizing bottles to a final headspace with 150 kPa of CO_2 gas (99.99%) [18]. No ZVI dosage was used as a negative control or positive control. A serum bottle with no ZVI dosage was supplied by pressurizing bottles to a final headspace with 150 kPa of CO_2 gas (99.99%) for negative control and 150 kPa of H_2/CO_2 gas mixture (80:20, v/v) [19] for positive control. After that, all bottles were incubated at 35 ± 2 ℃ in an incubator shaker (B_IPX1102C, Biobase, China) at 120 rpm. Each experiment was considered terminated when the substrates were consumed and when products became stable or reached a stationary stage.

The headspace pressure was measured using a manometer (digital pressure gauge, GD1010J, OCTA, Thailand) and was calculated as standard temperature and pressure (STP). In order to determine the components, 100 μL of gas samples in the headspace of each assay were monitored by CG-TCD (Shimadzu, GC-2014, Japan). The solid was dried and examined with scanning electron microscopy (SEM) coupled with energy-dispersive X-ray spectroscopy (EDS). The sludge was collected from each bottle at the end of the tests for extracellular polymeric substance (EPS) and next-generation sequencing.

The methane production data obtained in the experiment was fitted [20-21] to the modified Gompertz in Eq. (1). This model assumes that the CH_4 production rate corresponds to the reduction of CO_2 with H_2 from ZVI, making the equation suitable for CH_4 production in a batch experiment. The first-order model is given in Eq. (2) [22-23].

$$Y = A \cdot \exp\left\{-exp\left[\frac{R_m}{A} \cdot \exp(\lambda - t) + 1\right]\right\} \quad (1)$$

$$Y = A \cdot [1 - \exp(-k \cdot t)] \quad (2)$$

where Y is the cumulative CH_4 production (mmol); A is the maximum cumulative CH_4 (mmol); Rm is the CH_4 production peak rate (mmol/d); λ is the CH_4 production lag phase period (d); t is the reaction time (d); e is Euler' s function with a value of 2.718; k is the CH_4 production rate constant or the first-order reaction constant (d^{-1}).

2.4 Analytical techniques

For EPS extraction and analysis, a heat extraction method was followed, according to Morgan et al. [24]. Extraction of the Loosely bound (LB)-EPS and Tightly bound (TB)-EPS from the anaerobic sludge was conducted according to Li and Yang [25]. The sludge was resuspended in 15 mL of 0.05% NaCl solution and then diluted to its original volume of 50 mL. The NaCl solution for dilution was pre-heated up to 70 °C to ensure that the sludge suspension reached an immediate warm temperature of 50 °C. After that, the sludge suspension was sheared by a vortex mixer for 1 minute and centrifugation at 4000 xg for 10 minutes. The organic matter in the supernatant was readily extractable EPS and was regarded as the LB-EPS of the biomass.

For the extraction of the TB-EPS, the sludge pellet left in the centrifuge tube was resuspended in a 0.05% NaCl solution to its original volume of 50 mL. The sludge suspension was heated to 60 °C in a water bath for 30 minutes, and the sludge mixture was then centrifuged at 4000 xg for 15 minutes. The supernatant was collected and regarded as the sludge's TB-EPS extraction. Both the LB-EPS and TB-EPS extractions were analyzed for proteins and polysaccharides. Protein was analyzed by a UV/VIS spectrophotometer (G10S UV-Vis, Thermo Fisher Scientific, USA) following the modified Lowry method using bovine serum albumin as the standard. Polysaccharide content was determined with the phenol-sulphuric acid method using glucose as the standard.

For SEM-EDS analysis by TESCAN MIRA3 (Czech Republic), the images of the ZVI surface before and after anaerobic culture were taken at a voltage of 10 keV under a high vacuum mode of operation.

2.5 Next-generation sequencing for archaeal community structure

The next-generation sequencing (NGS) was used to monitor the archaeal diversities of inoculums before and after the anaerobic batch experiment. 1 mL liquid sludge samples were centrifuged and extracted DNA using a DNeasy PowerSoil Kit according to the manufacturer's instruction (Giagen, Germany). Extracted DNA was verified by electrophoresis in 1% agarose gel of 1x TAE (Tris/Acetic acid/EDTA, Bio-rad) stained with SRBRTM Safe DNA gel stain (Invitrogen, Life Technology) to check the integrity of the extracted DNA. The concentration of extracted DNA was quantified in duplicate using a

Qubit 4.0 fluorometer (Invitrogen, Life technology, USA) with a Qubit dsDNA BR assay kit (Thermo Fisher Scientific, USA). Purified DNA samples were sent to Novogene Co., Ltd (China) for next-generation sequencing.

Archaeal 16S rRNA sequences of V3-V4 regions were amplified using the primer pairs U519F [26] and 806R [27]. The PCR products were checked by Agilent 2100 (Agilent Technologies) and Q-PCR. The 288-bp insert DNA library of the DNA fragment was generated using the NEBNext® Ultra™ DNA Library Prep Kit for Illumina (NEB, USA). An Illumina HiSeq 4000 platform (Illumina Inc., USA) was used for 2 × 150 bp paired-end sequencing. Sequences with ⩾97% similarity were assigned to the same operational taxonomic units (OTUs). Paired-end reads were analyzed using FLASH (V1.2.7) software. Sequence analysis was performed with Uparse software (v7.0.1090).

3 Results and discussion

3.1 CO_2 bioconversion to CH_4 by anaerobic sludge and ZVI

For the first 11 days of incubation, the CO_2 in the headspace of the CO_2/ZVI at 16-64 g/L was not eliminated; only CO_2 in the headspace of the CO_2/ZVI at 96 g/L was eliminated with an initial rate of CO_2 depletion of 0.374 mmol/d (Figure 1). It is pointed out that as ZVI concentration increased, the rate of CO_2 conversion in the headspace occurred at a faster rate. H2 in the headspace of the CO_2/H_2 and the CO_2/ZVI of 96 g/L was fully consumed after day 4 of incubation, while none of H_2 was detected from the CO_2/ZVI of 16-64 g/L (data not shown). None of the H_2 accumulation in the headspace implied that H_2 and ZVI were the sole electron donors for converting CO_2 to CH_4, and anaerobic microorganisms adapted to these conditions faster.

The highest CH_4 production over time during the first 11 days was found in CO_2/H_2, followed by CO_2/ZVI at 96 g/L (Figure 2). However, the initial rate of CH_4 generation of the CO_2/ZVI at 96 g/L (0.129 mmol/d) was slightly higher than that of the CO_2/H_2 at 0.112 mmol/d. The CH_4 generation rates correlate with higher ZVI concentration. The presence of CH_4 in both CO_2/H_2 and CO_2/ZVI indicates that hydrogenotrophic methanogens can utilize both ZVI and H_2 as electron donors and convert CO_2 to CH_4 [20, 28-29].

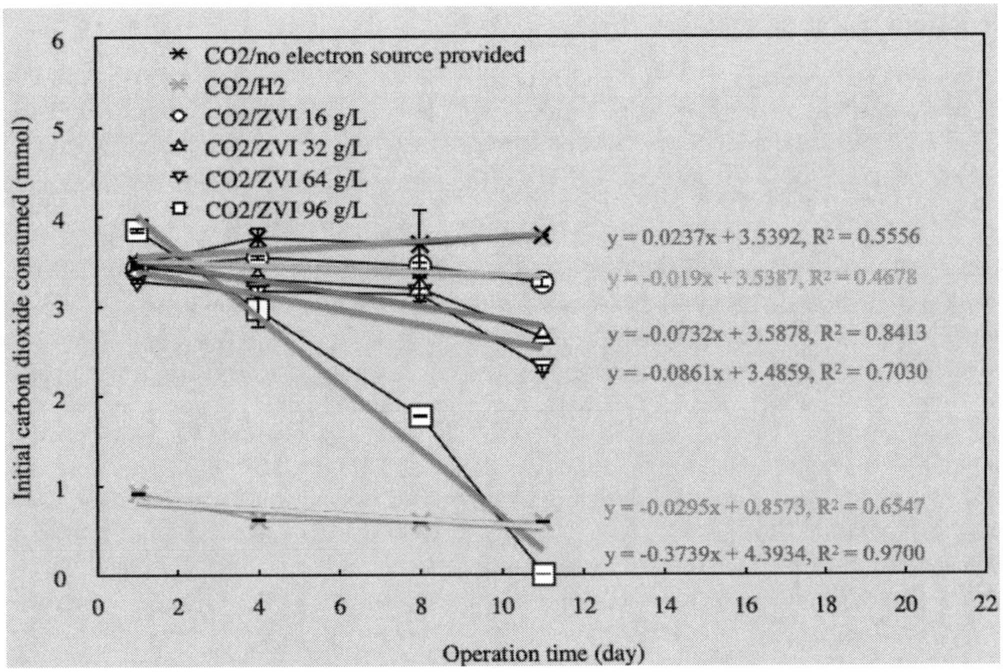

Fig. 1. Initial rate of CO_2 consumption in the headspace under various ZVI concentrations.

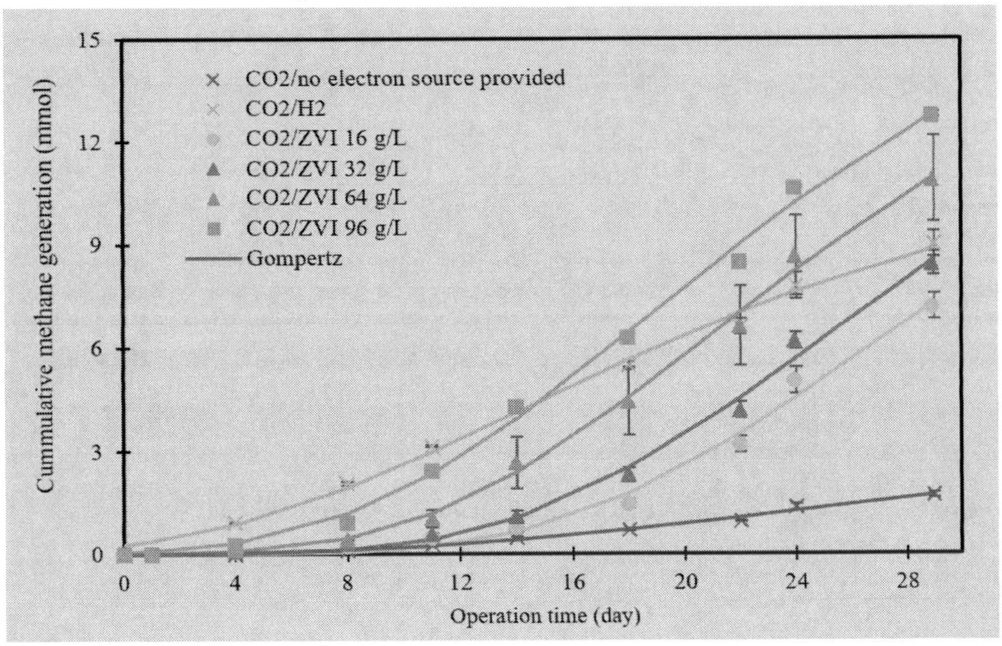

Fig. 2. Cumulative methane generation from CO_2 with H_2 and ZVI sources at different concentrations between experimental data and prediction by the modified Gompertz model. Points are experimental data, while lines represent model prediction.

3.2 Modified Gompertz model

The observed and predicted cumulative CH_4 production data of each batch test fitted to the modified Gompertz model is shown in Figure 2, and the kinetic parameters of the model are shown in Table 1. The cumulative CH_4 production of the CO_2, CO_2/H_2, and CO_2/ZVI at 16, 32, 64, and 96 g/L were: 1.70, 9.01, 7.20, 8.45, 10.91, and 12.78 mmol, respectively, for the observed data; 1.71, 8.87, 7.31, 8.57, 10.99, and 12.81 mmol, respectively, for the modified Gompertz model. Among the tested models, the modified Gompertz was magnificent for modeling CH_4 production in the CO_2/H_2. The fitting error of less than 10% between the observed and predicted CH4 yield data was considered an excellent fitting error in previous studies [30-31].

When ZVI concentration increased from 16 to 96 g/L, the lag phase (λ) decreased from 26 to 18 days. As ZVI increases, the decrease in λ is due to the high H_2 donor released from ZVI to promote faster CO_2 conversion to CH_4. This means that high ZVI dosages were suitable for methanogen adaptation and the conversion of CO_2 to CH_4. The kinetic constant observed for CO_2/ZVI was higher than that of CO_2/H_2. The highest kinetic constant of 0.1006 d^{-1} was obtained for the CO_2/ZVI at 64 g/L. The difference between predicted and experimental results for CO_2 with no electron source provided and CO_2/H_2 is 0.11 and 1.35%, respectively, while the CO_2 with a ZVI of 16-96 g/L ranges from 0.30-1.05%.

Table 1. Kinetic parameters of CH_4 production from conversion of CO_2 by anaerobic sludge with ZVI.

Parameters		CO_2 as a carbon source					
		No electron source	H_2 source	ZVI source at various concentrations (g/L)			
				16	32	64	96
CH_4 yield (mmol CH_4/mmol CO_2)	Experiment	0.230	1.218	0.973	1.142	1.475	1.728
	Modified Gompertz	0.231	1.199	0.988	1.158	1.485	1.732
R_m (mmol CH_4/d)		0.117	2.118	0.705	2.376	4.288	4.924
A (mmol CH_4)		1.289	31.087	42.957	46.426	46.211	50.150
λ (d)		24.39	14.235	26.28	25.01	20.76	18.53
k (d^{-1})	First-order	0.0724	0.0929	0.0960	0.0936	0.1006	0.0965
R^2		0.9913	0.9944	0.9903	0.9938	0.9968	0.9965
Difference between experiment and modified Gompertz (%)		0.11	1.35	1.05	1.14	0.72	0.30

3.3 EPS of anaerobic sludge conversion of CO_2 to CH_4

The EPS of anaerobic sludge from a batch experiment is listed in Table 2. Protein is regarded as a substance that is hard to extract. Polysaccharides and protein were the two main components of EPS. That is, there was more protein than polysaccharide in all extracted EPS, which confirmed previous studies that the predominant EPS material was protein [32]. The predominance of protein in the EPS may be due to a large quantity of exoenzymes [33]. It was found that the amount of LB-EPS was much greater than that of TB-EPS in both polysaccharides and proteins. Therefore, optimum ZVI dosage could help the formation of EPS, while too high a ZVI dosage may interrupt metabolite transport and reduce the surface charge.

Fe^{2+} is a chelating agent of EPS generation. A low concentration of ZVI can stimulate microbial self-protection mechanisms, and microbes secrete more EPS to prevent the introverted transfer of toxic materials [34-35]. Shen et al. [36] reported that Fe^{2+} enables to bond with EPS of sludge to form stable complexes, which is crucial to maintaining an intact structure of anaerobic sludge. In addition, the interaction between Fe^{2+} and EPS could help the formation of granulation [36]. It was believed that Fe^{2+} could decrease the surface charge of microorganisms by double-layer compression, which could strengthen the aggregation of microorganisms [37].

Table 2. The content of EPS in anaerobic sludge.

Sludge sample	Polysaccharides (mg/g SS)		Proteins (mg/g SS)	
	Loosely bound EPS	Tightly bound EPS	Loosely bound EPS	Tightly bound EPS
Original sludge	1.68 ± 0.00	0.62 ± 0.00	32.39 ± 1.58	16.27 ± 0.32
CO_2/no electron source provided	5.86 ± 0.15	2.28 ± 0.15	55.69 ± 0.47	29.83 ± 2.09
CO_2/H_2	7.08 ± 0.09	4.20 ± 2.28	63.94 ± 0.57	34.00 ± 0.95
CO_2/ZVI 16 g/L	7.06 ± 0.81	2.12 ± 0.17	62.13 ± 1.23	32.39 ± 0.38
CO_2/ZVI 32 g/L	4.74 ± 0.22	1.41 ± 0.03	52.53 ± 2.66	21.78 ± 0.76
CO_2/ZVI 64 g/L	6.10 ± 1.26	1.76 ± 0.12	57.97 ± 0.28	28.22 ± 3.23
CO_2/ZVI 96 g/L	2.93 ± 0.44	0.79 ± 0.41	40.31 ± 1.33	19.90 ± 0.38

3.4 Stimulated methanogenesis from CO_2 by ZVI

Figure 3 shows the abundance of archaeal communities in the anaerobic sludge at the end of the experiment. The archaeal community structures of the original sludge and the sludge samples taken from the CO_2, H_2/CO_2, and CO_2/ZVI were different. Among methanogens, the families Methanomethyliaceae and Bathyarchaeia were detected in the original sludge, and the sludge samples taken from CO_2 and H_2/CO_2 believed that this family might be involved with the CO_2 metabolic pathway. Methanomethyliaceae reported as methylotrophic H_2-dependent methylotrophic methanogens, play important roles in the carbon cycle of hydrocarbon-enriched subsurface ecosystems [38]. Bathyarchaeia, classified as a methylotrophic methanogen [39], could utilize various C1 substances and methylated compounds [40].

Methanothermobacteriaceae abundance was obviously increased in CO_2/ZVI (16–96 g/L), indicating that ZVI were highly beneficial to the enrichment of this family. When ZVI released Fe^{2+} to participate in the synthesis of hydrogenase, the genus Methanothermobacter, a member of the family Methanothermobacteriaceae, exhibited strong competitive adaptability. This genus was enriched, and methanogenesis was promoted [21]. Therefore, it indicated that Methanothermobacteriaceae adapted to the anaerobic environment faster due to the addition of ZVI.

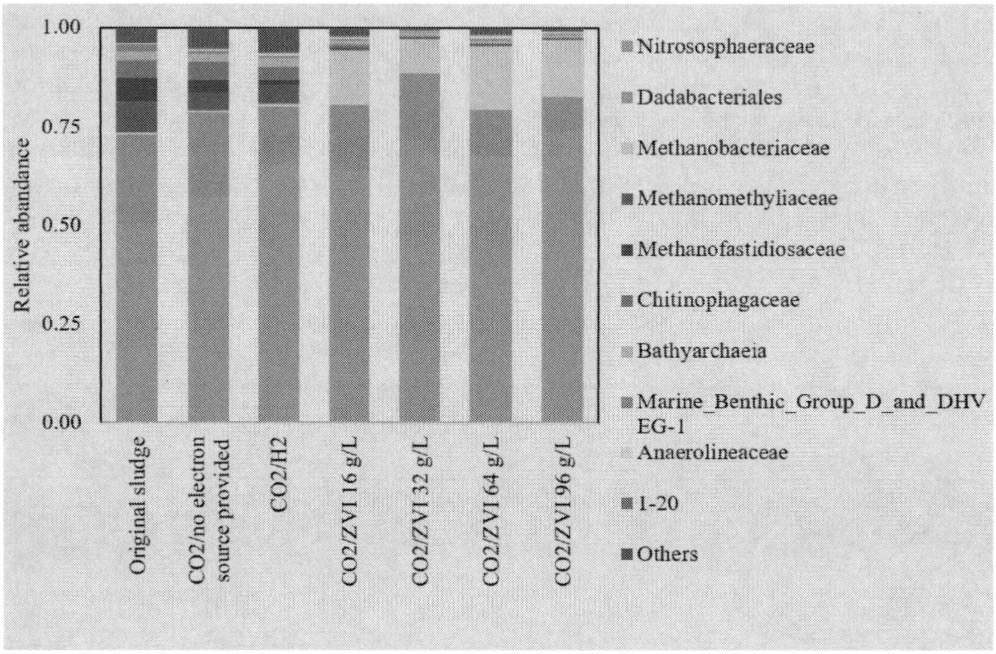

Fig. 3. Archaeal community abundance at the family level in the anaerobic sludge at various conditions.

3.5 Microbial surface morphology in the presence of ZVI

A high magnification of SEM in Figure 4 (a–c) showed that the sludge of CO_2 with no electron source provided had irregular clumps, cocci, and thin rod shapes. When the sludge was incubated with CO_2/ZVI 96 g/L, the blade-like crystals of ZVI are attached to the irregular clumps and cocci (Fig. 4d), and prismatic and needle-like crystals are attached to cocci and thin rod shapes (Fig. 4e-4f).

The EDS result showed the element composition profile. After anaerobic conditions, the sludge from CO_2 with no electron source provided consisted mainly of 50.93–57.86% C and 22.28–32.52% O, while the Fe element was the main component in the sludge of CO_2/ZVI 96 g/L, accounting for 65.01–72.61%. C and O were the main constituent elements of cells, biofilms, and metabolites [19].

Fig. 4. SEM and EDS (b-c, e-f) analysis of the sludge incubated under anaerobic conditions (a-c) without any electron donors and (d-f) ZVI powder 96 g/L filled with 150 kPa CO_2 in the headspace.

4. Conclusion

The amount of ZVI added to anaerobic incubation has an impact on the amount of CH_4 that is produced from CO_2 bioconversion. The modified Gompertz showed the best fit for cumulative CH_4 production from CO_2/H_2 and CO_2/ZVI. The addition of ZVI optimized methanogens, especially the family Methanobacteriaceae, which enhanced the conversion of CO_2 to CH_4 products. It indicated that ZVI had a positively impact on this family's favor.

This work (Grant No. RGNS 63-105) was supported by Office of the Permanent Secretary, Ministry of Higher Education, Science, Research and Innovation (OPS MHESI), Thailand Science Research and Innovation (TSRI) and Rajamangala University of Technology Rattanakosin (Fundamental Fund: fiscal year 2023 by National Science Research and Innovation Fund) (FRB6638/2566).

References

1. JI. Huertas, N. Giraldo, S. Izquierdo, *Mass transfer in chemical engineering* MarkoÅ J (Ed.) 134-150, ISBN: 978-953-307-619-5, InTech (2011)
2. F. Osorio, JC. Torres, *Renew. Energ.* **34**, 2164–2171 (2009)
3. DG. Mulat, F. Mosbæk, AJ. Ward, D. Polag, M. Greule, F. Keppler, JL. Nielsen, A. Feilberg, *Waste Manage.* http://dx.doi.org/10.1016/j.wasman.2017.05.054
4. M. Harasimowicz, P. Orluk, G. Zakrzewska-Trznadel, *J. Hazard. Mat.* **144,** 3, 698-702 (2007)
5. L. Appels, J. Baeyens, J. Degre`ve, R. Dewil, *Energ. Combust. Sci.* **34,** 755–81(2008)
6. P. Jajesniak, HEMO. Ali, TS. Wong, TS, *J. Bioprocess Biotech.* **4,** 155 (2014)
7. H. Kawaguchi, T. Sakuma, Y. Nakata, H. Kobayashi, K. Endo, K. Sato, *J. Biosci. Bioeng.* **110**, 1, 106-108 (2010)
8. K. Sato, H. Kawaguchi, H. Kobayashi, *Energ. Convers. Manage.* **66**, 343–350 (2013)
9. P. Frontera, A. Macario, M. Ferraro, P. Antonucci, *Catalysts* **7**, 12 (2017) doi:10.3390/catal7020059.
10. CV. Miguel, MA. Soria, A. Mendes, LM. Madeira, *J. Nat. Gas Sci. Eng.* **22** 1-8 (2015)
11. G. Luo, S. Johansson, K. Boe, L. Xie, Q. Zhou, I. Angelidaki, *Biotechnol. Bioeng.* **109**, 4, 1088-10894 (2012)

12. Y. Zhang, Y. Liu, Y. Jing, Z. Zhao, X. Quan, *J. Environ. Sci.* **24**, 4, 720-727 (2012)
13. Y. Feng, Y. Zhang, X. Quan, S. Chen, *Water Res.* **52**, 242-250 (2014)
14. V. Domrongpokkaphan, C. Phalakornkule, M. Khemkhao, *Int. J. Hydrogen Energy* **46**, 60, 30976-30987 (2021)
15. S. Karri, R. Sierra-Alvarez, JA. Field, *Chemosphere* **62**, 121–127 (2006)
16. J. Gonzalez-Estrella, R. Sierra-Alvarez, JA. Field, *J. Hazard. Material.* **260**, 278– 285 (2013)
17. RS. Tanner, *J. Microbiol. Meth.* **10**, 83–90 (1989)
18. I. Vyrides, M. Andronikou, A. Kyprianou, A. Modic, A. Filippeti, C. Yiakoumis, CG. Samanides, *J. CO_2 Util.* **27**, 415-422 (2018)
19. RS. Taubner, SK-MR Rittmann, *Front. Microbiol.* **7**, 532 (2016) doi: 10.3389/fmicb.2016.00532
20. AF. Hamzah, MH. Hamzah, HC. Man, NS. Jamali, SI. Siajam, MH. Ismail, *Heliyon* 9 (2023) e16791
21. J. Cheng, H. Dong, H. Zhang, L. Yuan, H. Li, L. Yue, J. Hua, J. Zhou, *Bioresour. Technol.* **314**, 123799 (2020) doi:10.1016/j.biortech.2020.123799
22. P. Neumann, A. Torres, FG. Fermoso, R. Borja, D. Jeison, *Int. Biodeterior. Biodegrad.* **100**, 85-88 (2015)
23. R. Bedoić, A. Špehar, J. Puljko, L. Čuček, B. Ćosić, T. Pukšec, N. Duić, *Renew. Sustain. Energy Rev.* **130**, 109951 (2020)
24. JW. Morgan, CF Forster, L. Evison, *Water Res.* **24**, 743–750 (1990)
25. XY. Li, SF. Yang, *Water Res.* **41**, 1022– 1030 (2007)
26. L. Øvreås, L. Forney, FL. Daae, V. Torsvik, *Appl. Environ. Microbiol.* **63**, 3367–3373 (1997)
27. K. Takai, K. Horikoshi, *Appl. Environ. Microbiol.* **66**, 5066–5072 (2000)
28. K. Mori, H. Tsurumaru, S. Harayama, *J. Biosci. Bioeng.* **110**, 4, 426–430 (2010) doi:10.1016/j.jbiosc.2010.04.012
29. T. Uchiyama, K. Ito, K. Mori, H. Tsurumaru, S. Harayama, *Environ. Microbiol.* **76**, 1783–1788 (2010)
30. O. Bakari, KN. Njau, C. Noubactep, *Case Stud. Chem. Environ. Eng.* **8**, 100377 (2023)
31. F. Raposo, R. Borja, MA. Martín, A. Martín, MA. De la Rubia, B. Rincón, *Chem. Eng. J.* **149**,1-3, 70-7 (2009) https://doi.org/10.1016/j.cej.2008.10.001
32. L. Huang, G. Liu, G. Dong, X. Wu, C. Wang, Y. Liu, *Chem. Eng. J.* **316**, 525–533 (2017)
33. B. Frølund, R. Palmgren, K. Keiding, PH. Nielsen, *Water res.* **30**, 8, 1749-1758 (1996)
34. ZG. He, YP. Yang, ZH. Shan, YH. Hu, H. Zhong, *Trans. Nonferrous Met. Soc. China.* **24**, 4, 1171-8 (2014)
35. Y. Wang, J. Qin, S. Zhou, X. Lin, L. Ye, C. Song, Y. Yan, *Water Res.* **73**, 252-64 (2015)
36. CF. Shen, N. Kosaric, R. Blaszczyk, *Water Res.* **27**, 1, 25–33 (1993)
37. HQ. Yu, HHP. Fang, JH. Tay, *Water Sci. Technol.* **41**, 199–205 (2000)
38. YF. Liu, J. Chen, ZL. Liu, ZW. Hou, B. Liang, LY. Wang, L. Zhou, LB. Shou, DD. Lin, SZ. Yang, JF. Liu, *Engineering* **24**, 264-75 (2023)

39. Z. Lyu, N. Shao, T. Akinyemi, WB. Whitman, *Current Biology* **28**, 13, 727–732 (2018) doi:10.1016/j.cub.2018.05.021
40. J. Hou, Y. Wang, P. Zhu, N. Yang, L. Liang, T. Yu, M. Niu, K. Konhauser, BJ. Woodcroft, F. Wang, *Sci. Adv.* **9**, 27 (2023) eadf5069. DOI:10.1126/sciadv.adf5069.

Anaerobic digestion of distillery wastewater and napier grass: characterization and process performance parameters

Louise Grace Avena[1*], *Michelle* Almendrala[1], *Rumina* Rodriguez[1], and *Kristine Gayle* Mojica[1]

[1] School of Chemical, Biological, and Materials Engineering and Sciences, Mapúa University, Intramuros, Manila, Philippines

Abstract. The BOD/COD ratio of napier grass (NG) + distillery wastewater (DWW) was found to be 0.4982, which reflected that the biodegradability of organic material is high and can be degraded and treated efficiently via biological processes. Moreover, further reductions in total solids (TS) and volatile solids (VS) were observed after thermal pretreatment at temperatures of 60°C and 100°C. VS/TS ratio also increased from 0.8767 to 0.9508 indicating a more significant proportion of organic matter are favorable for biogas production. This was evidenced by the resulting removal efficiencies ranging from 15% to 32% and 23% to 42%, for TS and VS, respectively. Overall, the study reported relevant effects of the application of thermochemical pretreatment in the process performance parameters in anaerobic digestion.

1 Introduction

Alcohol manufacturing companies produce vast amounts of wastewater that can be treated accordingly to recover valuable materials for biogas production. In 2015, only 10% of wastewater was treated in the Philippines, while 58% of the groundwater was contaminated; as the primary issue is the lack of sludge treatment and disposal facilities, untreated effluent is dumped into bodies of water [1]. Molasses-based ethanol distillery wastewater (DWW) is a byproduct of alcoholic fermentation and is essential for cost-effective biogas production. Around 10-15 L of DWW are believed to be created for every 1 liter of ethanol produced. When DWW is discharged into water streams, economic burden on businesses and environmental consequences are posed [2]. Effluent from sugar and alcohol industries today commonly undergoes anaerobic digestion (AD) not only for energy generation but also for solids removal and treatment.

* Corresponding author: lgsavena@mapua.edu.ph

The insufficiency of studies addressing the potential of the utilization of the by-products from sugarcane processing industries and for biofuel production implies the urgency for its research and development especially due to the increasing sugar demand. This study aims to determine substrate characteristics such as pH, biochemical oxygen demand (BOD), chemical oxygen demand (COD), total solids (TS) and volatile solids (VS) of the substrates and final digestate and evaluate the solids removal efficiency that could address the current problems with solids and wastewater management in ethanol distilleries in the Philippines. Moreover, this work also endeavors to determine AD process performance parameters such as TS reduction, VS reduction, and pH change, all of which are relevant in the future of anaerobic treatment systems. Ultimately, valuable information may be provided throughout the study regarding an improved waste-to-energy technique for DWW that will aid in recommending better pollution control strategies.

2 Methodology

2.1 Raw materials and inoculum

Fresh samples of napier grass (Pennisetum purpureum), also known as Elephant grass, were harvested from South Cotabato, while the DWW used in this study was collected from Absolut Distillers Inc. in Lian, Batangas. To prevent unwanted fermentation, the DWW sample was stored in a plastic container at a temperature of 4°C before the experiment. The nutrients required were sourced from the chemistry laboratory of Mapúa University in Intramuros, Manila. Before use, the fresh NG was blended. The substrates and micronutrients were diluted with Type III water, while the inoculum was prepared by combining cow manure and DWW in a 50:50 ratio, followed by seven days of inoculation before use.

2.2 Experimental setup and pretreatment

The batch AD setup followed the liquid displacement method during biogas collection [3] while all runs were subjected to chemical pretreatment using $Ca(OH)_2$, from which reactors R1 and R2 were focused on mono-digestion, R3 on co-digestion, and R4 and R5 on thermal pretreatments at temperatures of 60°C and 100°C, respectively.

2.3 Analytical methods

A 50:50 ratio by volume mixture of DWW and NG was mixed homogenously, wherein an 800 mL sample was sealed in a 1L HDPE bottle. Each sample were tested for biochemical oxygen demand (BOD) and chemical oxygen demand (COD) characterization (see Table 1).

Table 1. Test method used by NASAT Labs for feedstock characterization

Parameter	Test Method
BOD	5210-B: 5-Day BOD Test
COD	Closed reflux, Colorimetric Method

 https://doi.org/10.1051/e3sconf/202453003005

2.3.1 TS and VS

To measure TS and VS levels, the EPA Method 1684 was used. Samples placed on crucibles were then labeled, weighed, and heated at 105°C for 24 hours. Afterward, they were cooled in a desiccator before being weighed again. The sample is heated once more for an hour, cooled again, and weighed until the weight change is less than 4%.

$$TS = \frac{W_{final} - W_{dish}}{W_{initial} - W_{dish}} \tag{1}$$

To analyze volatile solids, it is necessary to ignite the dried residues in a furnace for a duration of two hours at a temperature of 550°C. After cooling in a desiccator, the residues must be weighed. This cycle of igniting, cooling, and weighing must be repeated until the weight difference is below 4%.

$$VS = \frac{(W_{final} - W_{ash}) - W_{dish}}{W_{final} - W_{dish}} \tag{2}$$

The mentioned procedures were carried out on the samples both before and after the onset of AD.

2.3.2 pH

Modifying the initial pH of samples before AD is critical to maintain digestion process stability, optimize microbial activity, and prevent inhibition. All pH values were adjusted within the range of 6.25 to 7.9 using 1M NaOH, creating an environment conducive to the effective functioning of methanogenic bacteria, as it significantly enhances biogas production and yield [4]. The sludge's pH levels were determined using a pH meter before and after digestion.

2.4 Removal efficiency

The advantage of the pre-treatment techniques was also measured by determining their removal efficiencies of TS and VS using Equation 3.

$$\text{Removal efficiency (\%)} = \frac{C_i - C_f}{C_i} \times 100 \tag{3}$$

where C_i is the initial concentration of the variable and C_f is its final concentration. This measure is an important indicator of organic matter decay which implies biogas production [5].

3 Results and Discussion

3.1 Feedstock characterization

The results of the feedstock characterization are shown in Table 2. The BOD:COD ratio is a metric used to assess the biodegradability of a substrate. A threshold value of 0.4 is indicative of easy biodegradability, while a value greater than 0.5 is considered most effective [6]. The BOD/COD ratio of NG + DWW was found to be 0.4982, the biodegradability of organic

 https://doi.org/10.1051/e3sconf/202453003005

material is considered high. In order to achieve greater efficacy, a value exceeding 0.5 is regarded as more favorable. As a result, NG underwent thermochemical pretreatment (TCP) in an effort to increase its suitability for the AD process and enhance its ability to biodegrade. The values of TS and VS evaluated in the substrate were within the acceptable range outlined in the existing literature as indicated in Table 2. A reduction in the values of TS and VS was detected after the TCP stage. The decrease in biomass size has been attributed to chemical reactions and alterations in the biomass structure induced by the breakdown of lignin in NG with Ca(OH)2. Increased cellulose and hemicellulose accessibility for subsequent microbial digestion was a factor in the reduction of solid composition as a whole [7,8]. Previous research has also reported comparable decreases in TS and VS [9,10].

Moreover, further reductions in TS and VS were observed after TP at temperatures of 60°C and 100°C, which is consistent with the results reported by Pumpoung et al. [11], who observed an increase in the removal of TS and VS as the temperature raised. In contrast, empirical evidence also suggests that TP does not always result in a statistically significant reduction in TS. This is because a significant proportion of lignocellulosic components continue to exist in a volatile state despite undergoing an increased dissolution process [12]. A notable disparity, conversely, was noted in the reduction of VS, which signifies that TP effectively degraded complex compounds into simpler molecules that were more readily digestible, resulting in lower VS levels than previously observed [13].

Nevertheless, after TCP, the VS/TS ratio increased from 0.8767 to 0.9508. As they are more susceptible to complete digestion, substrates with a higher VS/TS ratio—which indicates a more significant proportion of organic matter—are more favorable for biogas production [7,13]. In addition, the AD must maintain an optimal pH around neutral; the initial pH was adjusted as outlined in Table 4 (Appendix D). The pH value of NG alone was 7.93, while those of NG + DWW after pretreatment ranged from 6.67 to 7.04.

Table 2. Feedstock characterization.

Parameter	NG	DWW	NG*	DWW*	NG + DWW	NG (TCP)
TS (g/g_{sample})	0.1056	0.0769	0.3220[a], 0.3110[b], 0.8600[c], 0.1630[d]	0.0903[e], 0.0799[f], 0.1520[g]	0.0871	0.0903[+] 0.10259[++] 0.09254[+++]
VS (g/g_{sample})	0.0926	0.0649	0.8500[a], 0.2888[b], 0.8600[c], 0.1432[d]	0.0855[e], 0.0762[f], 0.8168[g]	0.0762	0.08586[+] 0.09056[++] 0.07998[+++]
VS/TS	0.8767	0.8449	-	-	0.8753	0.9508[+] 0.8827[++] 0.8643[+++]
pH	7.93	6.78	7.5[h]	6.52[f]	6.7	7.04[+], 7.23[++], 6.67[+++]
BOD (mg/L)	-	-	-	-	31,000	-
COD (mg/L)	-	-	-	-	62,228.78	-
BOD/COD	-	-	-	-	0.4982	-

* From literature
[14][a], [15][b], [16][c], [17][d], [18][e], [19][f], [20][g], [11][h], [+]60°C, [++]80°C, [+++]100°C

 https://doi.org/10.1051/e3sconf/202453003005

3.2 C/N Ratio

The production of biogas is significantly impacted by the carbon-to-nitrogen (C/N) ratio, given that carbon provides energy for microorganisms and nitrogen induces the formation of ammonia gas. An overly high C/N ratio can promote acidogenic bacteria over methanogenic bacteria, impeding biogas formation [21]. On the other hand, a very low C/N ratio encourages microbial development by causing rapid nitrogen consumption. For the production of biogas, a C/N ratio between 20 and 30:1 is ideal [22]. Previous research on the C/N ratios for DWW and NG is shown in Table 3.

Table 3. C/N Ratio of DWW and NG based on existing literatures.

	C/N Ratio	Reference
DWW	2.11	[23]
	3.75	[24]
	10.4	[25]
Napier Grass	39.35	[17]
	35	[26]
	30.62	[27]

NG has a high cellulose and xylan content that can be hydrolyzed to produce monomeric sugars applicable to microorganisms. Due to its high C/N ratio, it can be used as a proper compensatory element for substrates with lower C/N ratios and helps control the presence of inhibitory substances. C/N ratios of 8.22 for chicken manure and 39.35 for NG were recorded by Weerayutsil et al. [17], suggesting that mono-digestion based purely on manure did not produce the maximum amount of methane because an excessive amount of nitrogen inhibited the growth of methane-forming bacteria. On the other hand, methane generation was enhanced during co-digestion with materials with different C/N ratios while preserving system stability. Adjama et al. [28] investigated co-digestion using human waste and lignocellulosic feedstock (rice straw), and they found that a mixture of 50% rice straw and 50% human waste produced the most methane. Dube's research on pretreated napier grass and butcher shop waste showed that the best biogas yield was obtained with a 50:50 co-digestion ratio (70.3 Nml/g•VSadded) [29]. In light of these findings, the 50:50 DWW: NG ratio employed in this investigation is consistent with achieving a balanced C/N ratio in the substrate composition. The averaged values from existing literatures point to biogas production's ideal C/N ratio (20-30:1). This explains why the biogas output from DWW was the lowest (753.0 mL), the biogas output from NG alone was more significant (1253.0 mL), and the biogas output from DWW + NG were the highest (< R3 < R4 < R5).

3.3 Removal efficiency

The application of pretreatment procedures yielded noticeable reduction in TS and VS concentrations, as seen in Figure 1. The corresponding removal efficiencies observed ranged from 15% to 32% and 23% to 42%, respectively. Studies report that lignin and hemicellulose begin to solubilize at a temperature range of 150-180oC, paving way from easier degradation [30]. Nevertheless, the occurrence of inhibitors in such elevated temperatures could lead to

https://doi.org/10.1051/e3sconf/202453003005

decreased biogas yield and supplementary expenses. [31]. Hence, exploring the optimal exposure time for the low-temperature pretreatment (LTPT) of NG is recommended.

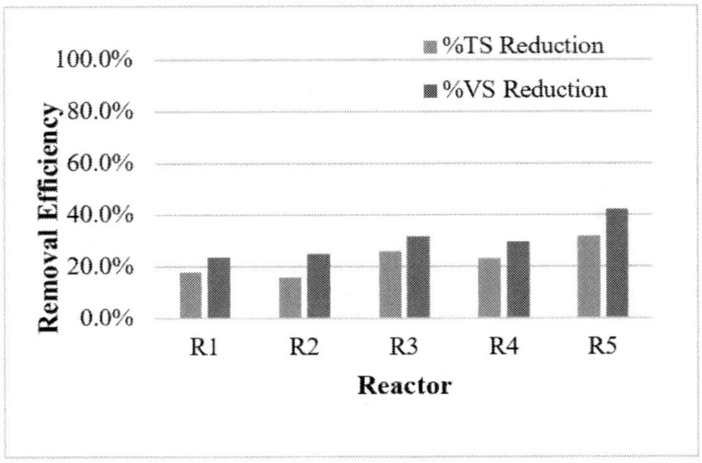

Fig. 1. Removal efficiency.

3.4 pH change

The digesters' initial and final pH values, as illustrated in Table 4 (Appendix D), are all situated between 6.25 and 7.9, thereby establishing an environment favorable for the efficient operation of methanogenic bacteria [4]. It is worth mentioning that the pH of R1 peaked at 7.93, whereas R5 displayed its lowest point at 6.67. Jayaraj et al. conducted a study wherein they observed that the degradation efficiency and biogas yield were significantly greater at pH 7, while the lowest values were documented at pH 5 [32]. The reduced biogas production can be attributed to the possible development of inhibitors, including furfural and soluble phenolic compounds, which are recognized to hinder biogas synthesis [33]. The influence of pH was more significant than the increase in organic acid yield. This is due to the inhibition of hydrolytic bacteria, which are accountable for the initial hydrolysis reaction during biogas production. As a result, the degradation of complex polymers into simpler monomers, which is necessary for the subsequent reaction in methanogenesis, was impeded [34]. One preventive measure suggested by prior research is to regulate the pH level between 4 and 7 to prevent the formation of inhibitory compounds [35]. The increased probability of producing inhibitory compounds, including volatile fatty acids (VFA) and ammonia, this finding is consistent with previous studies that have established a link between elevated pH levels, heightened ammonia toxicity, and the buildup of volatile fats [36].

The data in Figure 2 indicates that the final pH values for the NG treated with DWW fell within the acidic range, precisely at 5.86, 6.25, and 5.08. In contrast, the control (R1) maintained a pH of 7.44, considered normal. This result is consistent with prior investigations, in which NG-containing substrates tended to acidic final pH values, whereas the control group maintained a normal pH range [27]. The decrease in biogas generation observed in samples R2, R3, R4, and R5 can be ascribed to the pH dropping to toxic levels, specifically below pH 6.5. Such a low pH level impairs the sensitivity of microbial communities involved in AD. On the contrary, R1 maintains biogas production beyond the 28-day digestion period due to its pH remaining in close proximity to neutrality [37].

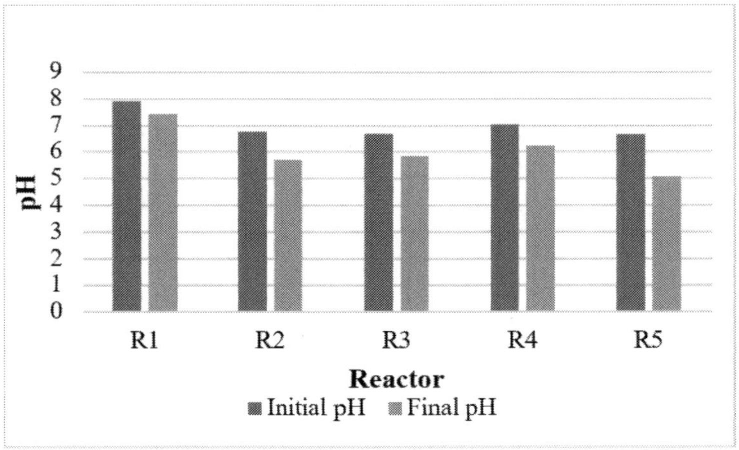

Fig. 2. pH values of digesters before and after AD.

4 Conclusion

This study examined the valorization of DWW and NG via AD for future energy generation and solid/wastewater management and treatment. Because of its higher C/N ratio, NG is an effective stabilizer for substrates like DWW with lower C/N ratios. The findings revealed that R4 and R5 experienced the most substantial decrease in VS. Moreover, the feedstock characterization indicated that NG + DWW had an easy biodegradability ratio, to further increase its effectiveness, there is a need for TCP to improve its suitability for AD. Furthermore, it was discovered that the utilization of TCP resulted in a more significant decrease in the percentage of VS, indicating its importance in facilitating subsequent microbial digestion. The digesters maintained consistent initial and final pH values between 6.25 and 7.9, which created a conducive environment for the effective growth and development of methanogenic microbes. Therefore, it was recommended that future research adopt continuous monitoring of pH levels and biogas production to evaluate the stability and sustainability of the AD process.

The authors would like to thank the School of Chemical, Biological, and Materials Engineering and Sciences, Mapúa University for their support on this study.

References

1. Wastewater Management and Sanitation in Asia | Asian Development Bank n.d. https://www.adb.org/features/wastewater-management-and-sanitation-numbers (accessed August 26, 2023).
2. Bhardwaj S, Das P. A Review: Advantages and Disadvantages of Biogas. International Research Journal of Engineering and Technology 2017.
3. Avena LG, Almendrala M, Caparanga A. Effects of thermal pretreatment and biotin supplementation on the anaerobic digestion of pineapple (Ananas comosus) wastes, kinetics of methane production, and statistical analysis. Energy Reports 2023;9:202–8. https://doi.org/10.1016/J.EGYR.2023.09.184.

4. Gebreeyessus GD, Mekonen A, Alemayehu E. A review on progresses and performances in distillery stillage management. J Clean Prod 2019;232:295–307. https://doi.org/10.1016/j.jclepro.2019.05.383.
5. Tariq M, Mehmood A, Abbas Y, Rukh S, Shah FA, Hassan A, et al. Digestate quality and biogas enhancement with laterite mineral and biochar: Performance and mechanism in anaerobic digestion. Renew Energy 2024;220:119703. https://doi.org/10.1016/J.RENENE.2023.119703.
6. Aguilar-Torrejón JA, Balderas-Hernández P, Roa-Morales G, Barrera-Díaz CE, Rodríguez-Torres I, Torres-Blancas T. Relationship, importance, and development of analytical techniques: COD, BOD, and, TOC in water—An overview through time. SN Appl Sci 2023;5. https://doi.org/10.1007/s42452-023-05318-7.
7. Avena LGS. Kinetic Study of Methane Production from Anaerobic Digestion of Pineapple (Ananas comosus) Wastes : Effects of Pretreatment and Biotin Supplementation by 2023.
8. Prapinagsorn W, Sittijunda S, Reungsang A. Co-digestion of napier grass and its silage with cow dung for methane production. Energies (Basel) 2017;10:1–20. https://doi.org/10.3390/en10101654.
9. Sanni SE, Akinrinola O, Ojima Yusuf E, Fagbiele OO, Agboola O. Chemical Kinetics of Alkaline Pretreatment of Napier Grass (Pennisetum purpureum) Prior Enzymatic Hydrolysis. The Open Chemical Engineering Journal 2018;12:36–56. https://doi.org/10.2174/1874123101812010036.
10. Kullavanijaya P, Chavalparit O. The effect of ensiling and alkaline pretreatment on anaerobic acidification of napier grass in the leached bed process. Environmental Engineering Research 2020;25:668–76. https://doi.org/10.4491/eer.2019.231.
11. Sp C, Grass N. C-17 Temperature Affects To the Potential of Biogas Production By Co-Fermentation From Chlorella sp. and Napier Grass. Proceeding of 23rd Tri-U International Joint Seminar and Symposium Bogor Agricultural University, Indonesia, 23-28 October 2016 2016.
12. Menon A, Ren F, Wang JY, Giannis A. Effect of pretreatment techniques on food waste solubilization and biogas production during thermophilic batch anaerobic digestion. J Mater Cycles Waste Manag 2016;18:222–30. https://doi.org/10.1007/s10163-015-0395-6.
13. Wang C, Hong F, Lu Y, Li X, Liu H. Improved biogas production and biodegradation of oilseed rape straw by using kitchen waste and duck droppings as co-substrates in two-phase anaerobic digestion. PLoS One 2017;12:1–19. https://doi.org/10.1371/journal.pone.0182361.
14. Kasulla S, Malik SJ, Yadav A, Kathpal G. Potential of Biogas Generation from Hybrid Napier Grass 2022;6:277–81.
15. Boonpiyo S, Sittijunda S, Reungsang A. Co-digestion of napier grass with food waste and napier silage with food waste for methane production. Energies (Basel) 2018;11. https://doi.org/10.3390/en11113200.
16. Sittijunda S. Biogas Production from Hydrolysate Napier Grass by Co-Digestion with Slaughterhouse Wastewater using Anaerobic Mixed Cultures. KKU Res J 2015;20:323–36.
17. Weerayutsil P, Khoyun U, Khuanmar K. Optimum Ratio of Chicken Manure and Napier Grass in Single Stage Anaerobic Co-digestion. Energy Procedia 2016;100:22–5. https://doi.org/10.1016/j.egypro.2016.10.141.
18. Ráduly B, Gyenge L, Szilveszter S, Kedves A, Crognale S. Treatment of corn ethanol distillery wastewater using two-stage anaerobic digestion. Water Science and Technology 2016;74:431–7. https://doi.org/10.2166/wst.2016.185.

19. Mikucka W, Zielinska M. Individual Phenolic Acids in Distillery Stillage Inhibit Its Biomethanization. Energies (Basel) 2022;15. https://doi.org/10.3390/en15155377.
20. Tansengco ML, Herrera DL, Tejano JC. Treatment of molasses-based distillery wastewater in a pilot-scale anaerobic sequencing batch reactor (ASBR). Electronic Journal of Biology 2016;12:367–73.
21. Choi Y, Ryu J, Lee SR. Influence of carbon type and carbon to nitrogen ratio on the biochemical methane potential, pH, and ammonia nitrogen in anaerobic digestion. J Anim Sci Technol 2020;62:74–83. https://doi.org/10.5187/JAST.2020.62.1.74.
22. Phuttaro C, Sawatdeenarunat C, Surendra KC, Boonsawang P, Chaiprapat S, Khanal SK. Anaerobic digestion of hydrothermally-pretreated lignocellulosic biomass: Influence of pretreatment temperatures, inhibitors and soluble organics on methane yield. Bioresour Technol 2019;284:128–38. https://doi.org/10.1016/j.biortech.2019.03.114.
23. Kumari P, Tanvi, Chaudhary S, Dhanker R, Goyal Sneh. Analysis of maturity parameters of distillery effluent and unchopped paddy straw through composting using microbial consortia. Pharma Innovation 2018;7:1–7.
24. Hakika DC, Sarto S, Mindaryani A, Hidayat M, Mufrodi Z. Detoxification of Distillery Wastewater by AOP Fenton for the Enhancement of Biogas Production. Jurnal Bahan Alam Terbarukan 2022;11:01–7. https://doi.org/10.15294/jbat.v11i1.35844.
25. BIO-DIESEL PRODUCTION FROM DISTILLERY WASTE WATER BY OLEAGINOUS YEASTS BY ENHANCING LIPID PRODUCTION USING BIOCHEMICAL 2020:2020.
26. Saitawee L, Hussaro K, Teekasap S, Cheamsawat N. Biogas proction from anaerobic co-digestion of cow dung and organic wastes (napier pak chong i and food waste) in Thailand: Temperature effect on biogas product. Am J Environ Sci 2014;10:129–39. https://doi.org/10.3844/ajessp.2014.129.139.
27. Haryanto A, Hasanudin U, Afrian C, Zulkarnaen I. Biogas production from anaerobic codigestion of cowdung and elephant grass (Pennisetum Purpureum) using batch digester. IOP Conf Ser Earth Environ Sci 2018;141. https://doi.org/10.1088/1755-1315/141/1/012011.
28. Adjama I, Derkyi NSA, Uba F, Akolgo GA, Opuko R. Anaerobic Co-Digestion of Human Feces with Rice Straw for Biogas Production: A Case Study in Sunyani. Modelling and Simulation in Engineering 2022;2022. https://doi.org/10.1155/2022/2608045.
29. Dube, N. N. Comparison of anaerobic digestion approaches using selected fibrous and non-fibrous organic waste. Cape Peninsula University of Technology. https://etd.cput.ac.za/handle/20.500.11838/3643.
30. Olatunji KO, Ahmed NA, Ogunkunle O. Optimization of biogas yield from lignocellulosic materials with different pretreatment methods: a review. Biotechnology for Biofuels 2021 14:1 2021;14:1–34. https://doi.org/10.1186/S13068-021-02012-X.
31. Wang M, Wang J, Li Y, Li Q, Li P, Luo L, et al. Low-Temperature Pretreatment of Biomass for Enhancing Biogas Production: A Review. Fermentation 2022, Vol 8, Page 562 2022;8:562. https://doi.org/10.3390/FERMENTATION8100562.
32. Jayaraj S, Deepanraj B, Velmurugan S. Study on the Effect of pH on Biogas Production from Food Waste by Anaerobic Digestion. The International Green Energy Confrence 2014;5:799–803.
33. Hendriks ATWM, Zeeman G. Pretreatments to enhance the digestibility of lignocellulosic biomass. Bioresour Technol 2009;100:10–8. https://doi.org/10.1016/j.biortech.2008.05.027.

34. Zainudeen MN, Kwarteng M, Nyamful A, Mohammed L, Mutala M. Effect of temperature and pH variation on anaerobic digestion for biogas production. Ghana Journal of Agricultural Science 2021;56:1–13. https://doi.org/10.4314/gjas.v56i2.1.
35. Banu J R, Sugitha S, Kavitha S, Kannah R Y, Merrylin J, Kumar G. Lignocellulosic Biomass Pretreatment for Enhanced Bioenergy Recovery: Effect of Lignocelluloses Recalcitrance and Enhancement Strategies. Front Energy Res 2021;9:1–17. https://doi.org/10.3389/fenrg.2021.646057.
36 Trosgård E, Granström K, Renström R. Small-scale biogas production in the province of Pampanga, Philippines Småskalig biogasproduktion i Pampangaprovinsen, Filippinerna 2015.
37. Bahira, Baki AS, Bello A. Effect of Varying pH on Biogas Generation using Cow Dungjournal/drjbb 2018;4:28–33.

Medical Waste Classification Using Convolutional Neural Network

Pensiri Akkajit[1,2*] and *Arsanchai* Sukkuea[3]

[1]Faculty of Technology and Environment, Prince of Songkla University, Phuket Campus, Phuket 83120 Thailand
[2]Waste Management for Sustainable Development, Center of Excellence on Hazardous Substance Management, Chulalongkorn University, Bangkok 10330
[3]School of Engineering and Technology, Walailak University 222 Thaiburi, Thasala, Nakhon Si Thammarat 80160 Thailand

Abstract. Medical waste disposal is a significant issue in developing countries like Thailand. It poses a persistent public health challenge as it leads to contamination of the environment and the spread of infectious diseases. This study aims to address this challenge by using a deep learning model to categorize different types of medical waste, including vials, masks, syringes, gloves, cotton, bandages, and IV tubes. Transfer learning method was employed to enhance the classification process. The study utilized the EfficientNet_b7 model and evaluated its performance based on accuracy, precision, recall, and F1 score. The results showed that with transfer learning, EfficientNet_b7 achieved a classification accuracy of 99% for both the training and testing datasets. Although there was a decline in accuracy, particularly for the syringe class, pretrained CNNs significantly improved the efficiency and accuracy of medical waste classification. Consequently, this proposed CNN model can serve as a viable alternative to conventional methods for classifying medical waste. By implementing these approaches, the efficiency of waste classification is improving, leading to a reduction in the costs associated with manual classification. This promotes sustainable waste management practices, which in turn contribute to the overall health of ecosystems and human well-being.

1 Introduction

Medical waste encompasses the waste generated within healthcare facilities, such as hospitals, clinics, blood banks, and laboratories. This waste includes items such as expired drugs, pharmaceutical products, swabs, syringes, needles, and sharp medical instruments [1]. This waste is comprised of two different categories, namely hazardous and non-hazardous. Out of the overall waste produced by healthcare activities, approximately 85% constitutes general non-hazardous waste and the remaining 15% is categorized as hazardous material, which may encompass infectious, chemical, or radioactive components [2]. Due to potential

* Corresponding author: pensiri.a@phuket.psu.ac.th.

contamination with blood, bodily fluids, and infectious materials, medical waste presents a significant environmental concern, carrying the risk of infection transmission. In order to protect the environment, medical waste management (MWM) is of importance which encompasses a range of activities essential for handling waste, starting from its generation, segregation, collection, transportation, treatment, and final disposal. In many developing counties, medical waste management suffers from inadequate regulation and a lack of operational standards [1]. The disposal of medical waste in these regions often involves problematic practices, which may include open dumping or incineration. Several limiting factors contribute to this situation, including insufficient funding, improper handling, and passing on the responsibility for discarding of the waste on to facility workers [3].

Thailand is a destination for medical tourism, particularly for procedures such as cosmetic surgeries, dental treatments, and cardiac care. It holds the top position in the list of medical tourism destinations, primarily attributed to its cost-competitive advantage compared to Malaysia and Singapore [4]. Before the COVID-19 pandemic, Thailand's healthcare waste (HCW) amounted to 0.68 kg per bed per day. In 2018, a total of 55,497 tons of infectious waste were generated, with the majority, specifically 49,898 tons of infectious waste, primarily treated through incineration facilities [4]. Phuket, the largest island in the southern part of Thailand, has reported the generation of approximately 1,200 kilograms of medical waste per day from hospitals and about 1.32 kilograms per day from clinics [5]. Waste that is indiscriminately dumped in undesignated places is a serious concern for residential communities. Mismanagement of this medical waste poses a significant risk factor for the spread of diseases and can lead to various adverse effects on the local communities. Akkajit et al. [1] assessed the knowledge, attitude, and practice (KAP) relating to the waste management among the healthcare workers in clinics at Phuket, Thailand from 172 clinics and showed significant and positive correlations between knowledge and attitude; and working experience and its duration as significant factors influencing good medical waste management practice.

Due to a lack of standardization, healthcare workers have difficulty sorting waste effectively. To accomplish this, we propose a deep learning-based classification method for medical wastes to automate the segregation of medical waste generated in healthcare center. Despite the remarkable accomplishments achieved by employing CNNs for classification and characterization tasks, the identification of medical waste remains challenging. Network Based on transfer learning (TL) for medical waste classification during the Covid-19 using various CNN architectures with TL was compared to the classification of medical waste images [6] with an accuracy of 99.40%, a sensitivity of 98.18%, and a specificity of 100% without overfitting. VGG19, InceptionV3, EfficientNetB7, and ResNet50 were built that can classify images into three categories, general, hazardous, and infectious which can primarily achieve above 95% accuracy and have performed remarkably well on test data [7]. A medical waste detection and classification system based on deep learning models (YOLOv5 and YOLO-resnet18) demonstrated a target detection model suitable for medical waste, achieving an accuracy of 90% for medical waste detection and 99% for classification [8].

Enhanced and comprehensive waste sorting practices can significantly increase overall recycling rates. This separation is more feasible in some industries due to the presence of specific types of waste. Medical waste, especially in healthcare center, is more difficult to separate than waste from other industries, since sharp objects and objects contaminated with infectious substances are common [9]. Herein, we investigated the applicability of CNNs to the classification of medical wastes using EfficientNet_b7 based on seven different classes included vial, mask, syringe, gloves, cotton, bandage, and IV tube. EfficientNets are a family of image classification models that achieve state-of-the-art accuracy and have been widely used in image classification. By fine-tuning the network parameters in each layer of the CNNs studied in this study, the classification effectiveness of CNN was enhanced. The proposed methodology presents several advantages over manual methods, including substantial time and effort reduction.

2 Methodologies

2.1 Dataset

In this section, the images were acquired from an open-source Roboflow [10]. The dataset consists of 2,026 images separated into seven classes of wastes included bandage, cotton, gloves, IV-tube, mask, syringe, and vial. Examples of the medical waste images used as a dataset are presented in Table 1.

Table 1. Image dataset for CNN-based medical waste classification.

Class	Example	Class	Example	Class	Example
Vial		Mask		IV tube	
Syringe	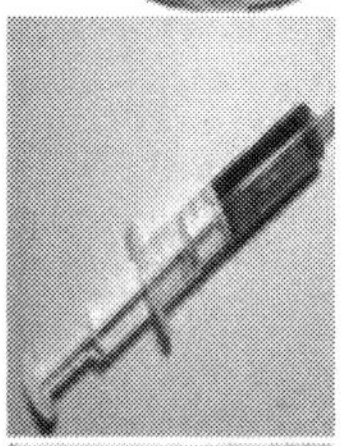	Gloves	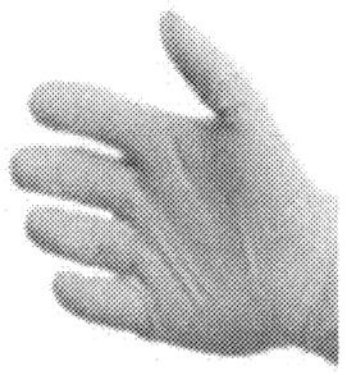	Cotton	
Bandage					

Figure 1 illustrates the methodology for CNN-based medical waste classification employed in this study including, model training, model testing, and model evaluation. This process was performed to identify and label the different types of wastes—vial, mask, syringe, gloves, cotton, bandage, and IV tube. During the annotation process, individual medical waste was marked, labelled, and assigned their corresponding class labels. The dataset was then divided into training and validation sets (80%) and testing sets (20%). Therefore, a total of 1,689 images were used for training and validation, while 337 images were allocated to the testing set. The training set was used to train the CNN, and the validation and testing sets were used to evaluate the performance of CNN model on unseen data.

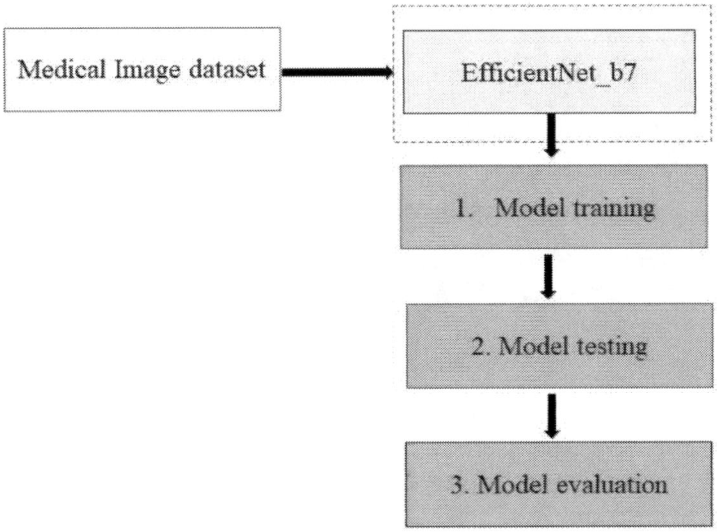

Fig. 1. Experimental design for classifying medical waste using EfficientNet_b7.

2.2 CNN training and validation

During CNN training, parameters are adjusted using labeled data to fine-tune the network's accuracy and performance. In addition, to tackle the issue of overfitting, an 80:20 data split between training and validation datasets and testing sets, coupled with a five-fold cross-validation technique was performed which repeated five times to ensure a comprehensive evaluation to help mitigate the impact of data variation across different splits, resulting in a robust evaluation.

2.3 Model evaluation

To assess the efficiency of CNNs in accurately classifying the different kinds of medical waste, the evaluation was based on the metrics, including accuracy, precision, recall, and the F1 score. To calculate the metrics, confusion matrix is used to calculate the evaluation metrics encompassing true positives (TP), false positives (FP), true negatives (TN), and false negatives (FN). The explanation and equations are followed [9,11,12].

3 Results and Discussions

3.1 CNN performance analysis

In this section, the performance of the proposed CNNs was assessed based on the metrics of accuracy, precision, recall, and F1 score to ensure their reliability and effectiveness in classifying the different kinds of medical waste. Tables 2 present a comparison of the model performance using the training and testing dataset. As can be seen, EfficientNet_b7 demonstrated the classification accuracy of 99% for both training and testing dataset. We found that the utilization of TL with pretrained CNNs substantially enhanced the efficiency and accuracy of medical waste classification. With the small image dataset in the testing set, however, the CNNs demonstrated the ability to distinguish different kinds of medical waste, thereby leading to the accurate in medical waste classification. The F1 score ranged from 98% to 100%, precision and recall ranged from 96 to 100% for the testing.

Table 2. CNN-based performance for medical waste classification.

Dataset	Class	precision	recall	F1 score	support	accuracy
Training and Validation	Bandage	1.00	1.00	1.00	225	0.99
	Cotton	1.00	1.00	1.00	244	
	Gloves	1.00	1.00	1.00	243	
	IV-tube	1.00	1.00	1.00	261	
	Mask	1.00	1.00	1.00	255	
	Syringe	0.96	1.00	0.98	215	
	Vial	1.00	0.96	0.98	246	
Testing	Bandage	1.00	1.00	1.00	42	0.99
	Cotton	1.00	1.00	1.00	42	
	Gloves	1.00	1.00	1.00	46	
	IV-tube	1.00	1.00	1.00	49	
	Mask	1.00	1.00	1.00	50	
	Syringe	0.98	1.00	0.99	47	
	Vial	1.00	0.98	0.99	61	

In addition, we compared the results of our study with other studies that utilized CNNs for image classification. For example, the SSD-MobileNet model was employed to identify medical waste, achieving a recognition accuracy of over 98.5% with an average recognition speed of 52 milliseconds [13]. A smart waste classification model using a hybrid CNN-LSTM approach with TL was developed using a dataset consisting of 27,027 images, including 17,005 organic waste and 10,025 recyclable waste images, to evaluate the performance of different models, including VGG-16, ResNet-34, ResNet-50, and AlexNet. The proposed model achieved the highest precision at 95.45% [14].

3.2 Confusion matrix

In addition, Figure 2 and 3 illustrate model prediction and confusion matrix of EfficientNet_b7 to determine how well they classified the seven classes. According to the results, the CNNs exhibited good discriminating capabilities to categorize seven classes of medical wastes. The classification accuracy shows a decline, especially for the syringe class, where it was misclassified. This led to a reduced classification accuracy for this class, as depicted in Figure 2, possibly due to its distinctive sharp shape characteristics.

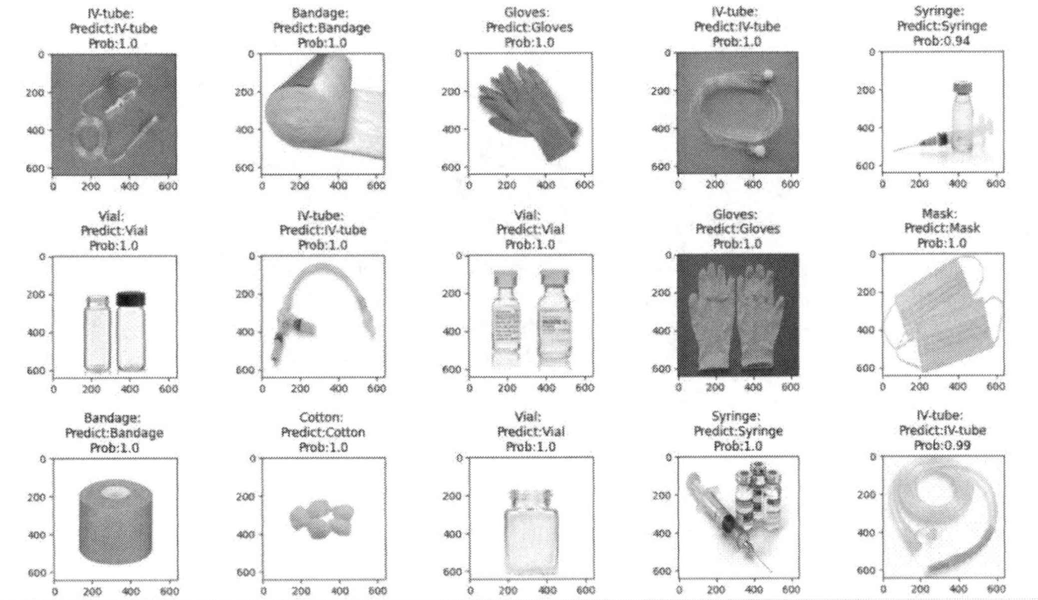

Fig. 2. Model prediction.

The multiclass classification performance was evaluated quantitatively on a normalized confusion matrix using the training and testing dataset (Fig. 3). According to the results the utilization of TL with pretrained CNNs substantially enhanced the efficiency and accuracy of medical waste classification. The CNNs demonstrated the ability to distinguish different kinds of medical waste.

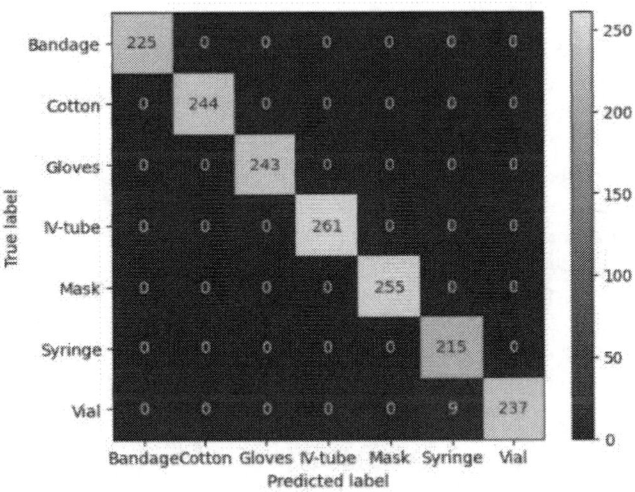

a) Training and Validation dataset

 https://doi.org/10.1051/e3sconf/202453004001

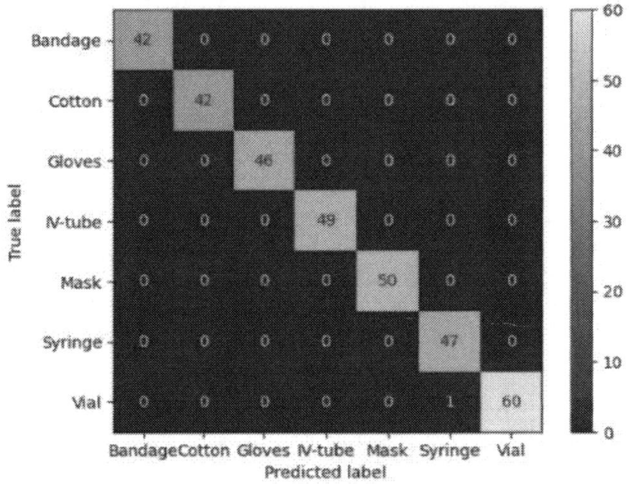

b) Testing dataset

Fig. 3. Confusion matrix for EfficientNet_b7.

4 Conclusions

This paper presents the development of a medical waste classification system based on deep learning model using the EfficientNet_b7 architecture, aiming to facilitate the segregation of medical waste. The image dataset was trained and optimized to create a classification model for medical waste using transfer learning. Our findings suggest that the utilization of transfer learning significantly improved the classification performance of the CNN model, even with a small dataset. Notably, the accuracy of medical waste classification reaches 99% for both the training and validation datasets, as well as the testing dataset. Our work represents a significant step forward in waste management through the creation and deployment of an intelligent system for medical waste classification. This innovation promises to substantially elevate the efficiency and accuracy of waste sorting processes, thereby contributing to the better management of medical waste.

Received funding to support activities promoting research and innovation from Hub of Waste Management for Sustainable Development (Center of Excellence on Hazardous Substance Management, Chulalongkorn University) and the National Research Council of Thailand (NRCT). The authors would like to acknowledge Prince of Songkla University. This project is funded by National Research Council of Thailand (NRCT): Contract number N42A660827 and Walailak University [grant number WU66249].

References

1. P. Akkajit, H. Romin, M. Assawadithalerd, J. Environ. Public Health, **8745472** (2020)
2. World Health Organization, Health-care waste. https://www.who.int/news-room/fact-sheets/detail/health-care-waste (2018)
3. J.M. Chisholm, R. Zamani, A.M. Negm, N. Said, M.M.A. Daiem, M. Dibaj, M. Akrami, Waste Manag Res. **39**, 9 (2021)
4. P.K. Srikanth, W. Srimanoi, P. Kashyap, C. Visvanathan, In: Ghosh, S.K., Agamuthu, P. (eds) Health Care Waste Management and COVID 19 Pandemic. Springer, Singapore (2022)

5. Phuket Provincial Health Office. Guidelines for infectious waste management in Phuket.http://www.phuketlocal.go.th/files/com_news_conference/20170529141158ryzuo.pdf (2017)
6. A. Wulansari, A. Setyanto, E.T. Luthfi, *A Convolutional Neural Network Based on Transfer Learning for Medical Waste Classification During Pandemic Covid-19*, in Proceedings of International Seminar on Intelligent Technology and Its Applications (ISITIA), Surabaya, Indonesia, 86-91 (2022)
7. M. Verma, A. Kumar, S. Kumar, *Medical Waste Classification using Deep Learning and Convolutional Neural Networks*, in Proceedings of 2022 IEEE Conference on Interdisciplinary Approaches in Technology and Management for Social Innovation (IATMSI), Gwalior, India, (2022)
8. Z. Cai, T. Bian, C. Bai, P. Li, C. Sun, *Intelligent medical waste detection and classification system based on machine vision*, in Proceedings of International Conference on Image, Signal Processing, and Pattern Recognition (ISPP 2023), 127070N (2023)
9. S. Asaeid, B. Hbalas, *Hospital Waste Sorting*, in Proceedings of the 11th student symposium on mechanical and manufacturing engineering. 25-26 August 2021, University Tun Hussin on Malaysia, Malaysia (2021)
10. B. Dwyer, J. Nelson, J. Solawetz, Roboflow (version 1.0) https://roboflow.com (2022)
11. P. Akkajit, A. Sukkuea, B. Thongnonghin, Ecol. Inform. **1023** (2023)
12. P. Akkajit, A. Sukkuea, *Rapid Classification of Microplastics by Using the Application of a Convolutional Neural Network,* in Proceedings of the 8th World Congress on Civil, Structural, and Environmental Engineering, CSEE'23, 29-31, March 2023, Lisbon, Portugal (2023)
13. X. Bian, Y. Chen, S. Wang, F. Cheng, H. Cao, IEEE Wireless Communications and Networking Conference Workshops (WCNCW), Nanjing, China, 1-6 (2021)
14. U.K. Lilhore, S. Simaiya, S. Dalal, R. Damaševičius, Multimed Tools Appl **83** (2023)

The performance of Adsorbing and Removing gaseous acetic acid by activate carbon

Chen-Fu Hsu[1],*Yu-Tsung* Chen[1], and *Chao-Heng* Tseng[2*]

[1]Researcher,National Taipei University of Technology,Institute of Environmental Engineering and Management Department,Taiwan
[2]Advisor,National Taipei University of Technology,Institute of EnvironmentalEngineering and Management Department,Taiwan

Abstract. Currently, gaseous molecular contaminants (acetic acid) is one of the main factors that cause defects in semiconductors and wafers , reduce product yields in high-tech process plants. Propylene Glycol Monomethyl Ethyl Ether Acetate (PGMEA) is a versatile solvent with many applications. Solvents used in the solder mask process in the manufacture of printed circuit boards (PCBs) .The diluent in photoresist and edge bead removers used in semiconductor and wafer fabrication is PGMEA, which forms acetate contaminants through an acid-catalyzed hydrolysis reaction. The concentration standard of acetic acid pollutants in clean rooms is 1ppb.In order to protect the expensive optical components, the acetic acid generated by PGMEA needs to be controlled.For this purpose, the activated carbon filter in the intake filter prevents acetic acid from contaminating the components and corroding the material. In this study, we used the commercially available activated carbon fabricated filter to adsorption of acetic acid. The effect of adsorption under different face velocity (0.03/0.06/0.09 m/s) and initial acetic acid concentrations (4/6/8/10 ppm) were measured according to the principle of ASHRAE 145.1standard. The experimental data are also evaluated for adsorption performance, adsorption capacity, kinetic model of acetic acid adsorption process, and isothermal model of acetic acid adsorption process.The results of the acetic acid adsorption experiment show that MM3000 activated carbon has the highest adsorption capacity and the longest adsorption saturation time than KPL activated carbon. The higher the inlet acetic acid concentration at a fixed wind speed, the shorter the saturation time and the higher the adsorption capacity; the faster the wind speed at a fixed acetic acid concentration, the shorter the saturation time and the higher the adsorption capacity.The Langmuir (R^2) of the MM3000 sample is higher than the Freundlich's coefficient of determination (R^2), which indicates that the adsorption phenomenon of the MM3000 sample is more suitable to be described by the Langmuir isothermal adsorption mode, which is the adsorption of a single molecular layer on a homogeneous surface. The Freundlich R^2 of the KPL sample is higher than the Langmuir's coefficient of determination (R^2), which means that the adsorption phenomenon of this KPL sample is more suitable to be described by the Freundlich isothermal adsorption mode, and it is a case of adsorption of multiple layers on an inhomogeneous surface.

[*] Corresponding author: tsengco@ntut.edu.tw

 https://doi.org/10.1051/e3sconf/202453004002

The adsorption of MM3000 activated carbon and KPL activated carbon is more suitable to be described by the proposed second-order adsorption kinetic model, which is a phenomenon of chemisorption.

Key words: Acetic acid, Activated carbon, Adsorption, Filter

1.Sources of Acetic Acid

Despite being a weak acid, acetic acid is corrosive, presenting itself as a colorless organic liquid compound with a distinctive acidic taste and pungent odor**[1][2]**.

1.1 Chemical factory

Acetic acid is considered a commodity chemical for a variety of end-user industries (textiles, fibers, pharmaceuticals, food, etc.). More than 65% of acetic acid production is involved in the production of polymers derived from vinyl acetate monomer (VAM) or cellulose acetate. The polyvinyl acetate produced is mainly used as a precursor for paints and coatings, or to produce plastics. VAM is one of the main additives in the polymer industry, used as an emulsifier, resin or surface coating, intermediate in the production of acrylic fibers and polymer wires. On the other hand, cellulose acetate is used in the production of cellulose acetate. Another prominent application is the production of acetic anhydride**[3].**

1.2 Film Stock Warehouses

The presence of acetic acid in film stock warehouses is related to the basic materials used in films. The base of these films is typically made from cellulose or acetyl cellulose, and residues of chemicals or solvents used in the film processing may contain acetic acid components. The base material of older film stocks, acetyl cellulose, may gradually release acetic acid under prolonged exposure to high temperature, humidity, or inadequate ventilation. Strict control of storage conditions and proper preservation of older film stocks are necessary to prevent the generation of acetic acid in film stock warehouses**[4]**.

1.3 Semiconductor Manufacturing Plants

Acetic acid is a major factor causing defects and reducing product yield in semiconductor and wafer manufacturing processes. Propylene glycol monomethyl ether acetate (PGMEA) is a multifunctional solvent used in these processes. PGMEA, through acid-catalyzed hydrolysis reactions, forms acetic acid pollutants. Cleanrooms have standards for total organic acid concentration (including acetic acid pollutants), set at 0.002 ppm. Sources of airborne molecular contaminants (AMCs) in cleanrooms include external gases, human emissions, manufacturing equipment, gases or vapors used in manufacturing processes, and building materials within the cleanroom environment. AMCs are categorized into five classes: acids (MA), bases (MB), condensables (MC), volatiles, and other unclassified substances**[5]**. AMCs can impact semiconductor process equipment performance through deposition. The photolithography area is particularly concerning, where acidic contamination and refractive organic compound contamination lead to the formation of crystal salts and optical haze, causing transmission losses and defects in the printing process**[6]**.

2. Activated carbon modification test

In this study, MM3000 activated carbon was modified with 40% phosphoric acid, and the activated carbon was placed in a 40% phosphoric acid solution for 1.5 hours. Wash the activated carbon with DI water until a constant pH is reached. Dry the activated carbon in an oven at 105°C for 1 hour and activate it at high temperature (420°C) for 1 hour. Table1. Adsorption capacity of carbon after being activated using different concentration of H_3PO_4 and KOH at room temperature**[7]**.

A filter surface wind speed parameter (0.03 m/s) and an acetic acid concentration parameter (10 ppm) were injected into the test cabin to conduct a continuous adsorption experiment, and the cabin was sandwiched in the middle filled with adsorbent (activated carbon approx. 37±1g) filter for adsorption. In order to understand the difference between modified activated carbon and unmodified activated carbon, this study used MM3000 activated carbon to adsorb acetic acid. The adsorption experimental values were substituted into the adsorption capacity formula to calculate the adsorption capacity per unit time, and the adsorption capacity and unit time were (hours) and efficiency versus unit time (hours) are plotted to present.

It can be found in (Table1) that the adsorption performance of MM3000 is even worse after being modified with phosphoric acid. The adsorption saturation time dropped significantly from 152 to 93 hours, and the adsorption capacity dropped significantly from 40.38 to 27.18 mg/g. It can be seen that using phosphoric acid to modify activated carbon is effective for this purpose. Studies have shown that the adsorption of acetic acid is not good. Because N-H appears in the functional group of MM3000 activated carbon, it is known that it has been modified. It is speculated that phosphoric acid is used for secondary modification, resulting in limited adsorption effect or worse adsorption performance.

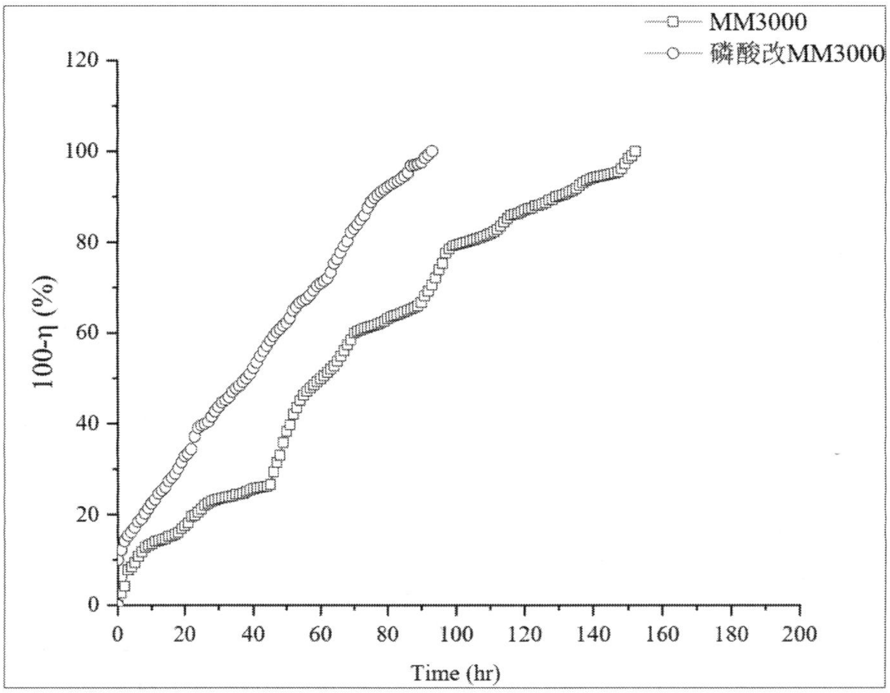

Fig.1. Breakthrough efficiency curve of MM3000 and phosphoric acid modified MM3000 activated carbon at surface wind speed of 0.03m/s and acetic acid concentration of 10ppm

https://doi.org/10.1051/e3sconf/202453004002

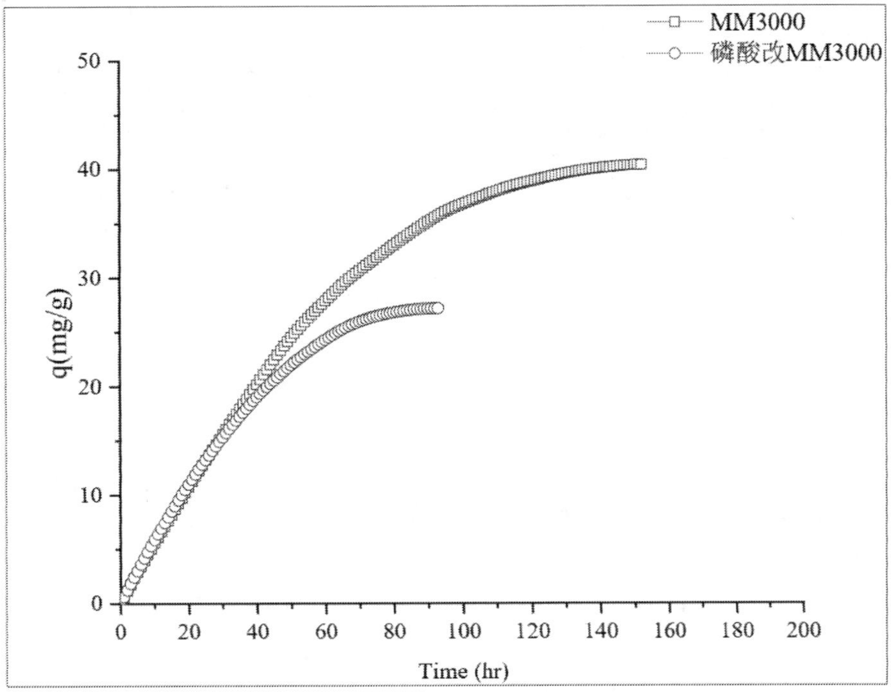

Fig.2.The adsorption capacity curve of MM3000 and phosphoric acid modified MM3000 at a surface wind speed of 0.03m/s and an acetic acid concentration of 10ppm

Table1.Comparison of adsorption time and adsorption capacity between MM3000 and phosphoric acid modified MM3000 activated carbon

Sample	Surface wind speed parameters (m/s)	Acetic acid concentratio n (ppm)	Adsorption saturation time (hr)	Adsorption capacity (mg/g)
MM3000	0.03	10	152	40.38
Phosphoric acid modification MM3000	0.03	10	93	27.18

3.Adsorption Modes

Isothermal adsorption mode**(Table2)**.Adsorption kinetic model formula**(Table3)**.
(Table4)shows the adsorption saturation time and adsorption amount of MM3000 activated carbon and KPL activated carbon filter adsorption experiments. It can be observed that the higher the acetic acid concentration of MM3000 activated carbon and KPL activated carbon under fixed surface wind speed parameters, the shorter the adsorption saturation time; at a fixed surface Under the acetic acid concentration parameter, the faster the surface wind speed, the shorter the adsorption saturation time; under the fixed acetic acid concentration parameter, the faster the surface wind speed, the greater the adsorption capacity; under the fixed surface

wind speed parameter, the higher the acetic acid concentration, the greater the adsorption capacity**[8]**.

(Table5)shows that the coefficient of determination (R2) of MM3000 Langmuir is higher than the Freundlich isothermal adsorption model, which means that the acetic acid adsorption phenomenon of this sample is more suitable to be described by Langmuir, which is biased towards the adsorption of a monolayer on a uniform surface.The coefficient of determination (R^2) of KPL Freundlich is higher than the Langmuir isothermal adsorption model, which means that the acetic acid adsorption phenomenon of this sample is more suitable to be described by Freundlich, which is biased towards the adsorption situation on the uneven surface**[9]**.

Among these three adsorption kinetics,**(Table6)**shows that the R^2 of the pseudo-second-order adsorption kinetic model of the MM3000 sample is higher than that of the pseudo-first-order model, and the trend of the pseudo-second-order adsorption kinetic model and experimental adsorption capacity of the adsorption capacity The lines are consistent. Closer to the pseudo-first-order intra-particle diffusion model; the adsorption situation of the MM3000 sample is more suitable to be described by the pseudo-second-order adsorption kinetic model, which is biased towards chemical adsorption. The trend of the results of this study is consistent with that of Huang Baihua (2018). Research on formaldehyde adsorption by raw carbon (wind speed m/s: 0.02, 0.05, 0.1; concentration ppm: 0.25, 0.5, 0.75, 0.9) and Ye Chengda (2019) active oxidation The trend of aluminum adsorbing formaldehyde is consistent (wind speed m/s: 0.1, 0.125, 0.15; concentration ppm: 0.75, 1, 1.25)**[8][9]**.

Table2. Isothermal adsorption mode formula

	Freundlich Isothermal adsorption mode		Langmuir Isothermal adsorption mode	
Definition	The earliest empirical formula of isothermal adsorption theory proposed, used to describe the adsorption situation of adsorbents on uneven surfaces.		Used to describe the adsorption situation of a monolayer. Each position on the surface of the adsorbent can only adsorb one molecule without affecting each other.	
Formula	$q_e = K_f C_e^{\frac{1}{n}}$ (1)	q_e:Equilibrium adsorption capacity(mg/g) K_f: Freundlich adsorption coefficient C_e:equilibrium concentration(mg/m^3) n:adsorption constant	$q_e = \frac{q_{max} K_L C_e}{1 + K_L C_e}$ (2)	q_e:Equilibrium adsorption capacity(mg/g) q_{max}:Maximum adsorption capacity(mg/g) K_L:Langmuir adsorption coefficient C_e:equilibrium concentration(mg/m^3)
Linear equation	$\log q_e = \log K_f + \frac{1}{n} \log C_e$ (3)		$\frac{C_e}{q_e} = \frac{1}{q_{max} K_L} + \left(\frac{1}{q_{max}}\right) C_e$ (4)	

Table3. Adsorption kinetic model formula

	Pseudo-first-order adsorption kinetic model	Pseudo-second-order adsorption kinetic model	Intra-particle diffusion pattern
Definition	Mainly used to describe the adsorption phenomenon and belongs to physical adsorption.	Mainly used to describe the adsorption phenomenon belongs to chemical adsorption	It is mainly assumed that all adsorption behaviors occur inside the particles. After the pollutants are adsorbed on the particle surface, they slowly diffuse into the pores of the particles. The adsorption behavior is affected by internal diffusion of particles and membrane diffusion.
Formula	$\frac{dq_t}{dt} = k_1(q_e - q_t)$ (5)	$\frac{dq_t}{dt} = k_2(q_e - q_t)$ (6)	$q_t = k_t \times t^{1/2}$ (7)
	k_1: First-order adsorption rate constant (1/hr) q_e: Quasi-equilibrium adsorption capacity(mg/g) q_t: Adsorption amount per unit time(mg/g)	K_2: Second-order adsorption rate constant (g/mg/hr) q_e: Quasi-equilibrium adsorption capacity(mg/g) q_t: Adsorption amount per unit time(mg/g)	q_t: Adsorption amount per unit time(mg/g) k_i: Internal particle diffusion rate constant(mg/g-hr$^{1/2}$)
Linear equation	$log(q_e - q_t) = -\frac{k_1}{2.303}t + logq_e$ (8)	$\frac{t}{q_t} = \frac{1}{k_2 q_e^2} + \frac{1}{q_e}t$ (9)	$q_t = k_i \times t^{1/2}$ (10)

Table4. Saturation time and adsorption capacity of MM3000 activated carbon and KPL activated carbon filter adsorption experiments

Surface wind speed parameters(m/s)	Acetic acid concentration parameters (ppm)	MM3000 activated carbon		KPL activated carbon	
		saturation time(hr)	Adsorption capacity (mg/g)	saturation time (hr)	Adsorption capacity (mg/g)
0.03	4	195	30.76	183	22.32
	6	180	34.88	173	27.47
	8	167	37.90	158	33.12
	10	152	40.38	143	38.06
0.06	4	167	49.78	161	35.89
	6	157	57.30	152	45.94
	8	148	61.58	139	55.64
	10	132	68.04	124	63.87
0.09	4	156	69.97	151	45.93
	6	144	78.99	138	57.09
	8	132	83.90	126	69.97
	10	117	91.25	111	83.63

Table5. Freundlich and Langmuir isothermal adsorption mode constants for MM3000 and KPL samples

Sample	wind speed(m/s)	Freundlich			Langmuir			
		K_f (mg/g) (m^3/mg)$^{1/n}$	n	R^2	q_{max}(mg/g)	K_L(m^3/mg)	R_L	R^2
MM3000 activated carbon	0.03	15.6495	3.3693	0.9995	51.0204	0.1508	0.0002~0.0005	0.9991
	0.06	28.5299	3.9825	0.9402	75.1880	0.2121	0.0002~0.0004	0.9965
	0.09	36.8129	3.5549	0.9922	113.6364	0.1580	0.0002~0.0005	0.9956
KPL activated carbon	0.03	5.8076	1.7100	0.9966	72.9927	0.0430	0.0008~0.0019	0.9763
	0.06	8.4528	1.5828	0.9997	135.1351	0.0361	0.0009~0.0023	0.9896
	0.09	10.1953	1.5363	0.9914	188.6792	0.0312	0.0010~0.0026	0.9313

https://doi.org/10.1051/e3sconf/202453004002

Table6. MM3000 activated carbon pseudo-first-order adsorption kinetic model, pseudo-second-order adsorption kinetic model and particle internal diffusion mode constants

Wind speed parameters(m/s)	Acetic acid concentration (ppm)	Experimental adsorption capacity (mg/g)	Pseudo-first-order adsorption kinetic model: k_1(min^{-1})	Pseudo-first-order adsorption kinetic model: $q_{e,1}$(mg/g)	Pseudo-first-order adsorption kinetic model: R_1^2	Pseudo-second-order adsorption kinetic model: k_2(g-^{1}min^{-1})	Pseudo-second-order adsorption kinetic model: $q_{e,2}$(mg/g)	Pseudo-second-order adsorption kinetic model: R_2^2	Internal particle diffusion pattern: ki (mg/g-min$^{1/2}$)	Internal particle diffusion pattern: qi(mg/g)	Internal particle diffusion pattern: R^2
0.03	4	30.76	0.028	84.32	0.4388	0.00003	32.54	0.9657	2.758	38.51	0.9894
	6	34.88	0.031	95.92	0.5153	0.00006	36.27	0.9913	3.210	43.06	0.9916
	8	37.90	0.035	100.12	0.7769	0.00007	40.42	0.9683	3.697	47.78	0.9858
	10	40.38	0.041	103.08	0.7110	0.00012	43.39	0.9602	4.028	49.66	0.9765
0.06	4	49.78	0.031	126.79	0.4164	0.00002	52.54	0.9584	4.824	62.34	0.9833
	6	57.30	0.032	139.07	0.5006	0.00004	58.82	0.9801	5.617	70.38	0.9902
	8	61.58	0.040	162.39	0.5522	0.00007	64.77	0.9866	6.174	75.11	0.9889
	10	68.04	0.050	191.26	0.6525	0.00012	71.80	0.9795	7.029	80.76	0.9741
0.09	4	69.97	0.041	250.36	0.5967	0.00002	74.76	0.9805	7.073	88.34	0.9893
	6	78.99	0.041	241.33	0.4452	0.00003	83.34	0.9855	8.223	98.68	0.99
	8	83.90	0.048	260.06	0.5962	0.00005	88.88	0.9837	9.044	103.90	0.9881
	10	91.25	0.060	293.31	0.6403	0.00009	97.11	0.9740	10.149	109.78	0.9775

4.Conclusion

4.1 Adsorption performance experiment

The MM3000 activated carbon sample has the largest adsorption capacity and the longest adsorption saturation time. It can be determined that MM3000 activated carbon has the best adsorption performance. The higher the inlet acetic acid concentration of the MM3000 sample and the KPL sample at a fixed wind speed, the shorter the saturation time; under a fixed acetic acid concentration The faster the wind speed, the shorter the saturation time; under a fixed wind speed, the higher the acetic acid concentration, the greater the adsorption capacity, and under a fixed acetic acid concentration, the faster the wind speed, the higher the adsorption capacity.

4.2 Isothermal adsorption mode

The results of the two isothermal adsorption modes show that the Langmuir R^2 of the MM3000 sample is higher than the determination coefficient (R^2) of the Freundlich isothermal adsorption mode, indicating that the adsorption phenomenon of this sample is more suitable to be described by the Langmuir isothermal adsorption mode, which prefers a monolayer on a uniform surface. The adsorption situation above. The Freundlich coefficient of determination (R^2) of the KPL sample is higher than the coefficient of determination (R^2) of the Langmuir isothermal adsorption model, which means that the adsorption phenomenon of this sample is more suitable to be described by the Freundlich isothermal adsorption model, which is biased toward the adsorption of multiple layers on an uneven surface.

4.3 adsorption kinetics model

Among these three adsorption kinetics, the results show that the R^2 of the pseudo-second-order adsorption kinetics model of the MM3000 and KPL samples is higher than that of the pseudo-first-order model, and the adsorption capacity of the two samples in the pseudo-second-order adsorption kinetics model is the same as the experimental adsorption capacity. The trend line is closer to the pseudo-first-order and particle internal diffusion models; the adsorption situation of the MM3000 sample is more suitable to be described by the pseudo-second-order adsorption kinetic model, which is biased toward chemical adsorption.

4.4 Advantage

The study of adsorbing acetic acid in indoor air quality under the best conditions and the best adsorption filter material composition will be of great help in improving indoor air quality.

References

1. Bo Chen, MengWang, Mingxiao Duan, Xiaotian Ma, Jinglan Hong, Fei Xie, Ruirui Zhang, Xiangzhi L.In search of key: protecting human health and the ecosystem from water pollution in China. Journal of Cleaner Production.2019. 228:101-111
2. Yingjun Liu, Pawel K. Miszta, Jianyin Xiong, Yilin Tian, Caleb Arata, Robert J. Webe, William W. Nazaroff, Allen H. Goldstein.Characterizing sources and emissions of volatile organic compounds in a northern California residence using space- and time-resolved measurements.Indoor Air.2019.29:630-644
3. Juan Luis Martín-Espejo, Jesús Gandara-Loe, José Antonio Odriozola, T.R. Reina, Laura Pastor-Pérez.Sustainable routes for acetic acid production: Traditional processes vs a low-carbon, biogas-based strategy.Science of the Total Environment,2022,80:3-6
4. N. S. Allen, M. Edge, J. H. Appleyard, T. S. Jewitt. Degradation of Historic Cellulose Triacetate Cinematographic Film: The Vinegar Syndrome.Polymer Degradation and Stability,1987,19： 379-387
5. C. R. Ayre, A. K. Mittal, J. F. Oullivan. The influence and measurement of airborne molecular contaminants in advanced semiconductor processing. Semiconductor Fabtech, 24th Edition, (2005)
6. G. Weineck, D. Zastera, A.J. Dallas, AMC control in photolithography: the pastdecade in review, Proc. SPIE 7638 (2010) 76383H
7. Arneli, Z F Safitri, AWPangestika, F Fauziah, V N Wahyuningrum, The influence of activating agents on the performance of rice husk-based carbon for sodium lauryl sulfate and chrome (Cr) metal adsorptions Materials Science and Engineering., 172 (2017) 012007
8. Huang Baihua, "Research on Indoor Formaldehyde Removal by Acidic Modified Activated Carbon", National Taipei University of Technology, 2017
9. Ye Chengda, "Research on Formaldehyde Removal of Activated Alumina Supported Platinum Metal", National Taipei University of Technology, 2019

Assessment of phosphorus flow of canteen food waste in educational institutions in Bangkok area

Sunatthiya Mokjatturas[1,3], *Sopa* Chinwetkitvanich[1], *Withida* Patthanaissaranukool[2,3], *Chongchin* Polprasert[1,3] *and Supawadee* Polprasert[2,3]

[1]Department of Sanitary Engineering, Faculty of Public Health, Mahidol University, Bangkok, 10400, Thailand
[2]Department of Environmental Health Sciences, Faculty of Public Health, Mahidol University, Bangkok, 10400, Thailand
[3]Center of Excellence on Environmental Health and Toxicology (EHT), OPS, MHESI, Bangkok, 10400, Thailand

Abstract. This study was to investigate the occurrence of food waste in educational institutions and their current food waste management. Three educational institutions in Bangkok area were selected, and sampling were conducted for three days in a row for each institution. The results showed that the amount of food waste ranging from 20 to 198 kg/day, accounting of 40 – 90% of total waste. Most of food waste was eventually ended up in landfills. The phosphorus proportion of uncooked food waste (UCFW) was in the range of 0.09 – 0.29 mg P/g, while that of cooked food waste (CFW) was in the range of 0.53 – 0.76 mg P/g. Total phosphorus mass entering the waste management system was ranging from 8 to 141 g P/d, while loss of phosphorus was in the range of 1.5 – 13.5 g P/d. Under the composting scenario used in this study, the yield of fertilizer-like product ranged from 0.3 to 37.2 tons per year resulting in calculated cost saving as high as 391 USD per year. From this study, food waste separation was highly recommended to improve assessment of phosphorus flow and recovery. Expectation of phosphorus recovery for agricultural sector benefits should be prioritized.

1 Introduction

An educational institution is a place for education and study activities provided by either public or private sectors. Besides, it is also a place where most people spend part of their lives since early childhood period (between 2 – 3 years of age). During compulsory education (first nine years of primary and secondary schooling), students spend time totally around 7,634 hours (or 848 hours per year) in schools/institutes [1]. Therefore, activities according to food consumption are inevitable, i.e., food preparation, food selling, leftover of food waste disposal, etc. Some students and personnels will have to deal with food consumption at least one meal a day, while some have to depend on up to two meals a day. Particularly, some colleges or universities arrange dormitory within their premises, resulting in food consumption activity for whole day. Wilkie et al. [2] mentioned that food waste was a majority component of the whole garbage (ranging between 47 to 58%) observed in three schools in Florida, accounting for food waste ratio of 52.2 g per student per day. In 2019, food waste generated from forty-six schools in the US approximately up to 530,000 tons. This amount could cause an environmental and economic impact equivalent to GHGs of 1.9 million metric tons of CO2 (MtCO2e), or 20.9 billion gallons of water, or 1.7 billion US dollars [3]. In case of college level, average of 57 g leftover was discarded in each tray [4], while 329.5 kg per day of food waste was observed in the Middle East [5]. Gustavsson et al. [6] reported the amount of food waste and loss of approximately 1.3 billion tons per year, or one-third of food production. In the view of management, Thailand Environment Institute (TEI) [7] reported the results of their survey that 'School' was mentioned as one of places recommended for food waste reduction promotion.

Currently, there are more than 12 million of students and more than 56 thousand of educational institutes, including more than 600 thousand of teachers and personnels in Thailand [8]. These students, teachers and educational personnels could spend time in their institutes up to 5 – 6 hours a day [9]. In 2022, Thailand generated 9.7 million tons of food waste, accounting for 38% of total solid waste. The component of this food waste was 39.5% of edible part and 60.5% of inedible part [10]. However, in consideration of food waste in educational institutes, the most proportion of them was edible [11]. In Thailand, food waste management still preferred landfill approach and sometimes open dumping (accounting for 57%), while the rest might go to incinerator or composting fertilizer [12]. In Bangkok area, there were 3,419 educational institutes with more than a million of related students [13]. The information of Bangkok solid waste during 2011 – 2020 illustrated that the proportion of food waste was in the range of 42 – 53%. Particularly, the higher amount of food waste of 4,323 tons per day was discovered in 2020 year. The survey results showed that the primary cause of food waste was excess food quantity per meal or leftover. Also, educational institutions were only sector where food waste management involves composting, while the others mostly applied landfill approach. Nowadays, Bangkok Metropolitan Administration (BMA) has a policy involving food waste utilization, especially, composting, effective microorganism (EM) liquid production, earthworms culturing, etc. [14].

As we know that food waste contains nutrients, particularly, nitrogen and phosphorus, which are essential for plantation. Therefore, the most appropriate attempt for food waste management would deal with biologically conversion of organic content in food waste, which fertilizer-like products were subsequently expected. Thailand had to import phosphorus fertilizer more than 200 thousand tons per year [15] because there are insufficient phosphorus reserves for domestic use. Unfortunately, phosphorus is an expendable element, which is expected to be a shortage in 50 – 100 years [16]. A phosphorus shortage in the near future will raise up a price of fertilizer, and consequently affect Thai agriculture investment cost. Hence, recovery of phosphorus from food waste is a matter of recent researches to find the best approach in bringing phosphorus back from waste, which should be further beneficial to agricultural sector. However, Thailand and Bangkok still do not have enough basic information on the mass flow of phosphorus in food waste system to create an integrated policy and appropriate approaches. For this reason, this study was to investigate the occurrence of food waste from the origin to the destination of waste management in educational institutions in Bangkok area of Thailand. Thus, a schematic mass flow of phosphorus in food waste system was synthesized and analysed. Loss of phosphorus would be identified, and guidelines for phosphorus recovery within food waste management would, then, be recommended.

2 Material and methods

Due to Bangkok is the capital of Thailand with dense populations, there are a vast variety of educational institutions. Therefore, there educational institutions (EI) in Bangkok area were selected to represent a variety of them. That is, the first one is a primary school (EI1) located in Bangkapi District, the second one is a technological college (EI2) located in Nongkhaem District, the third one is a governmental university (EI3) located in Ratchathewi District. Sampling was conducted for three days in a row for each educational institution, focusing on solid waste from a kitchen or a canteen. High proportion of food waste were expected from this source. Food wastes were identified into two categories; uncooked food waste (UCFW) and post-consumer food waste or cooked food waste (CFW). Characteristics of food waste, i.e, bulk density (BD), moisture content (MC), total solids (TS), volatile solids (VS), fixed solids (FS), total kjeldahl nitrogen (TKN), total phosphorus (TP), organic carbon content (OC) and C/N ratio were analysed according to Standard Methods [17–19]. In addition, waste weight data were recorded during sampling, and waste management of each educational institution were clarified in order to calculated phosphorus mass flow with mass balance concept. Then, phosphorus mass flow was illustrated to identified loss of phosphorus in food waste management system of each educational institutions. Eventually, suitable approaches for phosphorus recovery or reduction of phosphorus loss should be proposed. However, sorting food waste plays an important role prior further any approaches. Practical approaches in Thailand after sorting food waste were such as animal feed, composting and distributing to farmers' benefit, or intentionally anaerobic digestion to obtain high concentrations of phosphorus fertilizer crystals and methane gas as a renewable fuel.

3 Results and discussions

3.1 Solid waste generation and management

Solid waste data collected from a kitchen and a canteen of three designated educational institutions were described in Table 1. For the primary school (EI1), total solid waste was averagely observed as high as 221 kg/day, which contained high proportion of food waste of about 90%. Especially, most of their food waste was categorized as cooked food waste (CFW). This circumstance should be further found out within the institution, whether it was because of bad estimation in food production or bad quality of cooked food, resulting in lots of leftovers. However, this EI1 institution gave away their CFW for animal feeding instead of disposed to landfill approach. For the Technological college (EI2) and the public university (EI3), average solid wastes were observed of about 71 and 40 kg/day, respectively. Food waste proportions (UCFW+CFW) from canteen wastes were in the lower range of 40 – 50% than those of EI1, therefore, the higher proportions of general waste were observed. This should be due to these canteens of EI2 and EI3 prepared food for selling, thus, economical aspect would be their concern more than the primary school (EI1) did. Interestingly, higher recyclable waste proportion was found in EI3 institution (10.6%) than EI2 institution (4.6%). From the rough observation during three-day sampling, the public university (EI3) did sort recyclable waste more carefully than the technology college (EI2).

Table 1. The Solid Waste Generation.

Types of waste	Day (kg)			Average (kg)	SD (kg)	% of total waste
	1	**2**	**3**			
Educational institution 1 : School						
Uncooked food waste	0	30.25	18.65	16	15	7.36
Cooked food waste	264.15	237.70	45.35	182	119	82.38
General waste	10.85	20.70	16.85	16	5	7.29
Recyclable waste	8.30	0	11.40	7	6	2.97
Total	**283.30**	**288.65**	**92.25**	**221**	**112**	**100**
Educational institution 2 : College						
Uncooked food waste	14.06	4.37	6.18	8	5	11.48
Cooked food waste	16.10	22.10	25.33	21	5	29.64
General waste	34.60	43.31	38.52	39	4	54.33
Recyclable waste	3.00	3.25	3.50	3	0	4.55
Total	**67.76**	**73.03**	**73.54**	**71**	**3**	**100**
Educational institution 3 : University						
Uncooked food waste	5.00	15.15	20.50	14	8	33.57
Cooked food waste	4.50	3.95	11.20	7	4	16.23
General waste	16.00	15.00	17.00	16	1	39.64
Recyclable waste	4.00	3.90	4.90	4	1	10.57
Total	**29.50**	**38.00**	**53.60**	**40**	**12**	**100**

3.2 Characterization of uncooked and cooked food waste

Laboratory data results showed that density (BD) of UCFW from three institutions were in similar range of 221 – 284 kg/m^3, while CFW samples were observed much higher density in the range of 736 – 860 kg/m^3. Moisture contents of UCFW samples were in the range of 86 – 93%, which is higher than those of CFW samples (70 – 78%) because UCFW samples mostly consists of fresh vegetation parts. Total solid (TS) of UCFW samples were in the range of 73 – 140 mg/g, while volatile solid (VS) in UCFW samples were observed ranging from 64 to 93 %TS. For CFW samples, TS were found in the range of 219 – 270 mg/g, while their VS were ranging from 89 to 96 %TS. Volatile solid (VS) indicated that CFW contained more organic proportion that UCFW. For total phosphorus (TP) characteristic, UCFW samples were observed lower phosphorus contents (0.09 – 0.29 mg/g) than those of CFW (0.53 – 0.76 mg/g) as shown in Figure 1. Similarly, total kjeldahl nitrogen (TKN) data showed the higher range found in CFW samples (6.6 – 8.7 mg/g) than those in UCFW samples (1.9 – 3.0 mg/g). Organic carbon content (OC) of UCFW samples were in the ranges of 24 – 37%, while the OC of CFW samples were observed in the narrow range of 34 – 35%. However, C/N ratios of both UCFW and CFW samples were similar in the range of 10 – 13. This was understandable that cooked food waste composed of more edible part than uncooked food waste, therefore, higher phosphorus and nitrogen content and organic content were presented.

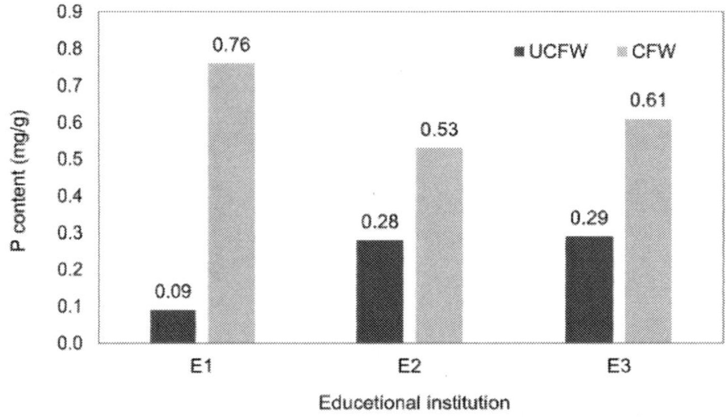

Fig. 1. P content in UCFW and CFW.

3.3 P flow in food waste

Mass balance of phosphorus in food waste system of three educational institutions were calculated and exhibited in Tables 2 to 4. For the primary school (EI1), Table 2 showed phosphorus content in UCFW and CFW samples were 0.09 and 0.76 mg/g, respectively. Therefore, total phosphorus mass in (UCFW and CFW) were 141 g/day or 28 kg/year (Note: 200 school days was used to calculate annual mass). Due to the primary school (EI1) gave away their CFW to animal feeding, which could be counted as recycled phosphorus approach, thus, phosphorus loss from food waste system of the primary school (EI1) was only 0.30 kg/year.

Table 2. Data for P flow in the primary school (EI1).

Flow	P Concentration (g/kg)	Total waste (kg/d)	P mass (g/d)	P mass (kg/year)
In				
UCFW	0.09	16	1.51	0.3
CFW	0.76	182	139	27.8
Total waste		**198**	**141**	**28.1**
Loss				
UCFW	0.09	16	1.51	0.3
CFW[a]	-	-	-	-

[a] Current situation of CFW in EI1 was recycled by animal feed.

For the technology college (EI2), phosphorus content in UCFW and CFW were 0.28 and 0.53 mg/g, respectively (Table 3), resulting in total phosphorus mass in of 13.5 g/day or 2.7 kg/year (Note: 200 school days was used to calculate annual mass). Loss of phosphorus from this EI2 was calculated from both of UCFW and CFW wastes (2.7 kg/year) because all food waste were disposed to landfill approach without any recycling of their food waste.

For the public university (EI3), phosphorus content in UCFW and CFW were 0.29 and 0.61 mg/g, respectively (Table 4). Total phosphorus mass in of this canteen were about 7.9 g/day or 1.6 kg/year (Note: 200 school days was used to calculate annual mass). Similar to EI2, loss of phosphorus was considered from both UCFW and CFW disposal to landfill without recycling, which was 1.6 kg/year.

Table 3. Data for P flow in the technology college (EI2).

Flow	P Concentration (g/kg)	Total waste (kg/d)	P mass (g/d)	P mass (kg/year)
In				
UCFW	0.28	8	2.30	0.5
CFW	0.53	21	11.22	2.2
Total waste		**29**	**13.52**	**2.7**
Loss				
UCFW	0.28	8	2.30	0.5
CFW	0.53	21	11.22	2.2

Table 4. Data for P flow in the public university (EI3).

Flow	P Concentration (g/kg)	Total waste (kg/d)	P mass (g/d)	P mass (kg/year)
In				
UCFW	0.29	14	3.93	0.8
CFW	0.61	7	4.00	0.8
Total waste		**21**	**7.93**	**1.6**
Loss				
UCFW	0.29	14	3.93	0.8
CFW	0.61	7	4.00	0.8

The flow of phosphorus in the food waste system shows that most of the phosphorus in food waste was lost in the range of 1 to 100%, or 1.51 to 13.52 g P per day of 0.3 to 2.7 kg P per year for this study (EI1, EI2 and EI3 cases). Schematic mass flow of phosphorus in this study was graphically shown in Figure 2. Though this seemed to be small amount of phosphorus for one source, there were more than four hundred educational institutions in Bangkok area. Therefore, food waste from educational institutions in Bangkok area could be interesting sources of phosphorus loss, which should be worth to apply waste sorting and phosphorus recycling.

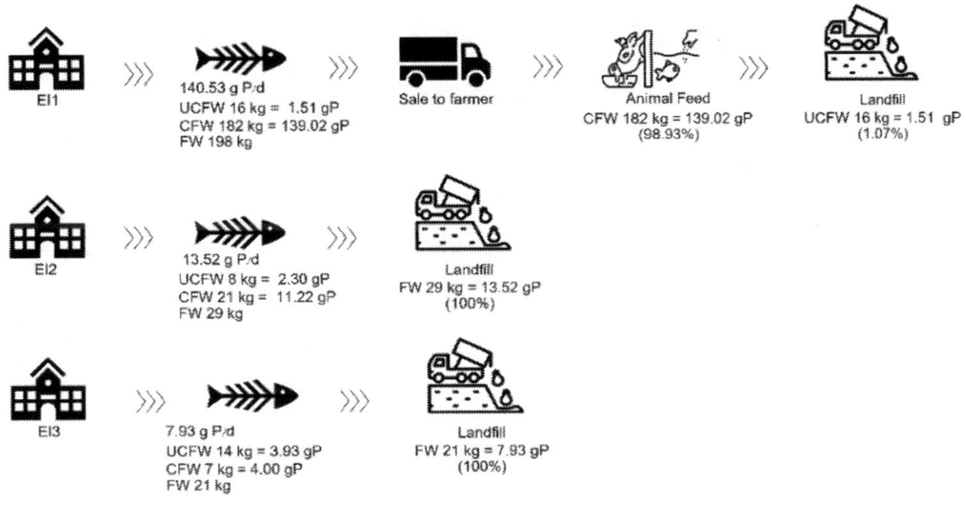

Fig. 2. P flow in 3 educational institutions.

3.4 Environmental and economic benefits for phosphorus recovery by composting approach

This section attempted to analysed some benefits, either environmentally or economically, from phosphorus recovery by composting process. Table 5 exhibited calculation results of cost saving if food waste was sent to composting process instead of landfill disposal. Due to cost of solid waste disposal into landfill by Bangkok Metropolitan Administration (BMA) was approximately 6.9 USD/m3 [14], waste weights from Tables 2 to 4 had to be converted into volumes by waste's densities. Then, cost of landfill disposal in case of all food wastes from three educational institutions going to landfill were shown in Table 5. Particularly, the expenditure of food waste (both UCFW and CFW) from the primary school (EI1) will be as high as 391 USD/year if they do not give away CFW for animal feeding. The other two (EI2 and EI3) will have to take the expenditure of solid waste disposal to landfill about 88 and 98 USD/year, respectively. This cost of landfill disposal would be reduced if all food waste went to composting process. In addition, the fertilizer-like product from composting process would also be benefit for agricultural purpose. Yields from composting process were also estimated and shown in Table 5, particularly, yield from CFW of the primary school (EI1) was as high as 31 tons per year.

Table 5. Feasibility Environmental and financial benefit assessment of phosphorus recovery from food waste using composting approach.

Items	The primary school (EI1)		The technology college (EI2)		The public university (EI3)	
	UCFW	CFW[a]	UCFW	CFW	UCFW	CFW
Cost of Solid Waste Collection and Disposal (USD)						
Waste volume (m^3/year)	11.3	45.7	7.1	5.7	12.7	1.6
Solid waste disposal cost (USD/m^3)[b]	6.9	6.9	6.9	6.9	6.9	6.9
Cost saving (USD/year)	**77**	**314**	**49**	**39**	**87**	**11**
Composting						
Total waste (kg/year)	3,200	36,400	1,600	4,200	2,800	1,400
Yield of compost (kg/year)[c]	**724**	**37,177**	**342**	**1,752**	**546**	**556**
P content (%)[d]	1.02	1.78	1.06	1.53	0.98	1.53

[a] Calculation of CFW in case of not give away to animal feeding.
[b] Solid waste disposal cost for sanitary landfill operated by BMA. [20]
[c] Composting of the mixture between food waste and green garbage [15] to obtain optimum C/N ratio is 25 [21], then, 11% of the mixture was counted as yield of organic fertilizer. [22]
[d] P content in organic fertilizer calculation [22].

4 Conclusion

Food wastes from kitchen and canteen generated in educational institutions in this study were quite substantial. The results showed that the proportion of food waste were in the range of 40 – 90%. This food waste, either uncooked or cooked food waste, contained nutrients both of nitrogen and phosphorus. If their food waste management was only landfill disposal, these nutrients were definitely counted as loss amount. The practical approach for phosphorus recovery such as composting will provide environmental and economic benefits. That is, landfill disposal cost would be reduced as much as 391 USD/year in case of the primary school (EI1), as well as, fertilizer-like product was yielded as high as 38 tons per year. However, composting will be success with the important condition of good sorting.

This research project is supported by Mahidol University, Grant no. NDFR 48/2564 and This research project is supported by National Research Council of Thailand (NRCT): NRCT5-RGJ63012-138. Besides, this research work is also supported in part by the grant from the Center of Excellence on Environmental Health and Toxicology (EHT), OPS, Ministry of Higher Education, Science, Research and Innovation.

References

1. The Organisation for Economic Co-operation and Development (OECD), *Education at a Glance 2023: OECD Indicators,* 352 (2023)
2. A. C. Wilkie, R. E. Graunke, C. Cornejo, Food Waste Auditing at Three Florida Schools, Sustainability **7**, 1370 (2015)
3. World Wildlife Fund (WWF), *FOOD WASTE WARRIORS A deep dive into food waste in US schools*, (2019)
4. K. J. Whitehair, C. W. Shanklin, L. A. Brannon, Written messages improve edible food waste behaviors in a university dining facility, J Acad Nutr Diet **113**, 63 (2013)
5. A. H. Abdelaal, G. McKay, H. R. Mackey, Food waste from a university campus in the Middle East: Drivers, composition, and resource recovery potential, Waste Manag. **98**, 14 (2019)
6. J. Gustavsson, C. Cederberg, U. Sonesson, R. Otterdijk, A. Meybeck, *Global food losses and food waste – Extent, causes and prevention,* 4 (2011)
7. Thailand Environment Institute (TEI), *REPORT ON CONSUMER INFORMATION FOR SUSTAINABLE CONSUMPTION AND PRODUCTION IN FOOD SUPPLY CHAIN IN BANGKOK,* 4 (2021)
8. Ministry of Education (MOE), Education statistic [Internet] 2023 [updated 2023; cited 2023 Nov 11] Available from: https://edustatistics.moe.go.th/
9. Department of Local Administration (DLA), *Manual for operating according to the policy of reducing study time and increasing learning time for educational institutions under local administrative organizations*, 15 (2017)
10. Pollution Control Department (PCD), *Thailand State of Pollution Report 2022,* 140 (2023)
11. Pollution Control Department (PCD), *Study of the composition of solid waste 2021,* 21 (2022)
12. Thailand Development Research Institute (TDRI), *Study on Food Loss and Waste Management in Thailand,* 1 (2019)
13. Bangkok Province Education Office (BPEO), *Educational information in Bangkok, academic year 2022,* 21 (2022)
14. The Department of Environment, Bangkok Metropolitan Administration (DOE), *Bangkok State of Environmental 2019-2020,* 103 (2021)
15. B. Thitanuwat, C. Polprasert, A. Englande, Quantification of phosphorus flows

throughout the consumption system of Bangkok Metropolis, Thailand Sci. Total Environ. **542**, 1106 (2016)

16. D. Cordell, J. O. Drangert, S. White, The story of phosphorus: Global food security and food for thought, Glob Environ Change. **19**, 292 (2009)
17. ASTM International, *Standard Test Method for Determination of the Composition of Unprocessed Municipal Solid Waste,* (2016)
18. American Public Health Association (APHA), *Standard Methods for the Examination of Water and Wastewater (20th ed.),* (1998)
19. A. Walkley, I. A. Black, An Examination of the Degtjareff Method for Determining Soil Organic Matter and a Proposed Modification of the Chromic Acid Titration Method, Soil Sci. **37**, 29 (1934)
20. The Department of Environment, Bangkok Metropolitan Administration (DOE), *Guidelines for collecting fees for sewage and solid waste management in Bangkok,* 8 (2021)
21. C. S. Akratos, A. G. Tekerlekopoulou, I. A. Vasiliadou, D. V. Vayenas, Olive Mill Waste: Recent Advances for Sustainable Management, 161 (2017)
22. Z. X. Keng, S. Chong, C. G. Ng et al., Community-scale composting for food waste: a life-cycle assessment-supported case study, J. Clean. Prod. **261**, 121220 (2020)

Cost-Benefit Analysis of Urban Green Infrastructures: A Case Study of the LiuChuan River Corridor, Taichung City, from Linsenliu Bridge to Minquanliu Bridge

Jen-Chih Chao[1*] *and Ji-Yuan* Lin[1, 2]

[1]Department of Landscape and Urban Design, Chaoyang University of Technology, Taichung 413310, Taiwan
[2]Department of Construction Engineering, Chaoyang University of Technology, Taichung 413310, Taiwan

Abstract. Green infrastructure has been shown to play the role of adaptation and adjustment under the effect of harsh climate. The creation of urban green infrastructure is an important indicator of sustainable urban development, and determining the cost-effectiveness of green infrastructure is a necessary step to help investment decisions. The whole economic worth of Taichung City's Liuchuan pollution cleanup and environmental improvement project is examined using cost-benefit analysis from three perspectives: investment value, tangible value, and intangible value. The study findings demonstrate that green infrastructure has long-term development value and can help the sustainable development of cities.

1 Introduction

Cities are increasingly expanding, putting more strain on land use, resources, and their inhabitants [1, 2]. The land area in many urban regions is predicted to increase dramatically by 2050, with cities housing nearly 68% of the world population [3]. Nevertheless, metropolitan areas house a sizable amount of the world's population, which is disproportionately affected by extreme weather occurrences. Urbanization has significantly aided economic development and increased the quality of life. At the same time, it brings with it a slew of environmental issues such as air pollution, flooding, and urban heat islands [4, 5]. As a result, addressing urban environmental concerns is an essential priority in order to ensure sustainable urban growth. Green infrastructure is rapidly being acknowledged as the foundation for cities' climate change adaptation [6]. Because it improves thermal comfort and helps regulate rainwater flow [7, 8]. A network of interconnected green spaces is referred to as green infrastructure [9]. Interconnected ecological features that provide various roles and bring multiple benefits [10]. It highlights that urban green space networks, such as hospitals, schools, and transportation, are not optional but rather essential for the high-quality growth of urban communities [11-13]. Furthermore, green infrastructure examines not only

* Corresponding author: jenchihchao@gmail.com

natural green places, but also the relevance of man-made green aspects [14]. Parks, land, trees, and other assets have traditionally been seen as critical components of urban green infrastructure [15]. These assets have been under increasing pressure in recent years as a result of development, budget cuts, and other factors, decreasing their current impact and frequently prohibiting future expansion [16]. There are numerous techniques available to handle green infrastructure design and performance evaluation issues, each with its own set of capabilities and constraints [17-19]. These techniques do not take into account cobenefits or cost/benefit impacts at many scales when determining whether green infrastructure initiatives are financially justified. As a result, more extensive cost/benefit analyses are required to better inform practitioners about the effects of their investments, which can lead to increased use of green infrastructure [20]. Cost-benefit analysis is a well-established method for weighing the advantages and disadvantages of various actions, and is an explicit aspect of the decision-making process. This article analyzes the Liuchuan pollution remediation and environmental improvement project in Taichung City as an example (phase 2-Linsenliu Bridge to Minquanliu Bridge) to discuss the costs and benefits of using green infrastructure for transformation.

2 Literature review

2.1 Green infrastructure

The early notions characterized green infrastructure as a full network of plants and open areas that serve numerous functions and bring multiple advantages to civilization [10]. Green infrastructure, defined in this way, might include parks, greenways, and green belts, and can often be linked back to Howard's Garden City idea [21]. Green infrastructure concepts are increasingly being used in cities to integrate various technological infrastructure with natural systems [22]. According to research, green infrastructure can help reduce flooding [23], improve water quality [24], reduce air pollution [25] etc. Green infrastructure provides numerous advantages.

2.2 Cost-benefit analysis

The basic definition of cost-benefit analysis is: from a social standpoint, to estimate the economic costs and benefits of a project using economic analysis methods in order to determine the project's adequacy and improve the efficiency of use of public sector resources and to optimize limited resources. configuration. Cost-benefit analysis is a popular economic analytical method for analyzing government public affairs such as public investment programs and public policies. The value of cost-benefit analysis is that it can consider the influence of development projects on all members of society, as opposed to typical private decisions, which only consider the impact on a single producer or consumer. A study of New York City's High Line, a train converted into a mile-long linear park, discovered that nearby communities faced rapid reconstruction and rising rental prices [26]. Although the benefits and functions of green infrastructure are ultimately determined by how green infrastructure is integrated into urban ecology and infrastructure systems, many of the effects of green infrastructure, both good and negative, are localized, and hence green infrastructure is important. Who is most affected is determined by the place [27]. The Center for Neighborhood Technology (CNT) found that municipal green infrastructure programs can deliver significant net benefits to communities by reducing capital costs of gray infrastructure by $120 million and providing more than $4 million in annual energy, air quality, and climate benefits [28]. The Forest Service conducted a cost-benefit analysis of street trees in

Midwestern communities. They discovered that the advantages of public street trees outweigh expenditures over a four-decade period. Small trees provided $160 in net benefits in 2005, while medium and big trees provided $640 and $2,320, respectively. However, the Forest Service analysis did not take discounts into account when calculating these advantages. Street trees prevent air pollution, save energy for heating and cooling, increase property value, reduce stormwater volume by intercepting rainfall, and save electricity [29]. If the net benefits of green infrastructure are positive, there is a compelling reason for municipalities to give incentives for private investment in green infrastructure, saving money, and producing better environmental outcomes.

3 Materials and Methods

3.1 Materials

In 2014, the Taichung City Government Water Conservancy Bureau completed the "Taichung City Liuchuan Pollution Remediation and Environmental Improvement Project" (first phase project), completed the interception and Chunghwa water purification site from Chongdeliu Bridge to Nantunliu Bridge and established a water purification plant on Zhongzhengliu Bridge. The segment from the bridge to the Minquanliu bridge is built using low-impact development approaches to restore the vitality of the Yanagawa River and improve the quality of life of residents by improving water quality and the environmental landscape. The second phase of the project aims to rebuild clean and diverse water areas in the canal section to achieve the goal of cleaning and beautifying drainage water quality, promoting the reuse of idle space, preserving the beauty of the ecological environment, promoting the ecological urban river, and transforming Liuchuan into a citizen ecological river in the Taichung metropolitan area. The main axis runs from Linsenliu Bridge to Sanminliu Bridge, and focuses on sewage interception project and low-impact development construction approaches. Through improved sewage interception facilities, extensive tree, shrub, and turf planting, and the installation of rain gardens, plant filter belts, and water pumps., permeable pavement, and other low-impact development measures to reduce the impact of point source and nonpoint source pollution on water quality, achieve 0 kilometers of severe pollution, improve the urban heat island effect, and prevent potential disasters caused by heavy rain runoff. Furthermore, this project portion strengthens the connection with the surrounding green belts by connecting Art Garden Road, Chonglun Park, and Lequn Park through landscape and environmental development, connecting Taichung City's blue and green belts, and providing citizens with a green tree. Figure 1 shows a leisure living environment that is shaded and suitable for ecological surroundings. The project has a total cost of about $12.4 billion and a length of 725 meters.

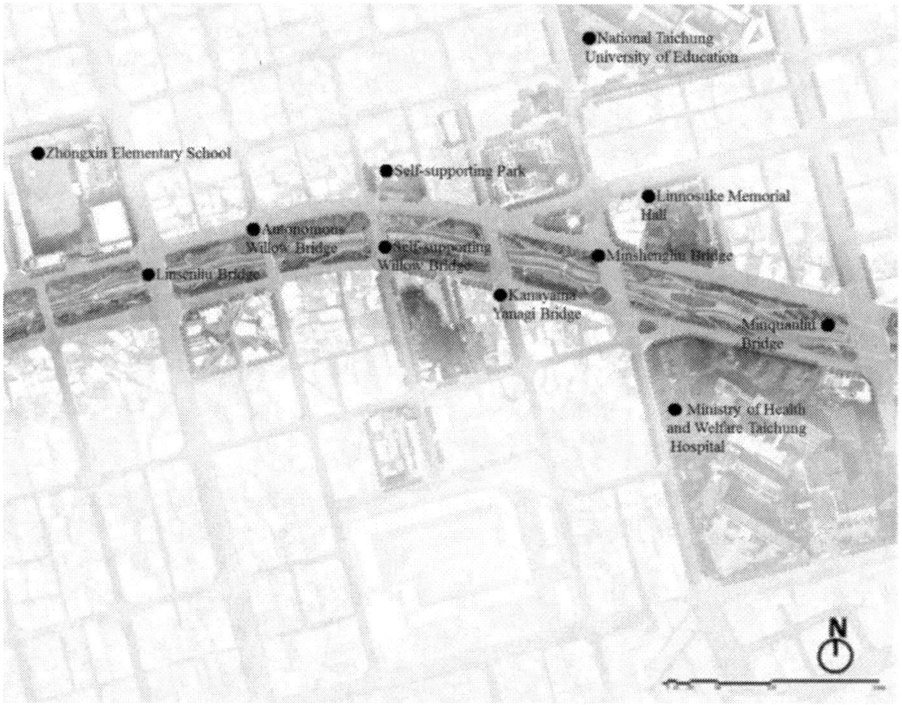

Fig. 1. Location map of the improvement plan for the section from Sanminliu Bridge to Linsenliu Bridge in Liuchuan Phase II.

3.2 Methods

The installation, maintenance, and opportunity costs were weighed against the benefits of avoiding stormwater runoff costs, reducing pollutants, and improving aesthetics. These costs and benefits will be spread out across the estimated life and examined using the Net Present Value calculation (calculation (1)). The values of B_t and C_t represent the benefits and expenses incurred in year t, respectively. The discount rate is i, and the net benefits are added together during the life cycle (n years). The real discount rate of 3.5% is used in all current value calculations. This ratio is appropriate for environmental projects with a 30-75-year life cycle[30].

Calculation of the net present value.

$$NPV = \sum_{t=0}^{n} \frac{(B_t - C_t)}{(1+i)^t} \qquad (1)$$

4 Results and Discussion

4.1 Results

Green infrastructure can bring numerous cost benefits, not only improving the urban environment, but also reaping long-term cost reductions. Based on a 40-year cost-benefit calculation, the annual benefit of greening is approximately \$51,000, and if greening saves water, electricity, and carbon taxes, the annual benefit to water resources is \$3,000. However, the annual benefit from carbon rights is approximately \$1.12 million. The total annual profit is approximately \$1.14 million, or a return of approximately 9%. Although the net present value is negative, the benefits may continue to accumulate over time. Here are some of the primary advantages:

Water safety: The 20-meter roads on both sides of the river from Liuchuan to Minquanliu Bridge to Linsenliu Bridge will be reduced by 8 meters each, according to the 102 Urban Planning Inspection. The drainage property and the green area will be exploited. The original concrete will be reduced on three sides in this project. The light changed to a soothing green belt that returned to the river and the people. High water is safer, low water is easier to maintain, and the frequency of maintenance for regular floods in two years is reduced from nine to one. Design the planting system and platforms at each level based on flood levels in different return periods and the flooding frequency in the 2-year and 25-year return periods in grade 10 to increase the river's flood-resistant capacity of the river and ecological green space.

Water Environment: During the planning and design phase, a comprehensive survey of local ecology and valuable old trees was conducted, ecological inspections were conducted, and eco-friendly measures such as avoidance and compensation were implemented to reduce cementation, increase green area, reduce heat island effects, and retain 40 trees (12 old trees)). Landscape design, 404 newly planted trees, multilayered planting design, increase green coverage area by 11,775 square meters, planting to attract butterflies and birds, improve the existing ecological environment, intercept sewage, purify water quality, and restore the river's ability to self-purify. Adopt low-impact development techniques such as rain gardens, plant filter strips, permeable pavements, infiltration ditches, infiltration dry wells, and water filling to improve nonpoint source runoff pollution through sedimentation, filtration, and adsorption, achieving the functions of microflood detention and rainwater reuse.

Water culture: a new kind of open urban waterfront space that integrates urban activities with cultural education and develops urban nodes. 1. The healing garden stimulates the five senses, i.e., the brain is more flexible to get therapeutic results when stimulated by vision, hearing, smell, touch, and taste. 2. Lin Zhisuke Plaza samples Lin Zhisuke's painted designs, simplifies, splits, and pastes them. It becomes a scene of people's activities on the facade; 3. The literary corridor extends the cultural context into the Yanagawa River via the resources of Taichung Literary Park, transforming it into a literary narrative landscape; In response to the Cultural City Midtown project, the Taichung Literary Promenade axis is created, connecting the flow of literary people in Taichung; 4. Three green sloped hills break the single line of the original straight canal, forming the undulating topography of the three hills, making the river bank have rhythmic height variations, and matching the plants to swing in the wind; Banyan trees 5 The park's old banyan trees, the ecosystem, and the banyan trees with place memories serve as a living space for inhabitants, a vast landscape wooden platform, and a platform for life and conversation. 6. The holiday market occupies the enlarged square space of the painted animation lane and serves as the market focal point. Crowds of people gather in the hinterland for relaxation and exchange, bringing more young life to the shoreline. 7. "Butterfly-attracting plants" are planted around the wilderness classroom pier, which primarily supplies food for butterfly larvae and nectar for adult butterflies, resulting in the ideal butterfly watching experience. Classroom with flowers. The humanistic space takes into account the demands of community and long-term care facility activities to provide a completely barrier-free place that can be used for bike trips, family and friend meetings, and educational visits. There are four barrier-free ramps, twenty barrier-free crossings, 45 barrier-free spots, and 1.450 meters of pedestrian space.

4.2 Discussion

Green infrastructure has several environmental, social, and economic benefits. Here are some of the main advantages:

Reduced operating and maintenance expenses: Green infrastructure generally has minimal operating and maintenance costs. Natural ecosystems such as wetlands or trees, for example, can naturally treat stormwater and lower running costs for stormwater management equipment. Furthermore, because they do not need large-scale energy consumption or infrastructure maintenance, green spaces and parks are often less expensive to establish and maintain [31]. Reduced energy costs: Green infrastructure may help communities save money on electricity. Cities with trees and green areas can reduce the urban heat island effect, lower summer cooling expenses, reduce heating and cooling demand, and offer natural shade [32-35]. Reduced water management expenses: Green infrastructure can help with stormwater management, flood risk reduction, and drainage system costs [34-37]. Rain gardens and ecological features, for example, can minimize stormwater runoff while also providing an opportunity for water recycling. Increased property values: Property prices are expected to increase in areas with surrounding green infrastructure. Local governments benefit from increased real estate tax income[26, 35]. Improved health: Green spaces and parks provide an opportunity for healthy and enjoyable activities, as well as helping minimize health-care expenses, such as stress and mental health problems [35, 37, 38]. Increased productivity: Green infrastructure can boost productivity since happy and healthy inhabitants and employees are more likely to work well. Businesses, schools, and other organizations will benefit from this [37, 39]. Reduce climate change impact costs: Green infrastructure contributes to climate change mitigation by reducing extreme weather occurrences and climate-related damage, which can lower disaster recovery costs [31, 35, 40]. In general, green infrastructure contributes to the achievement of sustainable development goals while also creating healthier, more attractive and more cost-effective urban and regional environments. Green infrastructure research focuses not only on cost reduction, but also on improving environmental, social, and economic advantages. These studies aid in the promotion of sustainable infrastructure building and management, as in well as the provision of more environmentally friendly and sustainable development models to cities and regions.

5 Conclusion

From the results of this study, it can be concluded that investing in green infrastructure is cost-effective. Even in the face of extreme urban climate conditions, green infrastructure can adapt to and mitigate urban flooding, heat island effects, and other disasters, with significant implications for sustainable urban development. Green infrastructure design and implementation reduce operating and maintenance costs while providing more environmentally friendly and sustainable solutions. This method promotes urban and infrastructural sustainability, while also benefiting society and the environment. While creating and maintaining green infrastructure may involve an initial investment, they can provide considerable long-term cost savings and help improve cities' sustainability, environmental, and social quality. As a result, to attain long-term overall cost effectiveness, urban planners and politicians frequently include green infrastructure into their urban development schemes. This approach aims to ensure that our cities can address current challenges while also providing additional advantages to society and the environment.

References

1. M. Luo and N.C. Lau, Geophysical Research Letters **46**, 6868-6877 (2019)
2. R. Ramyar, et al., Science of the Total Environment **703**, 135466 (2020)
3. C. He, et al., Nature Communications **12**, 4667 (2021)
4. X. Bai, et al., Annual review of environment and resources **42**, 215-240 (2017)
5. M. Cetin, Air Quality, Atmosphere & Health **12**, 1237-1249 (2019)
6. R. Ramyar, A. Ackerman, and D.M. Johnston, Cities **117**, 103316 (2021)

7. N. Kabisch, et al., Ecology and society **21**, (2016)
8. P. Li and Z.-H. Wang, Journal of Environmental Management **293**, 112963 (2021)
9. C. Bartesaghi Koc, P. Osmond, and A. Peters, Urban Ecosystems **20**, 15-35 (2017)
10. M.A. Benedict and E.T. McMahon, Renewable resources journal **20**, 12-17 (2002)
11. M.T.M. Frischenbruder and P. Pellegrino, Landscape and urban planning **76**, 67-78 (2006)
12. X. Li, L.C. Stringer, and M. Dallimer, Sustainable Cities and Society **80**, 103798 (2022)
13. S. Ronchi, A. Arcidiacono, and L. Pogliani, Sustainable Cities and Society **53**, 101907 (2020)
14. R. Hansen and S. Pauleit, Ambio **43**, 516-529 (2014)
15. S.L. Bell, et al., Geography Compass **13**, e12415 (2019)
16. J. Nam and N. Dempsey, Sustainability **11**, 4383 (2019)
17. A. Elliott and S.A. Trowsdale, Environmental modelling & software **22**, 394-405 (2007)
18. V. Jayasooriya and A. Ng, Water, Air, & Soil Pollution **225**, 1-20 (2014)
19. C. Zoppou, Environmental Modelling & Software **16**, 195-231 (2001)
20. V. Vandermeulen, et al., Landscape and urban planning **103**, 198-206 (2011)
21. A.M. Matsler, et al., Landscape and Urban Planning **214**, 104145 (2021)
22. Z.J. Grabowski, et al., Environmental Science & Policy **138**, 171-181 (2022)
23. C.G. Eger, D.G. Chandler, and C.T. Driscoll, Hydrological Processes **31**, 4492-4506 (2017)
24. H.E. Golden and N. Hoghooghi, Wiley Interdisciplinary Reviews: Water **5**, e1254 (2018)
25. T.A. Pugh, et al., Environmental science & technology **46**, 7692-7699 (2012)
26. D. Haase, et al., Habitat international **64**, 41-48 (2017)
27. K. Zhang and T.F.M. Chui, Science of the Total Environment **646**, 1219-1231 (2019)
28. T. Mittman and C. Kloss, (2015)
29. E.G. McPherson, et al., *Midwest community tree guide: benefits, costs, and strategic planting* (Pacific Southwest Research Station, Forest Service, US Department of Agriculture, 2006)
30. C. Almansa and J.M. Martínez-Paz, Science of the Total Environment **409**, 1305-1314 (2011)
31. O.J. Aladesote, Cost-Benefit Analysis of Green Infrastructure for Sustainable Stormwater Management of the Built Environment **97**, 13-13 (2022)
32. H. Pochee and I. Johnston, Building Services Engineering Research and Technology **38**, 728-756 (2017)
33. M.J. Nieuwenhuijsen, Annual Review of Public Health **42**, 317-328 (2021)
34. N. Yenil, et al., Journal of International Environmental Application and Science **12**, 140-146 (2017)
35. S.D. Pitman, C.B. Daniels, and M.E. Ely, Transactions of the Royal Society of South Australia **139**, 97-112 (2015)
36. D. Kim and S.-K. Song, Sustainability **11**, 3917 (2019)
37. A.D. Dunn, BC Envtl Aff L Rev **37**, 41 (2010)
38. K.J. Bowen and Y. Lynch, Current Opinion in Environmental Sustainability **25**, 90-95 (2017)
39. S. Cinderby and S. Bagwell, Area **50**, 126-135 (2018)
40. A.F. Samora-Arvela, et al., Journal of Spatial and Organizational Dynamics **5**, 176-188 (2017)

Sustainable Infrastructure Development in Indonesia: A Quantitative Evaluation of CO_2 Emission Reduction from Fly Ash-Cement Substitution in Ready-Mix Concrete

Rizka Lailatul Rohmah[1,2,*]

[1]Master Program in Occupational Safety and Health, Faculty of Public Health, Universitas Indonesia
[2]Department of Quality, Health, Safety, Environment and System, PT Adhi Persada Beton, Jakarta, Indonesia

Abstract. The construction industry is a major contributor to global CO_2 emissions, particularly through Portland cement production. In Indonesia, the utilization of fly ash in ready-mix concrete presents a promising approach to sustainable construction. This strategy aligns infrastructure development with environmental objectives by potentially reducing emissions and supporting global climate change mitigation efforts. This study investigates the environmental impact of incorporating fly ash into ready-mix concrete employed for the Yogyakarta-Bawen toll road project, with a specific focus on CO_2 emission reduction. Recognizing the critical role of sustainable construction practices in addressing climate change, this research aims to quantify the environmental benefits of substituting 40% of Portland cement with fly ash in FC 30 MPa ready-mix concrete. A comprehensive analysis revealed an 18% decrease in emissions associated with coarse aggregate and a 55% decrease from cement, despite a marginal increase from fine aggregate. Overall, this method achieved a significant 47% reduction in total CO_2 emissions, demonstrating the effectiveness of fly ash in enhancing the sustainability of construction materials. This substantial reduction underscores the potential of fly ash integration as a key strategy for mitigating the environmental impact of large-scale infrastructure projects.
Keywords: CO_2, Emission, Fly Ash, Ready-mix Concrete, Reduction

1 Introduction

The construction sector occupies a critical space at the nexus of economic development and environmental impact. Concrete, a fundamental building block for global infrastructure, also represents a significant source of carbon dioxide (CO_2) emissions [1]. Cement, the essential component of concrete, is a major contributor to global CO_2 emissions. Its annual production volume of approximately 4.1 billion metric tons accounts for 8-10% of worldwide emissions [2]. This figure positions the cement industry as the third-largest emitter of CO_2 globally, highlighting the pressing need for sustainable practices within this sector [3].

* Corresponding author: rizka.lailatul@ui.ac.id

Indonesia, a country with greenhouse gas emissions reaching 2.4 billion tonnes of CO_2 equivalent ($GtCO_2e$) in 2015 (representing 4.8% of global emissions), exemplifies the urgency for sustainable infrastructure development [4]. Recognizing this, Indonesia has made commitments under the Enhanced Nationally Determined Contributions (NDCs) to the UNFCCC Secretariat, aiming to increase emission reduction targets from 29% to 31.89% unconditionally and from 41% to 43.20% conditionally[5]. These commitments align with the country's National Medium-Term Development Plan 2020-2024 (RPJMN 2020-2024), which prioritizes environmental sustainability and climate change resilience alongside infrastructure development[6].

The Indonesian ready-mix concrete market, valued at US$ 459.5 million in 2022, is projected to grow at a compound annual growth rate (CAGR) exceeding 8%, reflecting the global trend of rising demand for ready-mix concrete[7]. This demand coincides with significant planned infrastructure investments. The Indonesian government aims to invest over USD 400 billion between 2020 and 2024, focusing on transportation, power generation, and the water industry[8].

In response to this growth, the Global Cement and Concrete Association (GCCA) has introduced sustainability guidelines, prompting the industry to explore alternative materials for decarbonizing concrete production[9]. One approach involves substituting conventional cement with industrial by-products like silica fume, fly ash, and ground granulated blast furnace slag[10]. Despite the global production of 777 million tons of fly ash in 2008, its utilization averages only 53.5% of the total output, indicating a substantial opportunity for increased use in concrete mixtures[11] [12].

The adoption of fly ash as a partial replacement for cement in ready-mix concrete presents a promising avenue for CO_2 emission reduction. Considering the cumulative CO_2 uptake of cement materials estimated at 22.9 Gt CO_2 from 1930 to 2021, which offsets 55.1% of the emissions from cement production over the same period, the potential for sustainable infrastructure development through material substitution is significant [13]. This quantitative evaluation aims to provide a comprehensive analysis of the impact of fly ash-cement substitution on CO_2 emission reduction in Indonesia's ready-mix concrete industry. This aligns with the country's environmental sustainability goals and contributes to the global effort to mitigate climate change impacts.

2 Objective and Methodology

The objective of this study is to critically assess the environmental impacts of adopting fly ash as a sustainability measure through its substitution for Portland cement in the concrete production of the Yogyakarta–Bawen toll road project. This evaluation aims to quantify the achieved reductions in CO_2 emissions, thereby elucidating the practical implications of incorporating sustainable construction materials and enriching the discourse on the efficacy of green construction methodologies in achieving sustainability goals in infrastructure projects. This study employs a quantitative research methodology to explore the implementation of fly ash as a sustainability measure in the Yogyakarta–Bawen toll road project, focusing on its partial substitution for Portland cement in concrete production. The methodology is structured around several pivotal steps:

2.1 Data Collection

This phase involved the meticulous gathering of comprehensive data pertaining to the concrete production process. Data collection encompassed two distinct periods: before and after the integration of fly ash as a partial substitute for Portland cement.

2.2 Emission Analysis

The study calculates CO_2 emissions for concrete production using both the conventional Portland cement methodology and the revised approach incorporating fly ash. Calculations will leverage established emission factors for each material, considering the volume of concrete produced.

2.3 Comparative Evaluation

The environmental outcomes of adopting fly ash as a sustainability measure are compared with those of the conventional approach. This comparison aims to quantify the environmental advantages, discussing the broader applicability of such sustainable practices in infrastructure development.

3 Inventory Data Collection

3.1 Project Information Overview

The batching plant located on Jl. Tempel-Seyegan in Margomulyo, Seyegan District, Sleman Regency, Yogyakarta Special Region, is established to fulfill the concrete demand for the National Strategic Project of the Yogyakarta–Bawen toll road construction. This strategic placement facilitates the logistical efficiency required for the sustained supply of concrete, instrumental in supporting the infrastructure project aimed at enhancing regional connectivity and transportation efficiency. Data collection for this study was conducted from August to December 2023.

3.2 Concrete Quantity Information

The project specifically utilizes Readymix FC'30 MPa concrete for the construction of pile caps and columns. The data collected reveals a monthly production volume of 8,100 cubic meters (m^3) for this type of concrete.

3.3 CO_2 Emission Factors for Raw Materials

The quantification of environmental impacts stemming from concrete production necessitates a rigorous analysis of CO_2 emission factors associated with all constituent raw materials. The emission factors for aggregates are derived from the established literature. [14] Additionally, the emission factor for cement is procured directly from SBI Cement Product, thus guaranteeing precision and direct applicability to the current study. The methodology for calculating cement's emission factors aligns with the "Cement CO2 and Energy Protocol, Version 3.1, CO_2 Emissions and Energy Inventory," a product of the collaborative efforts of the World Business Council for Sustainable Development's Cement Sustainability Initiative (CSI) and the European Cement Research Academy (ECRA). The comprehensive enumeration of emission factors for aggregates, fly ash, and cement is methodically presented in Table 1.

Table 1. CO_2 Emission Factor for Raw Materials

Material	Emission Factor (kg-CO2/kg)
Coarse Aggregate	0.0408
Fine Aggregate	0.0139
Fly Ash	0.027
Cement	0.712

3.4 Calculation Method of CO_2 Emissions

The methodology for calculating CO2 emissions within this study adheres to the rigorous mathematical framework established by Vázquez-Calle.[15] The foundational equation, denoted as Equation 1, encapsulates the aggregate CO_2 emissions attributable to the production of 1 m^3 of concrete, integrating the contributions from various materials. This equation is expressed as follows:

$$EF = W_c \times EF_c + W_{ca} \times EF_{ca} + W_{fa} \times EF_{fa} + EF_m \quad (1)$$

In this equation, W_c represents the mass of cement per cubic meter of concrete (kg), W_{ca} denotes the mass of coarse aggregate per cubic meter of concrete (kg), and W_{fa} signifies the mass of fine aggregate per cubic meter of concrete (kg). The variables EFc, EF_{ca}, and EF_{fa} refer to the emission factors for cement, coarse aggregate, and fine aggregate, respectively, measured in kilograms of CO_2 per ton of material. Lastly, EF_m stands for the emission factor of the mixer, quantified in kilograms of CO_2 per cubic meter of concrete. In this study, the emission factors for the batching plant operations were not considered due to the unavailability of specific data. The production of ready-mix concrete utilizes a batching plant with a capacity of 3 m^3 per batch.

4 Result and Discussion

This study conducts a meticulous quantitative evaluation of the CO_2 emission reductions achievable through the substitution of fly ash for Portland cement in the production of ready-mix concrete. By applying the specified formula (1), the study initially quantifies the emissions per cubic meter (m^3) for conventional ready-mix concrete, with the outcomes detailed in Table 2. These preliminary findings regarding CO2 emissions were further elaborated to encompass the total emissions derived from the aggregate monthly production volumes of concrete.

Table 2. Total CO_2 Emission from Ready-Mix Concrete FC 30 NFA

Material	Quantity (kg/m^3)	Total Emissions ($kg\text{-}CO_2/m^3$)	Total Emission Per Month ($ton\text{-}CO_2\text{-}e$)
Coarse Aggregate	920	37.54	304.04
Fine Aggregate	916	12.73	103.13
Cement	365	259.88	2105.03
Total Emission		310.15	2512.20

Table 3. Total CO_2 Emission from Ready-Mix Concrete FC 30 with 40% Fly Ash Substitution

Material	Quantity (kg/m^3)	Total Emissions ($kg\text{-}CO_2/m^3$)	Total Emission Per Month ($ton\text{-}CO_2\text{-}e$)
Coarse Aggregate	753	30.72	248.85
Fine Aggregate	961	13.36	108.20
Cement	164	116.77	945.82
Fly Ash	176	4.75	38.49
Total Emission		165.60	1341.36

Further analysis, as documented in Table 3, assesses the emissions per m_3 for concrete formulations incorporating fly ash as a sustainable alternative to Portland cement. This comparative analysis elucidates a pronounced reduction in CO_2 emissions attributable to the fly ash substitution strategy. Specifically, it was observed that traditional ready-mix concrete, leveraging Portland cement exclusively, accounted for 2512.20 tons of CO_2 emissions. In stark contrast, the adoption of fly ash in the concrete mix formulation resulted in a significant reduction, with total emissions being curtailed to 1341.36 tons of CO_2.

Table 4. Comparison of CO_2 Emission

Material	Total Emissions RMC FC 30 NFA (ton-CO2-e)	Total Emissions RMC FC 30 FA 40% (ton-CO2-e)	Reduction (ton-CO2-e)	Percentage (%)
Coarse Aggregate	304.04	248.85	55.19	18%
Fine Aggregate	103.13	108.20	+5.07	+5%
Cement	2105.03	945.82	1159.21	55%
Fly Ash	-	38.49	-	-
Total Emission Per Month	2512.20	1341.36	1170.84	47%
Total Emission Per Year	30146.42	16096.35	14050.08	47%

The analysis summarized in Table 4 reveals a significant reduction in CO_2 emissions through the application of fly ash as a sustainability measure in ready-mix concrete production. Adopting a 40% fly ash substitution resulted in an 18% decrease in emissions from coarse aggregate and a notable 55% reduction from cement use, highlighting the substantial environmental benefits of this approach. Although there was a slight increase in emissions from fine aggregate, the overall impact was a 47% reduction in total CO_2 emissions. This significant annual reduction of 14,050.08 tons of CO_2 highlights the efficacy of fly ash incorporation as a sustainability measure within the construction materials for the Yogyakarta–Bawen toll road project. This initiative underscores a pivotal contribution to environmental conservation, aligning with the project's sustainability goals. Moreover, it exemplifies a forward-thinking approach to infrastructure development in Indonesia, setting a benchmark for the adoption of green construction practices and substantially mitigating the carbon footprint associated with large-scale infrastructure projects.

5 Conclusion

This study presents a quantitative assessment of the environmental advantages achieved through the partial substitution of Portland cement with fly ash in ready-mix concrete production. The research focuses on the Yogyakarta-Bawen toll road project in Indonesia. A

40% fly ash substitution rate was implemented, leading to reductions in emissions of 18% and 55% for coarse aggregate and cement, respectively. While emissions from fine aggregate exhibited a slight increase, the overall outcome was a significant 47% decrease in total CO2 emissions. This translates to an annual reduction of 14,050.08 tons of CO2. These findings emphasize the environmental benefits associated with fly ash incorporation in concrete, providing a quantitative method for considerably lowering the carbon footprint of the construction industry.

References

1. E. A. Khalil. & M. N. AbouZeid, "Computation of the environmental performance of ready-mix concrete for reducing CO2 emissions: A case study in Egypt.," Energy Reports, **9**, 144-148, (2023).
2. L. Poudal & K. Adhikari, "Environmental sustainability in cement industry: An integrated approach for green and economical cement production.," Resources, Environment and Sustainability 4. Elsevier B.V, 100-124, (2021).
3. E. Mittelman, "The Cement Industry, One of the World's Largest CO2 Emitters, Pledges to Cut Greenhouse Gases," Yale Environment 360, (December 2018). [Online]. Available: https://e360.yale.edu/digest/the-cement-industry-one-of-the-worlds-largest-co2-emitters-pledges-to-cut-greenhouse-gases.
4. T. Harrisson, "The Carbon Brief Profile: Indonesia," Carbon Brief, (23 July 2021). [Online]. Available: https://www.carbonbrief.org/the-carbon-brief-profile-indonesia/.
5. Indonesia, "Enhanced Nationally Determined Contribution Republic Of Indonesia," (23 September 2022). [Online]. Available: https://unfccc.int/documents/615082.
6. Kementerian PPN, "Accelerating P4G's Partnerships and Indonesia's Leadership in Green Growth And 2030 Global Goals," (2021). [Online]. Available: http://greengrowth.bappenas.go.id/en/accelerating-p4gs-partnerships-and-indonesias-leadership-in-green-growth-and-2030-global-goals/.
7. GlobeNewswire, "Ready-Mix Concrete Market to Expand at 6.1% CAGR, Surpassing USD 782.9 million by 2031 | Analysis by Transparency Market Research, Inc.," (15 February 2024). [Online]. Available: https://www.globenewswire.com/news-release/2024/02/15/2829790/0/en/Ready-Mix-Concrete-Market-to-Expand-at-6-1-CAGR-Surpassing-USD-782-9-million-by-2031-Analysis-by-Transparency-Market-Research-Inc.html#:~:text=%2D%20The%20ready%2Dmix%20concrete%20market,t.
8. Mordor Intelligence, "Indonesia Ready-Mix Concrete Market Size & Share Analysis - Growth Trends & Forecasts (2024 - 2029," (2024). [Online]. Available: https://www.mordorintelligence.com/industry-reports/indonesia-ready-mix-concrete-market.
9. GCCA, "GCCA Sustainability Guidelines for co-processing Fuels and Raw Materials in Cement Manufacturing," (October 2019). [Online]. Available: https://gccassociation.org/sustainability-innovation/cement-best-practices-and-reporting/.
10. C. Orozco, S. Tangtermsirikul, T. Sugiyama, and S. Babel, "Examining the endpoint impacts, challenges, and opportunities of fly ash utilization for sustainable concrete construction," Scientific Reports, **13**, 1, 18254, (2023).

11. C. Herath, C. Gunasekara, D. W. Law, and S. Setunge, "Performance of high-volume fly ash concrete incorporating additives: A systematic literature review," Construction and Building Materials, **258**, 120606, (2020).

12. A. Arrigoni et al., "Life cycle greenhouse gas emissions of concrete containing supplementary cementitious materials: cut-off vs. substitution," Journal of Cleaner Production, **263**, 121465, (2020).

13. Z. Huang et al., "Global carbon uptake of cement carbonation accounts 1930–2021," Earth System Science Data, **15**, 11, 4947–4958, (2023).

14. L. K. Turner and F. G. Collins, "Carbon dioxide equivalent (CO2-e) emissions: A comparison between geopolymer and OPC cement concrete," Construction and Building Materials, **43**, 125–130, (Jun. 2013).

15. K. Vázquez-Calle, V. Guillén-Mena, and F. Quesada-Molina, "Analysis of the Embodied Energy and CO2 Emissions of Ready-Mixed Concrete: A Case Study in Cuenca, Ecuador," Materials, **15**, 14, 4896, (Jul. 2022).

Establishing the Potential Rice Loss Prediction Model of Climate and Nature Disaster Factors Based on Machine Learning Theory

Yun-Ju Hsu [1], *Yu-Keng* Chen [1*] and *Chao-Heng* Tseng [1]

[1]National Taipei University of Technology, Taipei, Taiwan

Abstract. The United Nations points out that extreme climate events are frequent and widespread in the 21st century and have become a global security issue. Artificial intelligence and machine learning have attracted much attention in environmental applications. This study aims at applying machine learning (ML) to rice disaster prediction, and uses SPSS to analyze environmental impact factors. After model training and evaluation, four models are provided, among which short-term prediction results show high accuracy on a single event, which are suitable for water damage, cold damage, and plant diseases and insect pests respectively. In terms of long-term prediction, using future meteorological prediction values to predict potential rice losses is better, especially within a specific time period. Ultimately, relevant units such as the Council of Agriculture or the Agri-Food and Food Administration can choose a suitable model based on different purposes (short-term or long-term forecasting).

Keywords: Climate change, artificial intelligence, machine learning, rice loss, Statistical Software for Social Sciences (SPSS)

1.Introduction

In recent years, the global economy has been booming, energy and fuels have been used extensively, and carbon dioxide emissions have increased rapidly, causing global warming to become increasingly serious. Increased surface temperatures have brought many negative impacts, such as melting icebergs, rising sea levels, and flooding of coastal areas [1] and natural disasters such as erosion [2]. The IPCC points out that extreme climates will become more frequent and widespread. Especially in the past two years, Climate Emergency has replaced Climate Change, and the world has declared a climate emergency [4, 5]. In order to respond promptly to the various climate crises brought about by the climate emergency,

* Corresponding author: abc283746@gmail.com

countries must adopt more progressive methods, such as artificial intelligence [6] and machine learning [7, 8].

Artificial intelligence (AI) is a technology that presents human intelligence through computer programs and is dedicated to solving common cognitive problems related to human intelligence, such as learning, problem solving, and pattern recognition. Provide data for AI to learn from past experiences and machine learning solutions. In recent years, AI has begun to be combined with the Internet of things (IoT) [13-15], allowing AI to analyze huge amounts of data faster, and the data can include Sources, places, objects and event information that have never been touched before.

Therefore, machine learning (ML) theory under the concept of artificial intelligence (AI) is used to build models. Machine learning models can solve the complexity and accuracy of big data [16-19], and predict the future climate and possibilities. Natural disasters suffered are set as impact parameters to predict rice losses in the agricultural industry chain, and predict long- and short-term impacts, and then introduce adaptation policy planning for the rice industry chain in response to climate change, so as to prepare rice-related industries for disasters. Advance prediction and preparation can reduce avoidable losses of people and property, and can also provide industries with early preventive measures against climate change disasters

2.Methods and materials

This study will use Statistical Software for Social Sciences (SPSS) to build the model. First, through two screening modes: independent sample T test and neural network-like, historical climate data, natural disaster data and agricultural related data from 1946 to 2019 were used to classify variables and screen important variables, and select the significant or important variables. Variables with higher reliability are selected, and then the model is trained using two different decision tree growth methods: Chi-square Automatic Interaction Detection (CHAID) and Classification and Regression Tree (CART), and the ROC analysis curve is used to determine whether the model has Predictive value.

The prediction part is divided into long and short-term predictions. The short-term prediction is the prediction month data, and the month with the greater rice loss among the five disaster items is selected as the prediction case, a total of five; the long-term prediction is the prediction year data, and the first part is the prediction for 2020. The actual 2020 observations and MOHC used HadGEM2-ES to simulate the future climate monthly data of IPCC and the TCCIP statistical downscaled monthly data prediction data of each county and city in the future to estimate the amount of potential rice losses in the future, and Compare the difference with the actual amount of rice losses in 2020; the second part is to predict the past three years (2017-2019). Meteorological factors and natural disaster factors from 2017 to 2019 are put into the best model to calculate the amount of potential rice losses. Forecast and compare the difference with the average of the actual rice loss amount in the three years from 2017 to 2019.

Finally, through the theory of machine learning, we completed the construction of a prediction model for the potential loss amount of rice affected by meteorological factors and natural disaster factors, and then judged whether the model was suitable for short-term prediction or long-term prediction, and its practical practicability. The research architecture is shown in Figure 3-1.

Data related to historical meteorology and natural disasters are taken from the Central Meteorological Administration's Climate Statistics Annual Report - Ground Data and data from the Statistics Network of the Ministry of Transport.

In terms of agricultural-related data, because the historical data collection period is relatively long, some data are unrecorded. In order to reduce the errors caused by missing values in the research data, this study uses Excel as the calculation tool to process the missing values, and based on Calculation of data in the "Council of Agriculture Statistical Annual Report" provided by the Council of Agriculture. If it is impossible to make an inference, assumptions will be made.

Since there may be more than one natural disaster in each month, each loss item also has more than one number. Therefore, this part of this study uses the average value for calculation. The data collected

in this study will be imported into SPSS for machine learning model construction.

(1) Dependent variable

The calculated amount of potential rice disaster loss is the part of the causal variable. Then, using "whether rice loss is caused" as the "category", it is judged

based on the amount of potential rice disaster loss. If there is a potential rice disaster loss amount, it is regarded as 1 ;If there is no potential rice disaster loss amount, it is regarded as 0.

(2) Independent variables

From the data related to meteorological factors and natural disaster factors from 1946 to 2019, the independent variables are summarized, including meteorological factors and natural disaster factors.

Table 1. Yield impact coefficient of rice disaster

NO.	Disaster type	Yield impact coefficient	references
1	Flooding injury	10%	[1]
2	Typhoon	17%	[2]
3	Chilling injury	3%	[3]
4	Drought injury	7%	[4]
4	Insect injury	15%	[5]

Table2. Independent variable

No.	variable name
X01	average temperature
X02	precipitation(mm)
X03	Maximum daily precipitation(mm)
X04	maximum wind speed(m/s)
X05	sunshine hours(hour)
X06	Relative humidity(%)

Among the natural disaster factors, in addition to the environmental impact on rice growth, sudden natural disasters, such as typhoons and heavy rains, will directly cause flooding and lodging of rice, increasing agricultural losses. The area of agricultural disaster losses caused by natural disasters from 1946 to 2009 was, in order, typhoon 10,145 thousand hectares (67%), heavy rain (13%), drought (10%), pests and diseases (5%), and cold damage (4%) and others (1%) [331]. This part is not far from the data collected by this study. However, this study only investigated the damage area of rice, so there are slight differences. This study collected rice from 1946 to 2019. Damaged area results

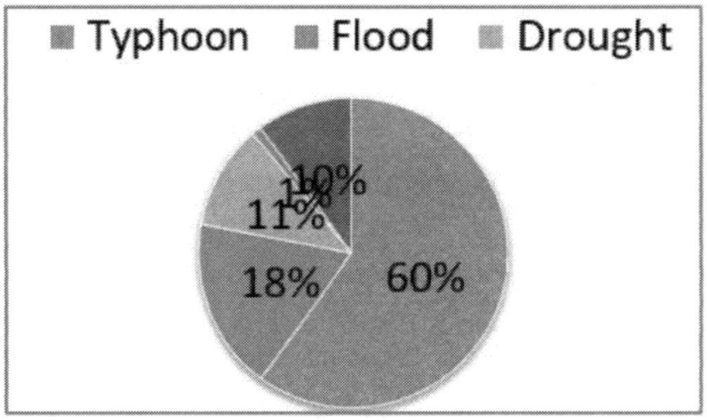

Fig1. Area of rice lost due to natural disasters from 1946 to 2019

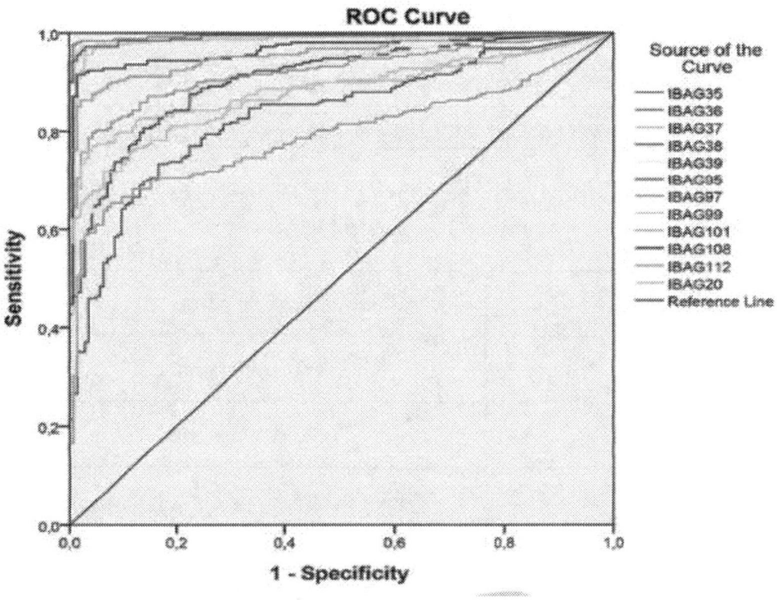

Fig2. ROC curve

3 Machine learning

After processing the data, the receiver uses the independent sample T test and the multi-layer perceptron (MLP) in the neural network (ANN) to screen important variables, and then puts the screened important variables into the decision tree.Carry out model training and evaluation, and finally use the Receiver Operating Characteristic (ROC) and the Area Under the Curve of ROC (AUC) to evaluate the accuracy and predictive value of the model to complete the construction of the model. . The software used is IBM SPSS 22.

Decision tree is used for training evaluation. The training decision tree growth method in this study is divided into two parts, namely Chi-square automatic interaction detection method (CHAID) and classification regression tree (CART). The model verification setting part mainly focuses on splitting samples. After different testing and training calibration ratios, the most appropriate training and calibration ratio is selected as the basis for data segmentation. This study uses whether rice losses are caused for decision tree model training, and trains with training-test ratios of 60:40, 70:30, and 80:20. The input independent variables are independent samples T-test and neural network-like screening. important variables.

In the decision tree segmentation sample verification part, SPSS will display the overall correct percentage. The presentation effect is equivalent to the concept of a confusion matrix. You can see the overall correct percentage during training and testing respectively, and choose a better sample segmentation ratio. Also evaluate which model algorithm is best

The model performance indicator is the Receiver Operating Characteristic (ROC), which has recently developed well in the fields of machine learning and data mining, and is also a relatively approachable diagnostic value indicator.

In signal detection theory, the ROC curve graphically presents the performance of a binary classifier system under a specific classification or threshold. The vertical axis (y-axis) of the graph is the true positive rate (TPR), also known as sensitivity; the horizontal axis (x-axis) is the false-positive rate (FPR)), expressed as 1 – specificity, while sensitivity is the probability of correctly judging the result as positive, and specificity is the probability of correctly judging the result as negative or negative.

4. Results and Discussion

The trends of meteorological factors and natural disaster factors from 1946 to 2019 were analyzed separately, and the relationship between the two factors for rice was discussed.

The trend charts of six meteorological factors from 1946 to 2019 are shown in Figure 4.1 to Figure 4.6

According to Figure 4-1, it can be found that the average temperature is increasing year by year, which is consistent with the global climate warming trend. In May 2020, the global surface temperature was 0.95°C higher than the 20th century average temperature, breaking the 140-year record. In Taiwan, the annual average temperature has continued to reach new highs in the past five years. From 2015 to 2019, it was among the top six with the highest average temperatures, and the annual average temperature after 2015 has never been lower than 24.2°C. In terms of global trends, in the past 30 years, the average temperature has increased by 0.2°C every 10 years. Since 1977, the global average temperature has been higher than the centennial climate value for 43 consecutive years, and the top eight hottest places in the past century have all occurred in nearly 10 years.

As can be seen from Figure 4-2, Taiwan's precipitation had stabilized until 2006, but it suddenly spiked in 2005. It is speculated that because 2003 to 2004 was the Year of the Holy Child, the precipitation in Taiwan was less than in previous years, and 2005 Year.

In Figure 4-3, the maximum daily rainfall trend from 1946 to 2019 fluctuates up and down with no definite pattern. The average value falls at 184.72 mm per day. According to the Central Meteorological Administration's definition of heavy rain, Taiwan almost There are often heavy rains, but because such climate conditions are suitable for the growth of rice, the monthly rainfall is about 100~200 mm.

In the maximum wind speed section of Figure 4-4, the overall range is large and irregular. The maximum wind speed falls on an average of 61.04 (m/s). In other words, the annual maximum wind speed in Taiwan causes a reduction of about 40% to 60% in rice yields. .

As for the relative humidity in Figure 4-5, the overall trend is stable and remains at around 76%.

Figure 4-6 For the sunshine hours, the average value falls at 145.57 (hr), and the overall trend is a slow decline. According to relevant studies, there has been a significant downward trend in sunshine rate since 1975. It is speculated that this is caused by the increase in the number of suspended particles in the world after the industrial revolution and the increase in

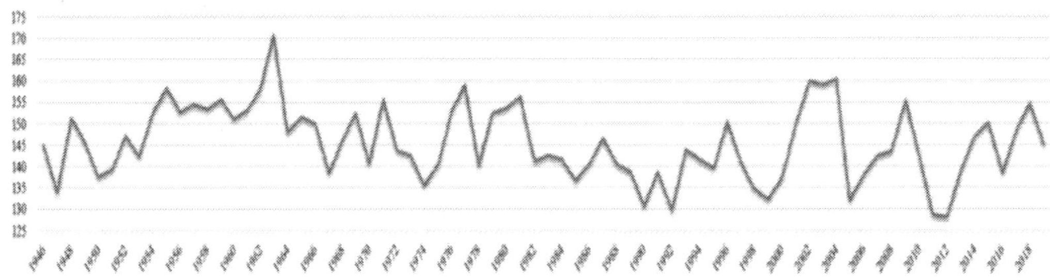

Fig3. PM 2.5 experiment of air purifiers with various air pollution removal mechanisms.

cloud condensation nuclei, resulting in an increase in cloud cover. Sunshine rate is defined as the ratio of actual sunshine hours to astronomical sunshine hours, expressed as a percentage. Various reasons point to the impact of suspended particles in the air, which will lead to a downward trend in sunshine rate and sunshine hours.. Comparison of PM 2.5 removal efficiency and CADR at the 60th minute of the

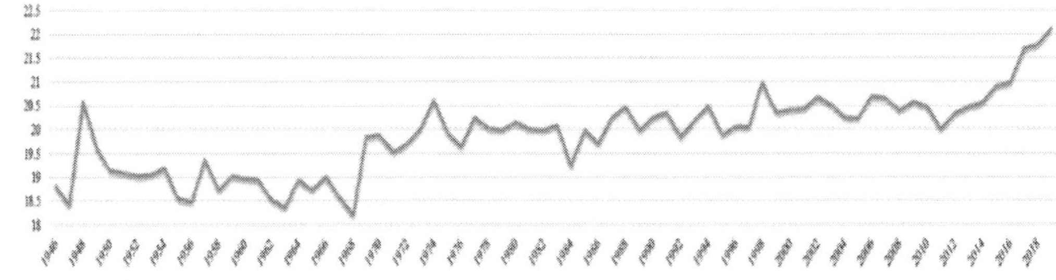

Fig4. Average temperature (°C) trend chart over the years

 https://doi.org/10.1051/e3sconf/202453004006

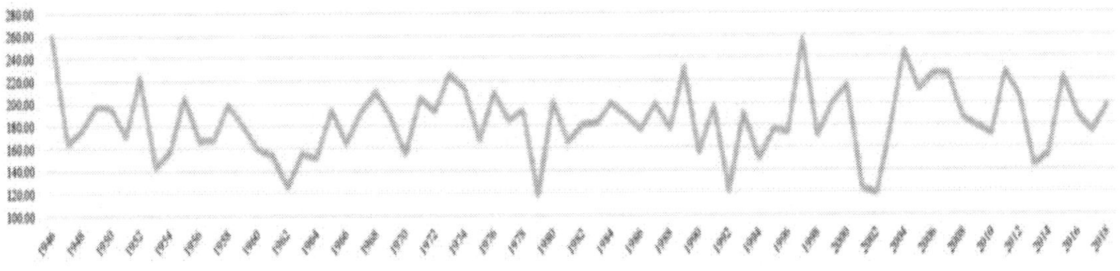

Fig5 . Trend chart of maximum daily rainfall (mm) over the years

The relationship between meteorological factors, natural disaster factors and rice

Calculate the total damaged area of rice from 1946 to 2019, the damaged area of rice in the first period, the damaged area of rice in the second period, the total potential loss amount of rice, the potential loss amount of rice in the first period and the potential loss amount of rice in the second period calculated in this study. Six natural

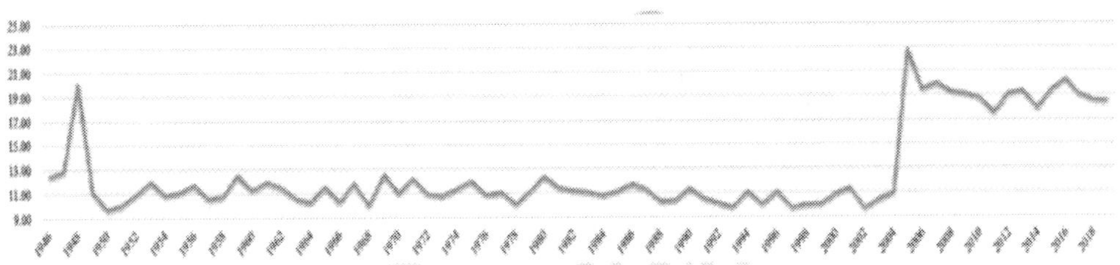

Fig6. Rainfall (mm) trend chart over the years

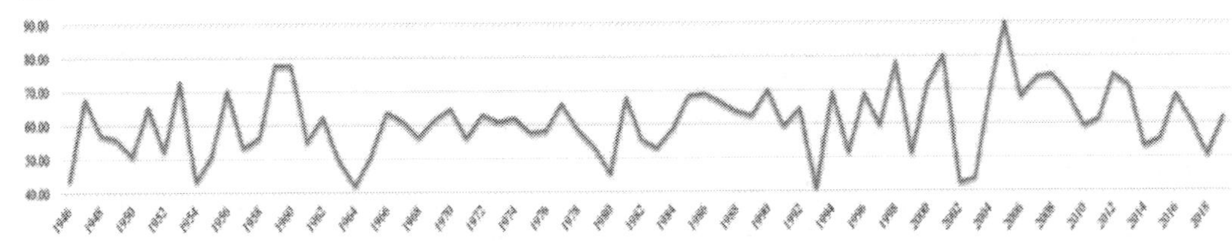

Fig7. Maximum wind speed (m/s) trend chart over the years

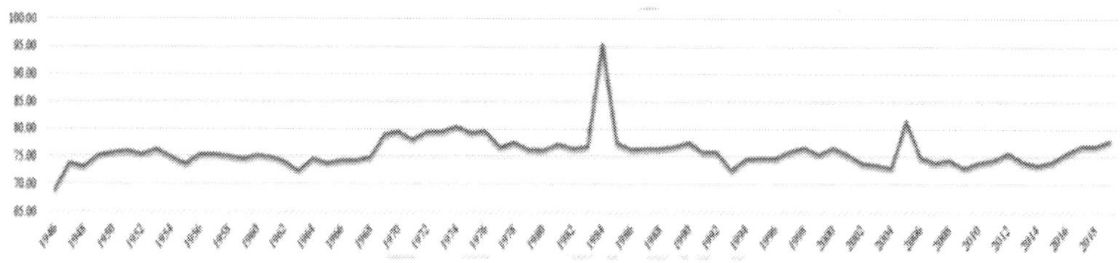

Fig8. Maximum wind speed (m/s) trend chart over the years

disaster factors and six meteorological factors were used to draw XY scatter diagrams, and the coefficient of determination of the linear trend (Coefficient of determination, recorded as R2)

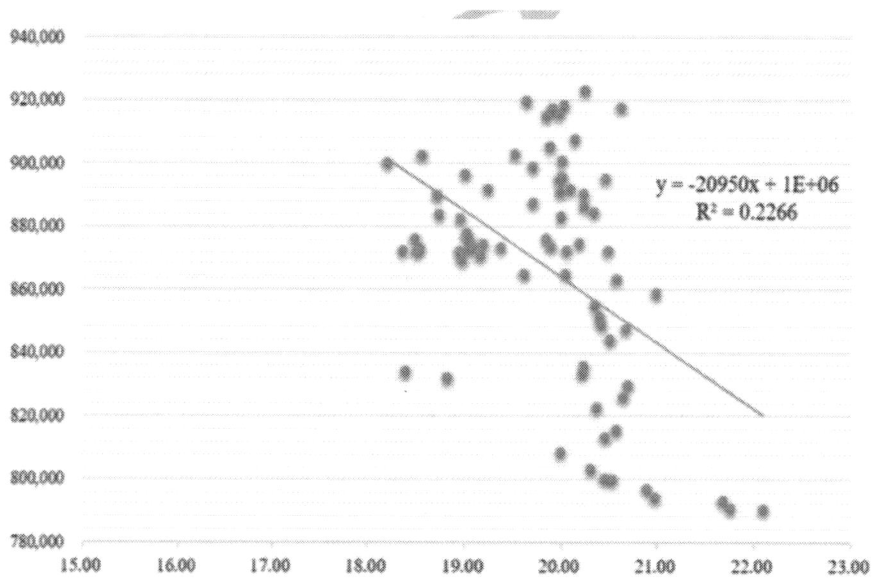

Fig 9. Relationship between total rice damaged area and average temperature

In the model construction, two growth methods of decision tree, chi-square automatic interaction detection method (CHAID) and classification and regression tree (CART) are used, coupled with two variable screening modes, independent sample T test and neural network, and three segmented sample ratios of 60:40, 70:30, and 80:20 for prediction. The results obtained are that the results of each model judged by the ROC curve and AUC have predictive value, and the models are respectively It has the best performance under the conditions of T-test + Decision Tree(CHAID_70:30), T-test + Decision Tree(CART_80:20), ANN + Decision Tree(CHAID_70:30) and ANN + Decision Tree(CART_70:30) Performance.

5.Conclusion

Use the machine learning (ML) software SPSS to do disaster prediction and climate simulation analysis of environmental impact factors on the industrial chain, and then make corresponding implementation strategies based on the analysis results to provide the industry with early preventive measures against climate change disasters and Address the development of climate adaptation policies, such as reducing crop impacts and establishing early warning system mechanisms.

Through the construction results of this rice potential loss prediction model, it can be seen that the amount of potential rice losses can be estimated by meteorological factors and natural disaster factors, and the meteorological prediction data of IPCC_HADGEM2-ES_RCP4.5 has also been applied to obtain quite good results, which is very accurate. In summary, it is successful to build a rice potential loss prediction model using machine learning theory, and it also represents the feasibility of introducing artificial intelligence into the agricultural industry chain.

Long-term monitoring, tracking management and early warning systems are the focus of future agricultural adjustment policies, integrating existing new technologies, such as using artificial intelligence tools for automated monitoring and intelligent management, to improve the resistance to climate crises caused by climate change. Ability to deal with adversity.

5.1Model research results

Using machine learning to build a potential rice loss prediction model: This study uses the decision tree algorithm in machine learning theory and historical data from 1946 to 2019 to combine two decision tree growth methods and two variable screening modes. And three different split sample ratios, model training and testing were carried out with SPSS software, and then the ROC curve and AUC area chart were used to judge the prediction results of the entire model. The overall results showed that each model had a considerable degree of accuracy and predictive value.

5.2Model training and evaluation

The judgment models of this study are respectively in T-test + Decision Tree (CHAID_70:30), T-test + Decision Tree (CART_80:20), ANN + Decision Tree (CHAID_70:30) and ANN + Decision Tree (CART_70:30). Best performance.

5.3 Directions in which the model can improve

In the IPCC_HADGEM2-ES_RCP4.5 prediction value processing part, this study is based on the historical proportions of each district, county and city. If the meteorological prediction values can be processed with more standard downscaling technology, the overall prediction results will be more accurate.

5.4 Model recommendations

This research model was built and trained based on the potential amount of rice losses, but the prediction results were not ideal. It is recommended that future research can use the actual amount of rice losses to train the model to reduce the error between it and the actual situation.

References

1. H.-W. Chuang, S.-R. Yang, T. Ou, and T.-P. Lin, "CLIMATIC ADAPTATION OF COASTAL COMMUNITIES ON THE SOUTHWEST OF TAIWAN," (in 英文), vol. 24, no. 6, pp. 1063-1069, 2016.
2. Joseph_Romm, Climate Change (What Everyon Need To Know). the United States of America by Oxford University Press, 2016.
3. T. C. a. C. Change, Climate emergency Pamphlet. The Campaign against Climate Change, 2011, pp. 10-13.
4. W. Ripple et al., "World Scientists' Warning of a Climate Emergency," BioScience, 11/26 2019.
5. J. MORTON, Don't Mention The Emergency 2018. Darebin Climate Action Now: Darebin Climate Action Now, 2018.
6. E.ManavalanandK.Jayakrishna,"AreviewofInternetofThings(IoT)embedded sustainable supply chain for industry 4.0 requirements," Computers & Industrial Engineering, vol. 127, pp. 925-953, 2019.
7. Hsu Ming-His, Lin Pay-Liam, Chen Su-Chin, Chen Sue-Huei, Fu Jin-Cheng, Yu Yi-Chiang, Chang Chih-Sheng, Wang An-Hsiang, Jang Jing-Huei, Huang Chen- Jia, Yeh Sen-Hae, Lee Wen-Cheng, Lin Sheng-Chi, Liu Che-Hsin, Chang Shin-Yi, Ke Ming-Chun, Chen Yi-Chen, Chen Yun-Ju, Chuang Ming-Jen,Ji Yandong and Li Long, "Progress in the application of machine learning algorithms in smart agriculture," (in Simplified Chinese), Journal of Tonghua Normal University, pp. 73-77, 2019.
8. Kang Zhejia, "Machine learning to explore the spatiotemporal relationship of PM2.5 concentration changes," Master, Bioenvironmental Systems Engineering, National Taiwan University, Institute of Bioenvironmental Systems Engineering, 2019.
9. Zhang Tingwei, "Manufacturers' drought adaptation behavior and its influencing factors under climate change - Taking Hsinchu Science Park as an example," 2018.
10. Wang Junhao and Tang Kaiwei, "European Union Agricultural Climate Zone Division and Assessment of Potential Impacts of Climate Change," Agricultural Extension Wenhui, vol. No. 56, pp. 267-280, 2011.
11. Zhou Yiling, "Research and Analysis on the Development and Current Situation of Environmentally Friendly Agriculture," pp. 1-9, 2017.
12. Shen Mingshi, "Measures taken by countries to ensure food security under climate change," PROSPECT & EXPLORATION, vol. Volume 9, Issue 3, pp. 82-98, 2011.
13. E.ManavalanandK.Jayakrishna,"AreviewofInternetofThings(IoT)embedded sustainable supply chain for industry 4.0 requirements," Computers & Industrial Engineering, vol. 127, pp. 925-953, 2019.
14. Guo Meijun and Zhang Ziying, "Common warning messages and the application of smart home Internet of Things," National Disaster Prevention and Rescue Science and Technology Center Disaster Prevention and Rescue Electronic News, vol. 166, 2019.
15. J. SNOW, "Modern problems require modern solutions: How can artificial intelligence stop climate change?," in National Geographic, ed, 2019.
16. H.-C.Li,Y.-C.Che,andM.-J.Guo, "The Development and Application of Taiwan Typhoon Loss Assessment System (TLAS)," 2013.

17. Hsu Ming-His, Lin Pay-Liam, Chen Su-Chin, Chen Sue-Huei, Fu Jin-Cheng, Yu Yi-Chiang, Chang Chih-Sheng, Wang An-Hsiang, Jang Jing-Huei, Huang Chen- Jia , Yeh Sen-Hae, Lee Wen-Cheng, Lin Sheng-Chi, Liu Che-Hsin, Chang Shin-Yi, Ke Ming-Chun, Chen Yi-Chen, Chen Yun-Ju, Chuang Ming-Jen, "Streamlining Report_ Typhoon Disaster Simulation and Early Warning Analysis," National Disaster Prevention and Rescue Science and Technology Center 2020.
18. Zhou Qiaoying, Wu Siyang, Chen Qiling, Guo Hongyu, Cai Huiwen, Wu Yongpei, Liao Dajing, Zhang Yaling, Zhang Shufen, Zhang Renyu, Chen Qinan, Liu Shihong, Chen Xinyi, "Application results of drone aerial images in crop damage identification," 2020.
19. M.-F. R. Lee, L.-J. Yang, and W.-Y. Kong, "Disaster response: swarm land, sea, air and underwater robots based on artificial intelligence," Science and Technology News, vol. 215, pp. 59- 73, 2018.

Simulation of Photovoltaic Water Pumping System in Sam Ratulangi University

Meita Rumbayan[1*], *Adinda F.* Nelwan[1], *David* Pang[1], *,Jefrey I.* Kindangen[1]and *Muhamad Dwisnanto* Putro[1]

[1]Faculty of Engineering, University of Sam Ratulangi, Manado, Indonesia

Abstract. The purpose of this study is to design and simulate a PV water pumping system at Sam Ratulangi University in Indonesia. The methods employed in this study consist of a literature review, design, and simulation using HOMER (Hybrid Optimization of Multiple Energy Resources) software. The comparison between PV water pumps and diesel-powered water pumps is analysed in terms of net present cost, cost of energy (COE), and emissions to meet the water requirements in the laboratory. The simulation results are based on two design scenarios for a solar photovoltaic water pumping system at a laboratory scale. In Scenario 1, the system has a daily load of 1.5 kWh/day, 0.52 kW PV modules, 4 batteries of 80 Ah each, and a 1 kW inverter. In Scenario 2, the system has a daily pump load of 2.5 kWh/day, 0.87 kW PV modules, 4 batteries of 80 Ah each, and a 1 kW inverter. The results indicate that the utilization of solar energy to power a PV water pumping system performs well and serves as a case study showcasing the use of solar energy. Further studies on the application of PV water pump should be implemented and tested to evaluate the system's performance.

1 Introduction

Photovoltaic water pump (PWP) system application can be a good option for renewable energy demonstration. It also give an example of green technology for further implementation. Besides that, the attraction of environment utilization of PWP system increase the beneficial of this technology application. The purpose of this research is to design and conduct techno-economic analysis for a solar pump that located in Sam Ratulangi university for demonstration site of solar energy utilization.

The PV water pump system is suitable for reducing Green House Gas that produced by diesel pump or electric grid that supply by fossil power plant. A PV system, or Photovoltaic system, is a highly suitable solution for reducing the Greenhouse Gas (GHG) emissions that are produced by diesel pumps or electric grids supplied by fossil fuel power plants. By harnessing the power of sunlight, PV systems generate electricity without relying on fossil fuels, thereby significantly mitigating the release of harmful GHG emissions into the atmosphere.

* Corresponding author: meitarumbayan@unsrat.ac.id

When compared to diesel pumps, PV systems offer numerous environmental benefits. Diesel pumps require the combustion of fossil fuels, which releases carbon dioxide (CO2), methane (CH4), and nitrous oxide (N2O) emissions. These emissions contribute to global warming and air pollution, leading to adverse effects on human health and the environment. In contrast, PV systems convert sunlight directly into electricity using photovoltaic cells, without any combustion process or associated emissions. Thus, using PV power instead of diesel pumps helps combat climate change by reducing the overall carbon footprint.

2 Literature Reviews

There are many studies has been conduct to demonstrate the utilization of optimization techniques in designing photovoltaic (PV) systems for water pumping and various other applications using HOMER (Hybrid Optimization of Multiple Energy Resources) software [1-4]. In [1] Salam et al designed a PV system to power the lighting load in a laboratory, while in [2] the system was designed for a typical house load in Oman. The authors employed HOMER software to determine the optimal design for each system. They reported that the cost of energy generated by the PV system in the laboratory application was calculated to be 0.561 USD per kilowatt-hour (kWh). Similarly, for the typical house load, the authors claimed a cost of energy of 0.389 USD/kWh.

In Kazem et al [5] presents the design and evaluation of a photovoltaic (PV) water pumping system in Sohar, Oman. The research focused on selecting the optimal components for the system, including PV modules, inverter, charger controller, and batteries specific to Oman. The daily energy load for the system was determined to be 2.22 kWh/day, with 0.84 kW of PV modules, 4 batteries (12V and 200 Ah), and a 0.8 kW inverter. The results of the study demonstrated that the PV system offered an attractive option with an optimum cost of energy at 0.309 USD per kilowatt-hour (kWh), as compared to the cost of energy from a diesel engine, which was reported to be 0.79 USD/kWh.

In [3] Girma et al conducted a study in Ethiopia where they designed and examined a hybrid PV/diesel generator system using HOMER software. The system was intended to supply a village with a daily energy consumption of 279 kWh. The optimized design indicated that 95% of the energy demand could be met by the PV system, while the remaining 5% would be generated by a diesel generator. The estimated cost of energy produced by the system was found to be 0.401 USD per kilowatt-hour (kWh). In [4], Mishra, and Singh present a proposal for the optimal design of a hybrid PV/biomass system, which was developed using HOMER software. The objective of the system was to provide energy to a village in India for two primary purposes: cooking and operating a water pumping system for irrigation. The authors aimed to determine the most efficient design configuration for the system in order to meet these energy demands effectively.

This study is the continuity techno economic analysis for the prototype that has been designed and developed [6]. The prototype, as initially conceptualized and crafted by Bakelli and colleagues in 2011, serves as the nucleus of the ongoing research endeavors. This study aims to deepen our understanding of the prototype's technological nuances, providing an in-depth exploration of its operational efficiency, durability, and adaptability to real-world scenarios. By scrutinizing the technical aspects of the prototype, we seek to uncover valuable insights that can inform future iterations or applications in related contexts.

The previous study by author aims to showcase the utilization of solar energy to empower island communities, indicating successful utilization of solar energy for water pumping. The authors suggest further studies to explore and scale up PV pump system applications in terms of technical and economic considerations [7]. The review work covers various aspects such as renewable energy source water pumping systems, designing methods, control strategies, and field performance of photovoltaic water pumping systems, as reported in [8]. Additionally, a review on solar photovoltaic water pumping systems discusses the scope and limitations of photovoltaic solar water pumping systems, describing the components and functioning of PV solar pumping systems. It also includes a review of research works by previous researchers and emphasizes the popular and ideal use of solar energy exploitation in water pumping applications [9].

Photovoltaic powered water pump system is a proven technology can be served for water requirements, irrigation and industrial application in both small and large size. The design of PV powered pump system depend on the requirement, solar availability and the budget. It is a challenge for the many researchers to design PV powered pump system based on the technical, economic and environmental analysis. In essence, the design of a PV powered Pump system represents a convergence of technology, economics, and environmental stewardship. The challenge lies in harmonizing these elements to create a system that not only meets the immediate water needs efficiently but also does so in a manner that is economically viable and environmentally responsible. Researchers engaged in the design of PV powered pump systems thus play a pivotal role in advancing sustainable water solutions, contributing to the ongoing global efforts to harness clean energy for essential applications.

3 Methods

The methods of simulation were used in this work consists of technical analysis and economic analysis based on HOMER software. HOMER is one the software application that used for analysing the feasibility of renewable energy system developed by National Renewable Energy Laboratory (NREL). HOMER (Hybrid Optimization of Multiple Energy Resources) software navigates the complexities of building cost effective and reliable hybrid microgrid.

The data is used about solar radiation and clearness index for the location of Manado taken from NASA database in HOMER and presented in Figure 1. As the solar radiation is high enough between 10:00 AM and 3:00 PM it is considered the PV system is generating electricity in this period to supply the 350 liter water storage in a tank.

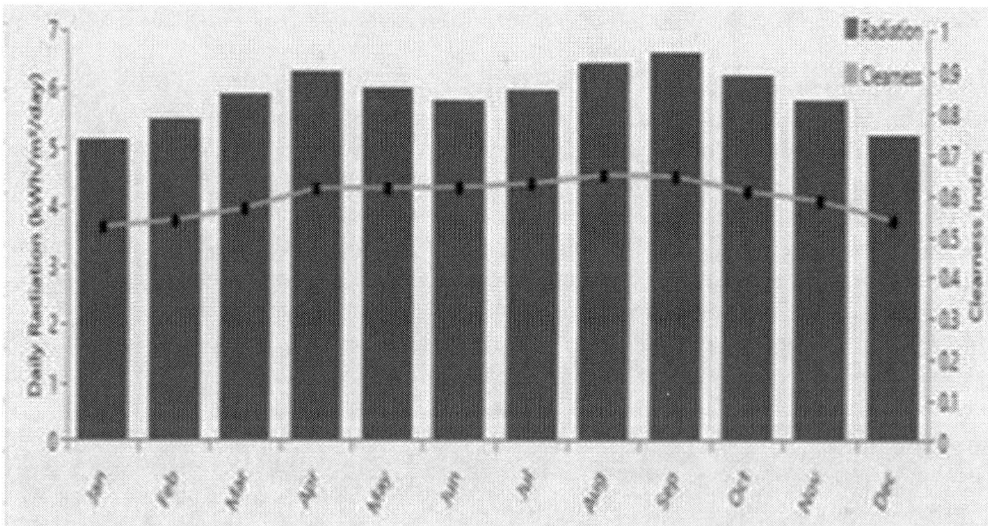

Fig. 1. The solar radiation and clearness index data for the location in this study

In this work, PV powered water pumping system has been designed for a toilet water supply in a laboratory that located at Faculty of Engineering, Sam Ratulangi University, Manado. The PV powered water pump system will be installed at laboratory for demonstration site. The proposed system in this study is comprised of several components, including a PV array with a limited number of modules, a charge controller, an inverter, batteries for energy storage, and additional components such as wires, protections, and module structures.

The PV water pump system is designed for 2 scenarios, i.e using 0.3 kW pump for scenario 1 and using 0.5 kW for scenario 2. The schematic diagram in HOMER model for the designed electric load to supply PV pump system is presented in Figure 2 and Figure 3 for scenario 1 and scenario 2 respectively.

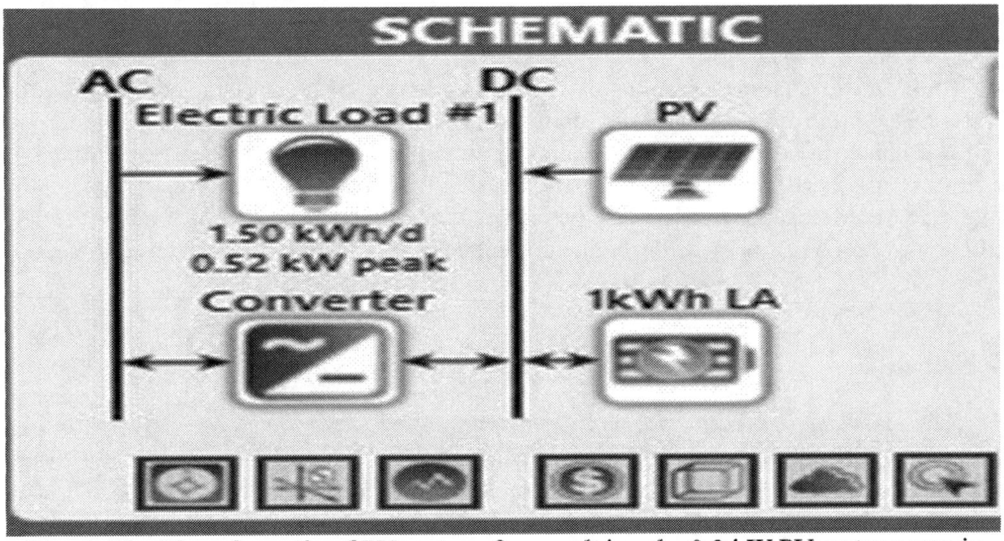

Fig. 2. The The schematic of PV system for supplying the 0.3 kW PV water pump in this work (scenario 1)

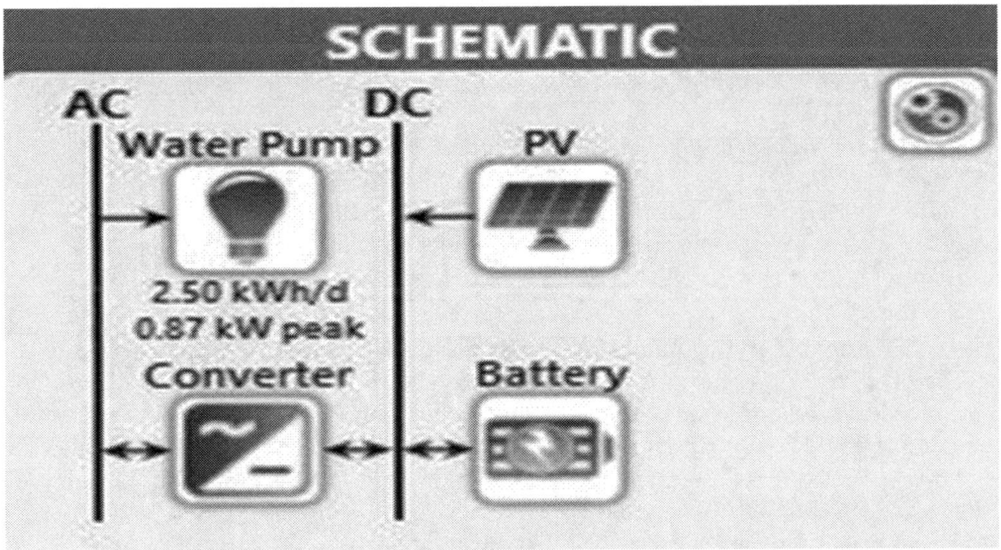

Fig. 3. The schematic of PV system for supplying the 0.5 kW of PV water pump in this work (scenario 2)

Figures 2 and 3 illustrate the intricately designed schematic of a cutting-edge photovoltaic powered water pump system meticulously crafted to fulfil the imperative role of supplying water to the designated demonstration site. This innovative system seamlessly integrates advanced photovoltaic technology to harness solar energy, presenting an eco-friendly and sustainable solution for the efficient and reliable provision of water resources at the demonstration site in Faculty of Engineering, Sam Ratulangi University.

At the heart of this innovation lies the seamless integration of advanced photovoltaic technology, a testament to its commitment to harnessing solar energy. The incorporation of cutting-edge photovoltaic elements underscores the system's commitment to sustainability, offering an eco-friendly alternative for meeting the water requirements of the demonstration site. This integration not only reduces the system's environmental impact but also positions it as a forward-thinking and responsible solution in the realm of water resource management.

This holistic approach emphasizes the convergence of technology and environmental consciousness, creating a symbiotic relationship between the innovative use of solar power and the imperative need for a reliable and efficient water supply. As a result, the system emerges as a beacon of sustainable engineering, providing not only a glimpse into the future of water management but also serving as a tangible example of how advancements in technology can be harnessed for the betterment of both humanity and the environment.

4 Results

The feasibility of the proposed PV water pumping system was designed and analyzed using HOMER. After simulating the system model in HOMER, the result shows that the greatest optimal result is achieved when the system is composed of 0.546 kW PV array, 4 batteries, and 0.8kW inverter. The economic analysis of the proposed model system find the total NPC is Rp. 16,800,000 (1,120 USD) and the cost of energy as 3,400 Rp/kWh equals 0.23 USD/kWh. The result is acceptable compare to other results were 0.561 USD/kWh, 0.389 USD/kWh [2], 0.309 USD/kWh [5], 0.401 USD/ kWh [6], 0.21-0.304 USD/kWh [10] and 0.327-0.361 [11]. The result of techno economical study is presented in Table 1.

Table 1. The result of techno economical study between scenario 1 and scenario 2

	Scenario 1	Scenario 2
Cost of energy (Rp/kWh)	3400	2370
Net Present Cost (Rp)	16,800,000	27,000,000

To compare the PV system results with the diesel generator to choose the best and costless system, This comparative study analysis shows that a size of 0.3 kW and 0.5 kW generator has been estimated with capital, operating, and total net cost being Rp 3,000,000 (equal to 300 USD), 200 (USD/yr), and 8,185 USD, respectively. The cost of Energy (CoE) is found to be Rp. 3212 per kWh (0.21 USD/kWh), which is considered to be high as compared with CoE of the proposed PV system.

The schematic of 0.3 kW and 0.5 kW diesel generator plant for powering the water pump as shown in Figure 4 and Figure 5 respectively.

https://doi.org/10.1051/e3sconf/202453005001

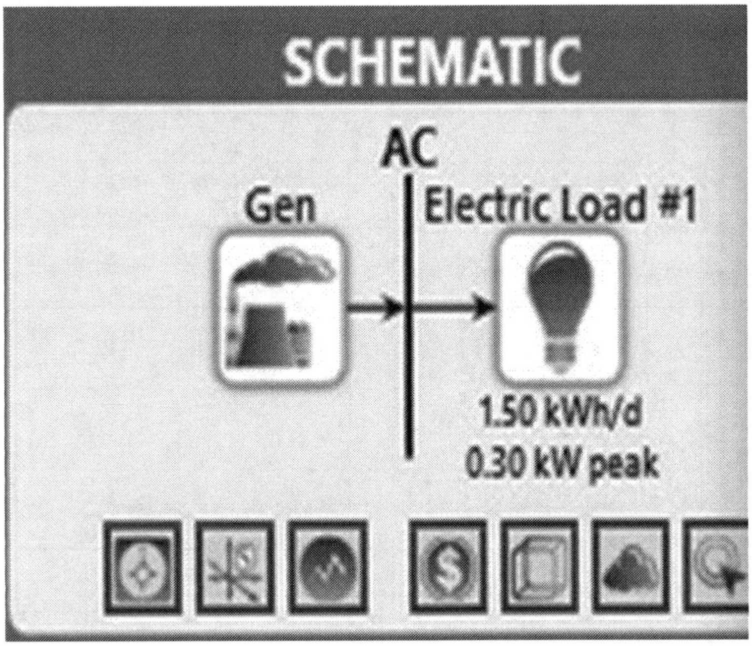

Fig. 4. The schematic of 0.3 kW diesel plant for supplying water pump pump 0.3 kW (scenario 1)

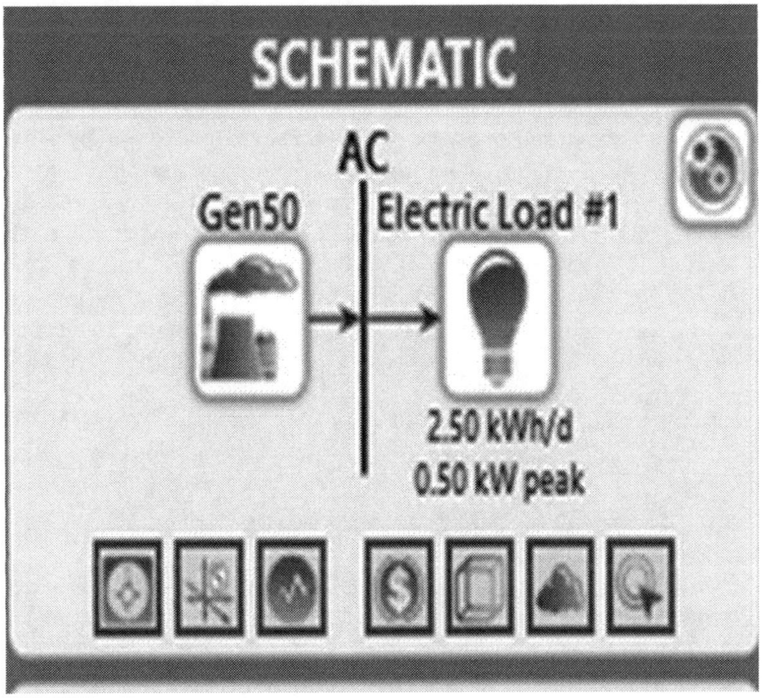

Fig. 5. The schematic of PV system for supplying the 0.5 kW of PV water pump in this work (scenario 2)

The amount of greenhouse gas emission produced by the diesel generator for 0.3 kW and 0.5 kW is presented in Table 2.

Table 2. The amount of greenhouse gas emission produced by the diesel generator for 0.3 kW and 0.5 kW

Pollutant	Emission (kg/year)	Emission (kg/year)
Carbon dioxide	426	810
Carbon monoxide	2.69	5.05
Unburned hydrocarbons	0.117	0.223
Particulate matter	0.0163	0.03
Sulfur dioxide	1.04	1.98
Nitrogen oxides	2.52	4.75

5 Conclusions

The technical and economic analysis of PV powered water pump is found feasible to implement. In this study water pumping system for a laboratory in Sam Ratulangi university has been designed. For scenario 1, the system has a daily load of 1.5 kWh/ day, 0.52 kW PV modules, 4 batteries 80 Ah and 1 kW inverter. For scenario 2, the system has a daily pump load of 2.5 kWh/day, 0.87 kW PV modules, 4 batteries 80 Ah and 1 kW inverter .

The economic analysis in term of Cost of Energy (COE) and Net Present Cost (NPC) have been calculated for scenario 1 and scenario 2. The COE is Rp 3400 per kWh for scenario 1, while Rp 2370 per kWh for scenario 2. For the NPC is Rp. 16,800,000 for scenario 1 while Rp 27,000,000 for scenario 2.

The environment analysis shows that substituting diesel power pump system by PV powered pump will reduce the greenhouse gas (GHG) emission 425 kg/year of CO2, 2.69 kg/year of CO, 0.117 kg/year of HC, 0.0163 kg/year of particulate matter, 1.04 kg/year of SO2, and 2.52 kg/year of NOx.

Photovoltaic systems provide a sustainable and environmentally friendly alternative to diesel pumps and fossil fuel-based electric grids. By harnessing the power of the sun, these systems enable the production of clean, renewable electricity, effectively reducing GHG emissions associated with the combustion of fossil fuels. Implementing PV systems on a wider scale can play a crucial role in mitigating climate change and transitioning towards a more sustainable and low-carbon future.

In essence, this research encapsulates more than just a design and simulation; it encapsulates the thoughtful planning and strategic placement of PV panels within the Faculty of Engineering at Sam Ratulangi University. It serves as a visual testament to the university's commitment to sustainable energy practices and offers insights into the optimized utilization of solar resources within the educational institution's infrastructure.

Further research and practical implementation are necessary to explore innovative approaches, optimize system performance, and expand the adoption of PV water pumps in small-scale settings. The continuity of research is integral for bridging the gap between theoretical conceptualization and practical implementation. This study represents a forward-looking exploration, building upon established foundations to contribute to the evolution of the prototype's trajectory. It provides valuable insights that can inform future developments, applications, and decisions within the realm of technological innovation and economic viability.

Authors wishing to acknowledge research funding support from Sam Ratulangi University through PNBP 2023.

References

1. M.A Salam, A. Aziz, A.H.A Alwaeli, H.A Kazem, Optimal sizing of photovoltaic systems using HOMER for Sohar, Oman. International Journal of Renewable Energy Research,Vol.3, no. 2, pp. 301–307 (2013)
2. H.A Kazem, A.A. Alkurwi, M.M Alabdul Salam, and A.H.A Alwaeli, Levelized electricity cost for photovoltaic system in Sohar-Oman. Proceedings of the IEEE 8th International Conference and Exhibition on Ecological Vehicles and Renewable Energies (EVER '13), Monte-Carlo, Monaco (2013)
3. Z. Girma, Hybrid renewable energy design for rural electrification in Ethiopia. Journal of Energy Technologies and Policy, vol. 3, no. 13, pp. 38–52 (2013)
4. Mishra, R., and Singh, S., 2013. Sustainable energy plan for a village in punjab for self energy generation. International Journal of Renewable Energy Research, vol. 3, no. 3, pp. 640–646.
5. H.A.Kazem, A.H.A. Al-Waeli,, A.H.K Al-Kabi, A. Al-Mamari, Technoeconomical assessment of optimum design for photovoltaic water pumping system for rural area in Oman. International Journal of Photoenergy (2015)
6. Y. Bakelli, A.H Arab , and B. Azoui, Optimal sizing of photovoltaic pumping system with water tank storage using LPSP concept. Solar Energy, vol. 85, no. 2, pp. 288–294. (2011)
7. M. Rumbayan, D. Mamahit, F. Kambey, B. Maluegha, G. Mangindaan, H. Jody, Design and installation of a solar water pumping system in Bunaken Island of Indonesia. AIP Conference Proceeding, Volume 2694, Issue 1.(2023)
8. V. C Sontake., V. R Kalamkar., Solar photovoltaic water pumping system - A comprehensive review. Renewable and Sustainable Energy Reviews, Elsevier, vol. 59(C), pages 1038-1 (2016)
9. M. R Dhagel, V.S Girnale, C.P Patil, Review on solar photovoltaic water pumping system. International Journal for Scientific Research & Development, Vol. 5, Issue 01, (2017)
10. A.H Al-Badi, M.H Albadi, A.M. Al-Lawati, A.S. Malik, Economic perspective of PV electricity in Oman. Energy, vol. 36, no. 1, pp. 226–232 (2011)
11. A.H Al-Badi, M AL-Toobi, S. AL-Harthy, Z. Al-Hosni, A. .Al-Harthy, Hybrid systems for decentralized power generation in Oman. International Journal of Sustainable Energy, vol.31, no. 6, pp. 411–421 (2012)

Quantifying Fracture Dynamics in Clean Energy: A Novel Fractal Perspective

Dayu Ye [1,2,3,*], *Guannan Liu* [1,2] *and Gao Feng* [1,2]

[1]State Key Laboratory for Geomechanics and Deep Underground Engineering, China University of Mining and Technology, Xuzhou 221116, Jiangsu, China
[2]Mechanics and Civil Engineering Institute, China University of Mining and Technology, Xuzhou 221116, Jiangsu, China
[3]WA School of Mines: Minerals, Energy and Chemical Engineering, Curtin University, Kalgoorlie, WA 6430, Australia

Abstract. Geothermal energy, as an emergent source of power, has consistently been a focal point of scholarly investigation both domestically and internationally, with a particular emphasis on the prediction and assessment of its extraction efficiency. In common geothermal extraction projects, the seepage behavior of water within the fractures for the geothermal reservoirs significantly impacts the efficiency of resource extraction. This study introduces an interdisciplinary fractal model for geothermal extraction. An enhanced fractal theory, tailored for geothermal reservoirs, is proposed, employing four innovative fractal parameters—fracture fractal dimension, fracture tortuosity fractal dimension, fracture roughness parameter, and maximum fracture length—to quantitatively characterize the fracture structure. This refined fractal theory is applied to geothermal reservoir extraction under the complex thermal-hydro-mechanical coupling. The findings indicate that the proposed structural parameters effectively characterize the micro-macro interactions during the geothermal extraction process. Significant evolution of these fractal parameters is observed throughout the extraction process. Furthermore, there is an inverse relationship between geothermal extraction efficiency and two key fractal parameters—fracture tortuosity fractal dimension and fracture roughness parameter—while the reservoir stress is directly proportional to these parameters.

1 Introduction

As a renewable energy source, geothermal energy, characterized by its stability, longevity, and renewability, is widely acknowledged as a crucial direction for future energy development. In recent years, the exploration and application of geothermal resources have garnered considerable attention and research both domestically and internationally [1-3]. Factors such as the geological-hydrological conditions of the geothermal reservoir, the depth of geothermal extraction, and parameters of the injection and production wells significantly

[*] Corresponding author: theoye@cumt.edu.cn

influence the efficiency of heat extraction. Additionally, the multiphysical field coupling effects in geothermal reservoirs, such as reservoir stress and deformation, porosity, stratum depth, temperature, and the fracturing behavior of geothermal strata [4-6], also significantly impact the efficiency of geothermal extraction.

An increasing number of scholars are delving into the multifactorial coupling effects during the geothermal extraction process [6-8]. These effects encompass the multiphysical phenomena of thermal-hydro-mechanical coupling, including dynamic studies of physical-mechanical parameters like porosity, permeability, reservoir temperature, and water extraction temperature. This lays the groundwork for deriving various types of multiphysical field analysis models [9-11]. However, despite these advances in geothermal extraction research, quantitative nonlinear analyses of the reservoir's fracture network evolution under the disturbances of geothermal extraction and the coupled thermal-fluid-solid interactions remain challenging. Traditional methods relying on CT, digital rock reconstruction, and discrete fracture network studies for geothermal reservoirs involve significant computational efforts, typically focused on micro-nano scales. Furthermore, traditional fractal theory-based analytical methods struggle to integrate the effects of extraction disturbances and complex multiphysical field couplings [12-14].

As the primary conduit for fluid migration, the nonlinear evolution of the fracture structure in geothermal reservoirs directly impacts seepage behavior. Under the influences of extraction disturbances and the thermal-fluid-solid coupling in the reservoir, temperature changes induced by extraction alter the intensity of the reservoir's thermal expansion effect. This, in turn, leads to variations in reservoir stress and results in deformations of varying degrees. Deformations in geothermal reservoirs, such as contraction or expansion, cause the fracture network structure to undergo nonlinear evolution. The multifaceted coupled effects described above directly influence the efficiency of heat extraction. Therefore, quantitatively exploring the nonlinear evolution of the fracture network in geothermal reservoirs during extraction and elucidating the impact of this evolution on heat extraction efficiency is of paramount importance.

To address the industry bottleneck of achieving precise micro-macro quantitative analysis during geothermal extraction under multifactorial coupled influences, this study proposes an interdisciplinary geothermal extraction model. This study introduces an enhanced fractal theory suitable for geothermal reservoirs, utilizing four innovative fractal parameters (fracture fractal dimension, fracture tortuosity fractal dimension, fracture roughness parameter, and maximum fracture length) to quantitatively characterize the fracture structure. Applying this enhanced fractal theory to typical geothermal reservoirs, this model explores geothermal reservoir extraction under complex thermal-fluid-solid coupling. Subsequently, the quantitative evolution characteristics of fracture number, tortuosity, and size during extraction is investigated and the impact of fracture roughness and tortuosity on reservoir stress and extraction efficiency is analyzed. The findings of this study are anticipated to provide a scientific basis for the rational utilization and development of geothermal energy, contributing significantly to sustainable development, ensuring stability in energy supply, and enhancing energy efficiency.

2 Fractal Modeling of Geothermal Extraction under Multi-Field Effect

2.1 Fractal Characterization of Rough Fracture Networks in Geothermal Reservoir Extraction

The number of fractures whose lengths in the interval $\left[l, l+dl\right]$ can be obtained by Eqs. (1) and (2):

$$-dN\left(l\right) = Nf\left(l\right)dl = N\delta_f l_{\min}^{\delta_f} l^{-(\delta_f+1)} dl \tag{1}$$

In most cases, the aperture of natural fractures satisfies the following relationship:

$$a = \eta_2 l^{\omega} \tag{2}$$

where a is the fracture aperture, η_2 is the proportional coefficient for shale fracture length, and ω is the power-law index of fracture aperture.

Based on the classical cubic law, the fluid flow through a fracture is:

$$q = \frac{a^3 l \Delta p_f}{12\mu L_f} \tag{3}$$

where μ is the dynamic viscosity of the fluid and L_f is a fracture length.

And the gas flow model considering the distribution of fracture spatial location is given by:

$$q = \frac{a^3 l\left(1-\cos^2\theta_1 \sin^2\theta_2\right)\Delta p}{12\mu L_f} \tag{4}$$

where θ_1 is the natural fracture azimuth angle and θ_2 is the dip angle.

where $r = l_{\min} / l_{\max}$ is defined as the fracture length ratio.

Combining Eqs. (1)- (4), the effective permeability for the shale is:

$$\kappa_f = \frac{N\cdot\varepsilon\cdot D_s \eta_2^3 l_{\min}^{3\omega+1}\left(1-\cos^2\theta_1 \sin^2\theta_2\right)\Delta p}{12A\cdot\left(3D_T+1-D_s\right)}\cdot\left[r^{Ds-(3\omega+1)}-1\right] \tag{5}$$

2.2 Evolution of Reservoir Stress During Geothermal Extraction

Under the combined influence of stress deformation caused by extraction disturbances, the nonlinear evolution of hydraulic pressure due to extraction water seepage, and the temperature of the geothermal reservoir, the reservoir stress control equation can be expressed as [15]:

$$\sigma_{ij} = 2G\varepsilon_{ij} + \frac{2G\upsilon}{1-2\upsilon}\varepsilon_{kk}\delta_{ij} - K\alpha_T T_{ij} \tag{6}$$

where σ_{ij} is the stress tensor influenced by both water seepage and extraction disturbances, G is the shear modulus of the geothermal reservoir, ν is the rock's Poisson's ratio, α_T is the rock's thermal expansion coefficient, and T_{ij} is the temperature tensor of the geothermal reservoir.

For the extraction of geothermal resources during water seepage, the strain in the rock is the sum of strains caused by temperature, stress, and seepage pressure, namely:

$$\varepsilon_{i,j} = \frac{1}{2G}\sigma_{ij} - \left(\frac{1}{6G} - \frac{1}{9K}\right)\sigma_{kk}\delta_{ij} + \frac{\alpha_T \Delta T}{3}\delta_{ij} + \frac{\alpha}{3K}\Delta p \delta_{ij} \tag{7}$$

$$Gu_{i,ij} + \frac{G}{1-2v}u_{j,ji} - ap_j - K\alpha_T T_j + F_i = 0 \tag{8}$$

2.3 Water Seepage through Rough Fractures During Extraction

During geothermal extraction, the evolution equation for porosity, influenced by extraction disturbances and stress changes, is as follows [16]:

$$\phi = 1 - (1-\phi_0)\exp\left[\frac{1}{K_s(p-p_0) + \alpha_T(T-T_0) + (\varepsilon_v - \varepsilon_{v0})}\right] \tag{9}$$

$$\frac{\partial \phi}{\partial t} = -(1-\phi_0)\left(\frac{1}{\kappa}\frac{\partial p}{\partial t} + \alpha_T\frac{\partial T}{\partial t} - \frac{\partial \varepsilon_v}{\partial t}\right)\exp\left(\frac{p-p_0}{\kappa} + \alpha_T(T-T_0) - (\varepsilon_v - \varepsilon_{v0})\right) \tag{10}$$

Therefore, by correlating the above equations with Equation (10), the governing equation for groundwater seepage can be established as:

$$H\frac{\partial p}{\partial t} - \nabla\left(\frac{\kappa}{u}\rho_w \nabla p\right) = H\alpha_T\frac{\partial T}{\partial t} - H\kappa\frac{\partial \varepsilon_v}{\partial t} + Q_z \tag{11}$$

2.4 Water Seepage through Rough Fractures During Extraction

Neglecting the thermal filtration effect, the total heat flux during geothermal extraction comprises two components: thermal conduction and fluid thermal convection, expressed as:

$$\dot{q}_T = -\lambda_M \nabla T + \rho_w C_g \dot{q}_g (T + T_{ar}) \tag{12}$$

where C_g is the thermal conduction constant of the rock layer and λ_M is the heat convection coefficient. The energy conservation equation for heat flow and the relationship with specific heat capacity is as follows:

$$\frac{\partial[(\rho C)_M (T+T_{ar})]}{\partial t} + (T+T_{ar})K_g\alpha_g\nabla\dot{q}_g + (T+T_{ar})K\alpha_T\frac{\partial \varepsilon_v}{\partial t} = -\nabla\dot{q}_T \tag{13}$$

$$(\rho C)_M = \phi(\rho_g C_g) + (1-\phi)(\rho_s C_s) \tag{14}$$

where $(\rho C)_M$ is the specific heat capacity of the geothermal reservoir, and C_g and C_s represent the thermal constants of water and rock during the seepage process, respectively.

3 Validation of the Proposed Fractal Model

To ascertain the accuracy of the fractal multi-field coupling model, this study utilized field geothermal extraction data from studies by Wang et al. and Koh et al. [17,18]. The simulation area, depicted in Figure 1, encompasses the main simulation parameters and extraction data of the geothermal reservoir as shown in Table 1.

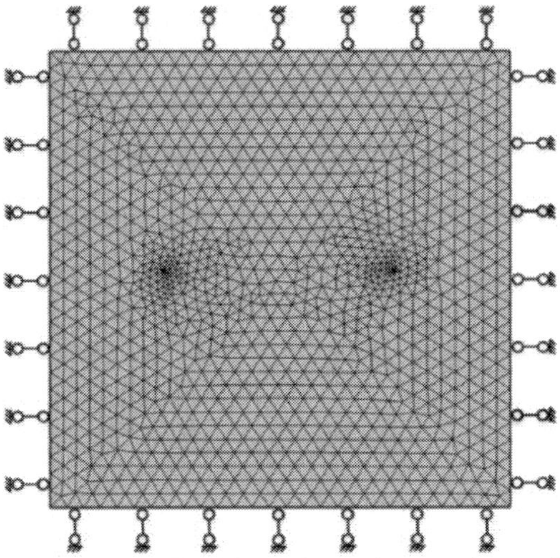

Fig.1. Schematic of Field Geothermal Extraction Project.

Table 1. Primary Data of Field Geothermal Extraction.

Parameter	**Value**
Young's modulus of rock, E (GPa)	24
Poisson’ s ratio of rock, v	0.189
Heat conductivity, $\lambda(W/(m \cdot K))$	6.0×10^{-6}
Constant pressure heat capacity, $C_p(J/(kg \cdot K))$	900
Density of the reservoir, ρ_r(kg/m-3)	2700
Solid bulk modulus, V(GPa)	48.2
Biot’s coefficient, α	0.49
Hydrostatic fluid pressure, N(MPa)	10.0
Tension strength, R_m(MPa)	12.1
Density of fluids, ρ_f(kg/m-3)	1000
Rock porosity, $\emptyset$	0.143
Dynamic viscosity, μ(Pa·s)	3.0×10^{-4}
Coefficient of compressibility, $a(1/\text{Pa})$	5.31×10^{-9}

As demonstrated in Figure 2(a), the calculated pore pressure, radial stress, and tangential stress show good agreement with experimental results, indicating model consistency with calculations based on steady-state heat transfer. Furthermore, Figure 2(b) compares the comprehensive simulation results of geothermal extraction under the influence of multifactorial coupling with actual field project data. This comparison substantiates the accuracy of this model.

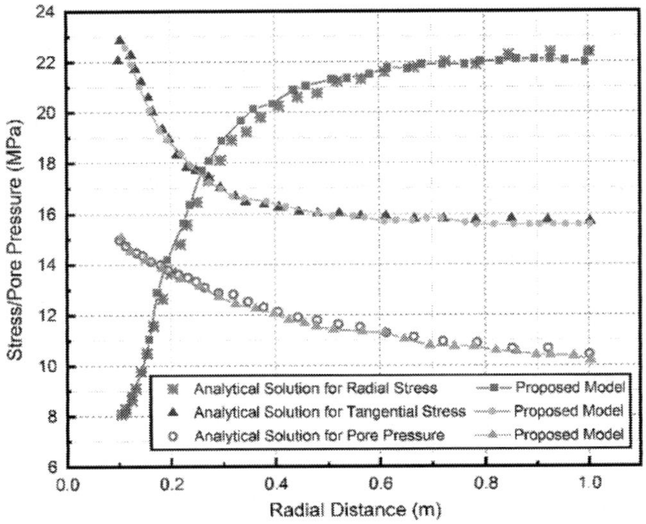

(a)

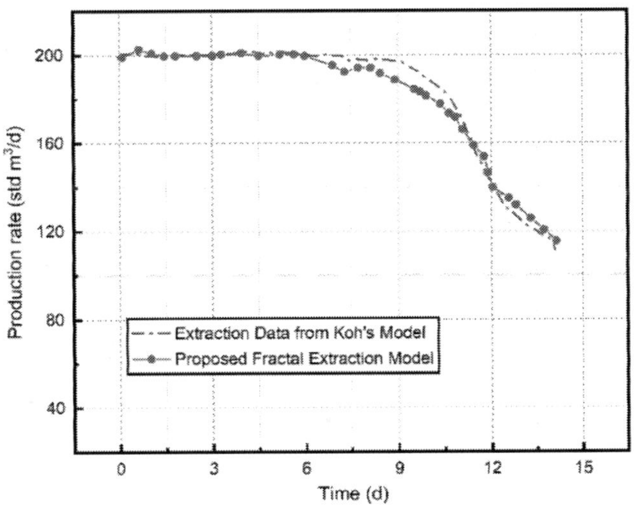

(b)

Fig. 2. Model Accuracy Validation

4 Results and discussion

Building on the above results, this section explores the contributions of different degrees of fracture curvature and fracture roughness to geothermal reservoir stress and production well water temperature. Figure 3 shows the contribution of different fracture roughness parameters to geothermal reservoir stress. As Figure 3 indicates, reservoir stress consistently increases under varying levels of fracture roughness. Additionally, at the same extraction time, reservoir stress is inversely proportional to the fracture roughness parameter. A larger ε, representing rougher fracture surfaces, impedes the flow of injected cold water. Under multi-field coupling, more difficult fluid flow leads to reduced water pressure-induced effective stress, resulting in lower reservoir stress. When increases from 0.06 to 0.10, the maximum reduction in reservoir stress is 5.39%.

Figure 4 illustrates the effect of different fracture roughness on production well water temperature. As seen in Figure 4, the production well water temperature consistently decreases over different extraction periods. At the same extraction time, a greater fracture roughness parameter results in higher water temperatures. As previously discussed, a larger ε denotes rougher geothermal reservoir fractures, making water seepage more challenging. Consequently, with slower water seepage, more heat is extracted, raising the production well water temperature as increases. When the fracture roughness parameter increases from 0.06 to 0.10, the production well water temperature at 80 days of extraction increases by 1.78°C.

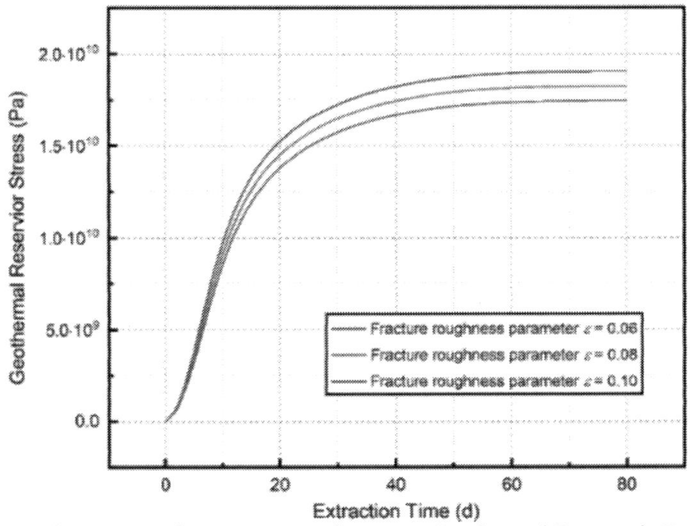

Fig. 3. The Impact of Fracture Roughness on Geothermal Reservoir Stress.

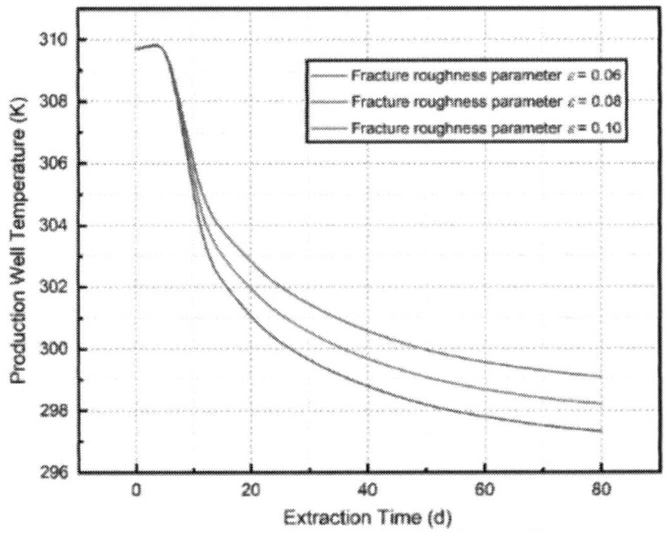

Fig. 4. The Impact of Fracture Roughness on Production Well Water Temperature.

Furthermore, the influence of fracture curvature on geothermal reservoir stress and production well water temperature is also investigated. As shown in Figures 5 and 6, in the fractal model proposed in this study, the degree of fracture curvature in the geothermal reservoir is quantitatively represented by the tortuosity fractal dimension D_T. As stated in Section 2, a smaller D_T indicates a more complex fracture network. Figure 5 shows the impact of different fracture tortuosity fractal dimensions on reservoir stress. It is evident that at the same extraction time, a larger fracture tortuosity fractal dimension leads to greater geothermal reservoir stress. A larger D_T implies a simpler fracture structure, thereby facilitating water seepage. Therefore, more water is injected into the geothermal reservoir from the injection well over the same period, correlating D_T positively with reservoir stress. When D_T increases from 1.2 to 1.6, the maximum increase in geothermal reservoir stress is 10.12%.

The influence of fracture tortuosity fractal dimension on production well water temperature is depicted in Figure 6. As shown, at the same extraction time, the production well water temperature is inversely proportional to the fracture tortuosity fractal dimension D_T. As mentioned, a smaller D_T represents a more complexly curved fracture structure. This complexity hinders water seepage and extraction, thus prolonging seepage time and increasing the amount of heat extracted. Consequently, the production well water temperature increases. When D_T increases from 1.2 to 1.6, the production well water temperature at 80 days of extraction increases by 2.39°C.

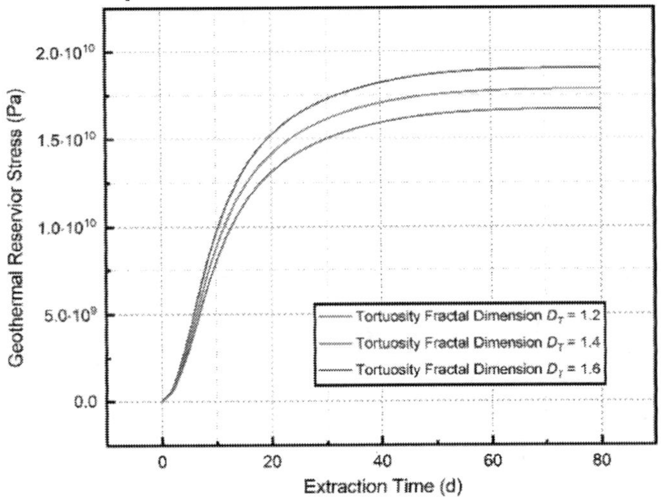

Fig. 5. The Impact of Fracture Tortuosity Fractal Dimension on Geothermal Reservoir Stress.

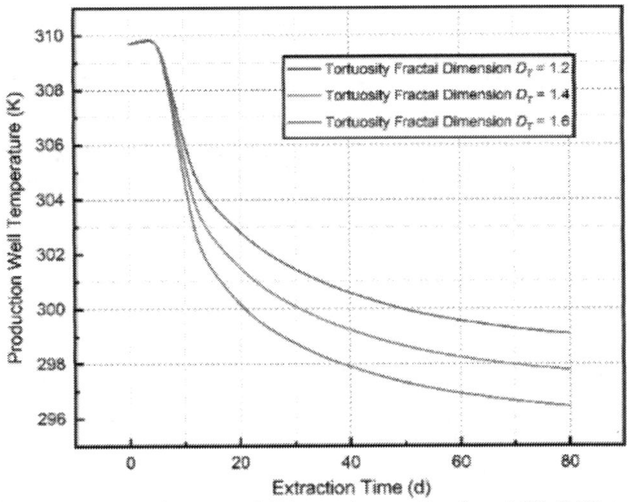

Fig.6. Impact of Fracture Tortuosity Fractal Dimension on Produced Well Water Temperature.

5 Conclusion

This study aims to address the industry challenge of quantitatively and comprehensively evaluating the impact of nonlinear evolution in reservoir fracture behavior on the safety and extraction efficiency of geothermal projects. Employing fractal theory adapted for porous media, this study developed an interdisciplinary geothermal extraction fractal model. This model incorporates four innovative fractal parameters (fracture fractal dimension D_S, fracture tortuosity fractal dimension D_T, fracture roughness parameter ε, maximum fracture length l_{max}) to quantitatively characterize the fracture structure, applying the refined fractal theory to geothermal reservoir extraction under complex thermal-fluid-solid coupling. The key findings are as follows:

(1) The fractal geothermal extraction model proposed in this study effectively characterizes the evolution of fracture network distribution, curvature, roughness, and length in geothermal projects. The model's computational results align closely with data from actual geothermal extraction engineering.

(2) Under the influence of coupled multiphysical fields, more complex fracture structures can be quantitatively represented by smaller fracture tortuosity fractal dimension D_T; rougher fracture surfaces can be quantitatively characterized by larger fracture roughness parameters ε.

(3) The fracture tortuosity fractal dimension and fracture roughness parameter significantly influence the reservoir stress and produced well water temperature during geothermal extraction. For instance, when ε increases from 0.06 to 0.10, the maximum reservoir stress decreases by 5.39%, and the produced well water temperature after 80 days of extraction increases by 1.78°C; when D_T, increases from 1.2 to 1.6, the maximum increase in geothermal reservoir stress is 10.12%, and the produced well water temperature after 80 days increases by 2.39°C.

References

1. Liu Y, Wang G, Guo X, et al. A Joint Method Based on Geochemistry and Magnetotelluric Sounding for Exploring Geothermal Resources in Sedimentary Basins and Its Application[J]. Water, 2022, 14(20): 3299.
2. Cant J L., Siratovich P A., Cole J W., et al. Matrix permeability of reservoir rocks, Ngatamariki geothermal field, Taupo Volcanic Zone, New Zealand. Geothermal Energy, 2018, 6(1):2.
3. Liu C, Li Y, Cao S, et al. Effects of Seawater Recharge on the Formation of Geothermal Resources in Coastal Areas and Their Mechanisms: A Case Study of Xiamen City, Fujian Province, China[J]. Frontiers in Earth Science, 2022, 10: 872620.
4. Mayo S., Josh M., Nesterets Y., et al. Quantitative micro-porosity characterization using synchrotron micro-CT and xenon K-edge subtraction in sandstones, carbonates, shales and coal. Fuel, 2015, 154(aug.15):167-173.
5. Singh M, Mahmoodpour S, Bär K, et al. Heat Extraction in Geothermal Systems with Variable Thermo-Poroelastic Fracture Apertures[J]. Geotechnics, 2023, 3(4): 1196-1206.
6. Zhang C, Zhang H, Song R, et al. Simulation of thermo-hydro-mechanical coupled heat extraction process of geothermal reservoir under varied geological conditions with CO2 as working fluid[J]. Journal of Hydrology, 2023: 130549.
7. Chen Y., Ma G., Wang H., et al. Optimizing heat mining strategies in a fractured geothermal reservoir considering fracture deformation effects. Renewable Energy, 2020, 148:326-337.

8. Ghassemi A., Zhou X. A three-dimensional thermo-poroelastic model for fracture response to injection/extraction in enhanced geothermal systems. Geothermics, 2011, 40(1):39-49.
9. Li C, Guan Y, Wang X, et al. Experimental and numerical studies on heat transfer characteristics of vertical deep-buried U-bend pipe to supply heat in buildings with geothermal energy. Energy, 2018,142: 689-701.
10. Zhang J, Zhao M, Wang G, et al. Evaluation of heat extraction performance of multi-well injection enhanced geothermal system[J]. Applied Thermal Engineering, 2022, 201: 117808.
11. Liu Z, Wu M, Zhou H, et al. Performance evaluation of enhanced geothermal systems with intermittent thermal extraction for sustainable energy production[J]. Journal of Cleaner Production, 2023: 139954.
12. T. Huang, X. Guo and F. Chen, Modeling transient flow behavior of a multiscale triple porosity model for shale gas reservoirs, Journal of Natural Gas Science & Engineering. 23 (2015) 33-46.
13. P. Au, J. Liu and Y. Leong, Yield stress and microstructure of washed oxide suspensions at the isoelectric point: experimental and model fractal structure, Rheologica Acta. 55 (2016) 847-856.
14. Y. Xia et al., A NEW METHOD FOR CALCULATING FRACTAL DIMENSIONS OF POROUS MEDIA BASED ON PORE SIZE DISTRIBUTION, Fractals-Complex Geometry Patterns and Scaling in Nature and Society. 26(1) (2018) 1850006.
15. I. Palmer, J. Mansoori, How permeability depends on stress and pore pressure in coalbeds: a new model, Soc. Petrol Eng. Res. Eng. 12 (1998) 539–544.
16. Mandelbrot, B. Benoit, The fractal geometry of nature, Am. J. Phys. 51 (3) (1998) 468.
17. Wang Y, Dusseault MB. A coupled conductive–convective thermo-poroelastic solution and implications for wellbore stability[J]. Journal of Petroleum Science & Engineering, 2003, 38(3):187-198.
18. Ahmad, Ghassemi, and, et al. Effects of heat extraction on fracture aperture: A poro–thermoelastic analysis[J]. Geothermics, 2008.

Enhancing Photovoltaic Thermal System Efficiency Using Micro-Channel Container and Nanoparticle Composited Phase Change Material: Effect of Dimensionless Water Flow

Wisanu Phukaokaew[1], *Amnart Suksri*[2,3], and *Tanakorn Wongwuttanasatian*[2,4,*]

[1]Department of Energy Engineering, Faculty of Engineering, Khon Kaen University, Khon Kaen 40002, Thailand
[2]Centre for Alternative Energy Research and Development, Khon Kaen University, Khon Kaen 40002, Thailand
[3]Department of Electrical Engineering, Faculty of Engineering, Khon Kaen University, Khon Kaen 40002, Thailand
[4]Department of Mechanical Engineering, Faculty of Engineering, Khon Kaen University, Khon Kaen 40002, Thailand

Abstract. The photovoltaic thermal (PV/T) hybrid system has attracted significant interest from researchers in the solar field. These systems efficiently harness solar radiation, converting it into a usable source of energy and making use of any excess heat generated in the panel. In this study, PCM is utilized as a medium for combining with nanoparticles. Nanoparticle composited phase change materials (nc-PCMs) are created by mixing lauric acid (LA) with magnesium oxide (MgO) and containing them in a micro-channel container to enhance their thermal conductivity. The testing conducted on the water at various dimensionless Reynolds number (Re) of 1100, 2200, 3300, 4400, and 5500. Research studies have discovered that the presence of a water inlet reduces the temperature of the PV module, resulting in an enhancement of the overall efficiency. According to the results, it has been concluded that the best Re for a water system is Re = 5500. The enhanced photovoltaic (PV) modules boosted the electrical efficiency by 16.01% compared to the reference module, leading to an overall efficiency of 54.91%.

Keywords: PVT system, Nanoparticles, PV cooling, Thermal management, Micro-channel

1. Introduction

The energy produced by the sun is widely recognized as a popular form of clean energy. Furthermore, it is worth noting that solar energy offers various economic advantages, including the minimum maintenance costs. Moreover, an important aspect of solar energy is its cost-effectiveness, since it is essentially free [1, 2]. Solar photovoltaic technology is

*Corresponding author: tanwon@kku.ac.th

widely recognized as one of the most efficient renewable energy technologies now in use [3]. A photovoltaic (PV) module is a technological apparatus that employs the principles of photovoltaics to transform solar energy into electrical energy. Currently, this technology has become increasingly cost-effective. The market demand for PV modules is experiencing significant and widespread growth. One limitation associated with the utilization of PV modules is the reduction in power generation efficiency that occurs when the temperature of the PV module exceeds a certain threshold. Phase change materials (PCMs) have been employed for the purpose of regulating the operational temperature of PV systems [4]. These materials are capable of storing thermal energy through the process of latent heat storage, wherein a phase transition occurs while the temperature remains constant. The consideration of a phase change material's latent heat of fusion is generally necessary when choosing a suitable PCM for the purpose of reducing the temperature of a PV module. However, it is worth noting that numerous phase change materials (PCMs) exhibit significant latent heat while simultaneously possessing a relatively poor thermal conductivity, as indicated by previous studies [5, 6]. The improvement of PCM thermal conductivity and the reduction of PV module system temperature have been the subject of a multitude of scholarly publications. By incorporating a metal characterized by high thermal conductivity into a PCM, it is possible to augment heat dissipation capacities. Despite increasing the cooling effect and heat transfer properties, the incorporation of thermal additives into a system, results in a reduction in the latent heat of fusion and an increase in the total weight [7, 8].

In addition, a number of intriguing studies have been conducted to enhance the reduction in temperature of PV modules. Especially, utilizing various PCMs and structures of PCM's container to enhance thermal conductivity and facilitate heat transmission between the PCM and the PV module. For example, three types of stainless steel PCM container i.e. grooved-, tubed-, and finned-container, were studied to investigate the characteristics of cooling performance of PV modules. It was found that the finned-container showed the highest cooling by 6.1 °C and increased electrical efficiency by 5.3% [9]. The utilization of aluminium mesh (sheet) was implemented within the PCM container's core in order to facilitate the dispersion of heat in a horizontal orientation of the PCM, as indicated by reference [10]. The study revealed that the incorporation of a mesh structure resulted in enhanced cooling of the photovoltaic system. Govind S. Menon et al. [11] examined an unglazed photovoltaic thermal (PVT) system using a serpentine coil sheet and tube thermal absorber setup using water and copper oxide-based nanofluid. The PVT system cooled with various nanofluid adding, and different water flows had 14–17% higher electrical efficiencies compared with the reference case. Recently, PVT/Water and PVT/TEG/nanofluid were studied in the year 2023. It was found that the PVT/Water and PVT/TEG/nanofluid setups had 25.1% and 41.2% temperature drops and improved by 5.8% and 8.5%, respectively, in electrical efficiency compared to the reference PV panel [12]. Although, there have been several studies focused on water flow in tubes of PVT system by varying volume flow rates of the water, the flow rates were in terms of kg/s or litre/min which can be applied in a specific case and cannot be used for other different cases. Thus, the water flowrate is converted into the dimensionless (Reynolds number, Re) in this investigation to provide a benchmark for future similar case studies.

On the basis of the above brief literature review, in addition to enhancing the thermal conductivity of a used PCM, it is of interest to create micro-channel containers for PV cooling. Lauric acid added with magnesium oxide (MgO) nanoparticles, was selected as the PCM in this investigation because its melting point (41-44 °C) is suitable the ambient temperature (T_{amb}=38 °C) in Thailand. The mixture is termed nanoparticle composited PCM (nc-PCM). After that, the nc-PCM is fully filled into the micro-channel container. Inside the micro-channel container, 8 water tubes have been inserted to transfer heat from the nc-PCM to the water tubes. As a result, the increases in water temperature can be expected. In this study, five different levels of Re of the water inside the tubes, namely 1100, 2200, 3300, 4400, and 5500, are tested. A micro-channel PCM container is installed at the back of the PV module. The best possible dimensionless is determined by taking into consideration both the electrical efficiency and the thermal efficiency. Incorporating nc-PCM and a micro-channel structure give a higher thermal conductivity and heat transfer in the container. It is expected that this special designed container could enhance PV temperature reduction and, as a result, improve the efficacy of the system.

2. Methodology

2.1 Materials

In this research, nanoparticle composited PCM (nc-PCM) was synthesized using the amalgamation of lauric acid (a type of organic fatty acid, denoted as LA) with MgO (a nanoparticle composed of metal oxide). Naturally occurring as periclase, magnesium oxide is a hygroscopic white solid mineral and a source of magnesium with the chemical formula MgO. The particle size is 125 μm. Based on the findings derived from my previous research, the incorporation of lauric acid (LA), an organic fatty acid, was employed to form nc-PCM by combining it with nano metal-oxide particles (MgO). The process led to the creation of nc-PCMs with different amounts of LA and metal-oxide particles, namely 99:1, 98:2, 94:6, 92:8, and 90:10. In summary, the best effective nc-PCM is LA: MgO in a 94:6 ratio. This composition has exhibited a significant increase in thermal conductivity, with enhancements of around 65.96% for referencing systems employing a 20-watt PV [13].

Fig. 1 illustrates the PVT assembly being studied, which consists of a microchannel in a container serving as the thermal enhancement mechanism and nc-PCM functioning as the thermal management substance. The PVT system comprises a PV module that is affixed to a rectangular container made of stainless steel, which serves as the container for nc-PCM. The choice of stainless steel was made due to its economical production cost and its favorable thermal conductivity, which facilitate efficient transmission of heat from the photovoltaic cells to the PCM section. Micro-channels were integrated into the PCM container in order to augment the available heat transfer area and improve the heat dissipation capabilities of the system. During daytime operation, heat from the PV module is absorbed by the PCM. The PCM then transfers the heat to water in a water tube, allowing for its discharge to the external environment. Additionally, high-temperature water offers other advantages, including the creation of residential hot water.

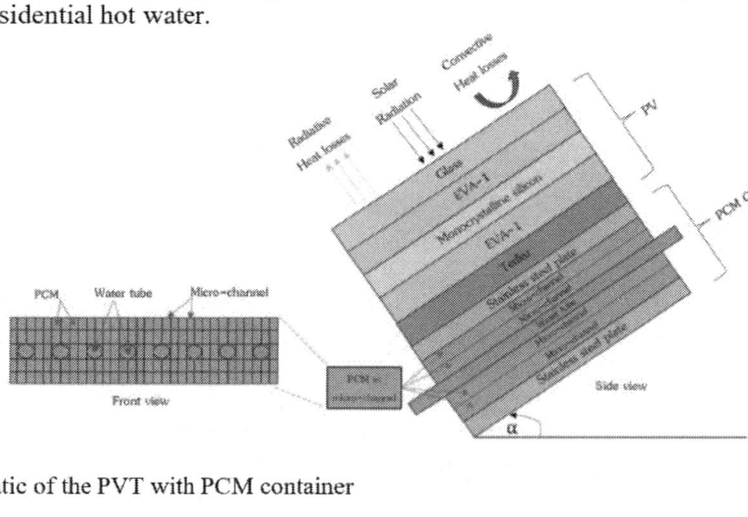

Fig. 1. Schematic of the PVT with PCM container

2.2 Experimental setup

The experiments were conducted in outside settings to compare the performance of a reference PV module with a PV module containing nc-PCM within a container. This study utilizes monocrystalline PV modules with a power output of 100 W. The temperatures are measured only through the utilization of K-type thermocouples. The resistive load (13.5 Ω) is connected to the system, and measurements are taken for the maximum voltage (Vmax) and maximum current (Imax). The KIPP&ZONEN pyranometer is employed to quantify solar irradiance throughout the course of a day at the Center of Alternative Energy Research and Development, Khon Kaen University in Thailand. The power, electrical efficiency (η_{el}), and thermal efficiency (η_{th}) of the PV modules under investigation are determined by employing equations (1) - (4) in order to establish the correlation between the temperature of the PV module and its electrical power profile. The angle evaluated for the PV tilt was 15°, and the PV modules were positioned with their surfaces oriented towards the south.

$$P = VI \tag{1}$$

$$\eta_{el} = \frac{P}{IR \; x \; A} \; x100\% \tag{2}$$

$$\eta_{th} = \frac{\dot{Q}_U}{IR \; x \; A} \; x100\% \tag{3}$$

$$\dot{Q}_U = \left(\dot{m}c_p\right)_w (T_{fo} - T_{fi}) \tag{4}$$

Where: P is the power output in watt (W), V is voltage in volts (V), I is current in amperes (A), IR is the solar irradiance in watt per square meter (W/m^2), A is area of the PV module in square meter (m^2), $\dot{Q}_U$ is useful heat rate from PVT module in kilojoule per second (kJ/s), $\dot{m}$ is mass flowrate in kilogram per second (kg/s), c_p is specific heat in kilojoule per kilogram-degree Celsius (kJ/kg°C), $T_{fo}$is temperature outlet in degree Celsius (°C), and T_{fi} is the temperature inlet in degree Celsius (°C).

Furthermore, the flowrate of the water input (volume flowrate) is transformed into the Reynolds number (Re) or dimensionless quantity to establish a standardized metric for future investigations. The flowrate of the water inflow and the Reynolds number are determined by employing equations (5) and (6), respectively. The overall efficiency is the sum of the electrical efficiency and the thermal efficiency, as shown in equation (7). The experimental setup and equipment utilized for the measurement of parameters are shown in Figs. 2–4.

$$Q_{vf} = vA_t \tag{5}$$

$$Re = \frac{\rho v D}{\mu} \tag{6}$$

$$\eta_{overall} = \eta_{el} + \eta_{th} \tag{7}$$

Where: Q_{vf} is volume flowrate in cubic meter per second (m^3/s), v is velocity in meter per second (m/s), A_t is area of tube in square meter (m^2), ρ is density of the fluid in kilogram per cubic meter (kg/m^3), D is dimeter of tube in meter (m), and μ is dynamic viscosity in Pascal-second (Pa.s).

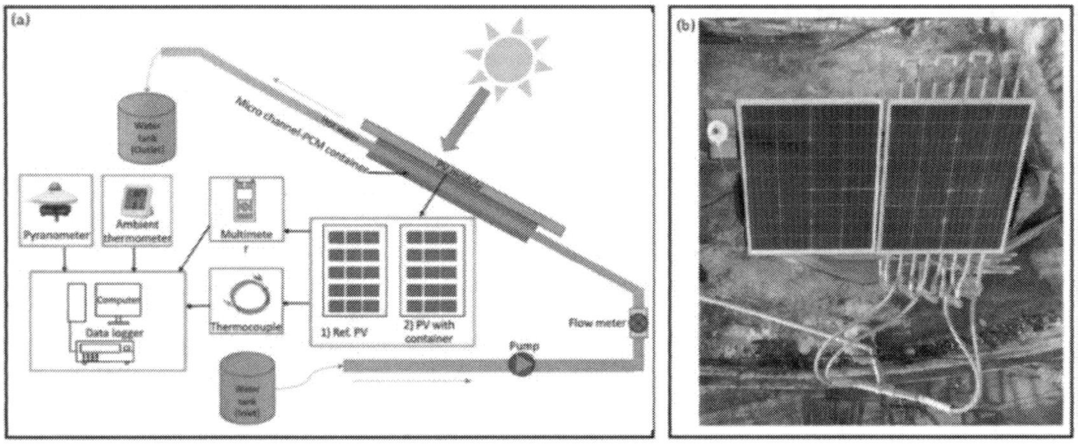

Fig. 2. (a) Schematic of experimental setup (b) Real experiment setup

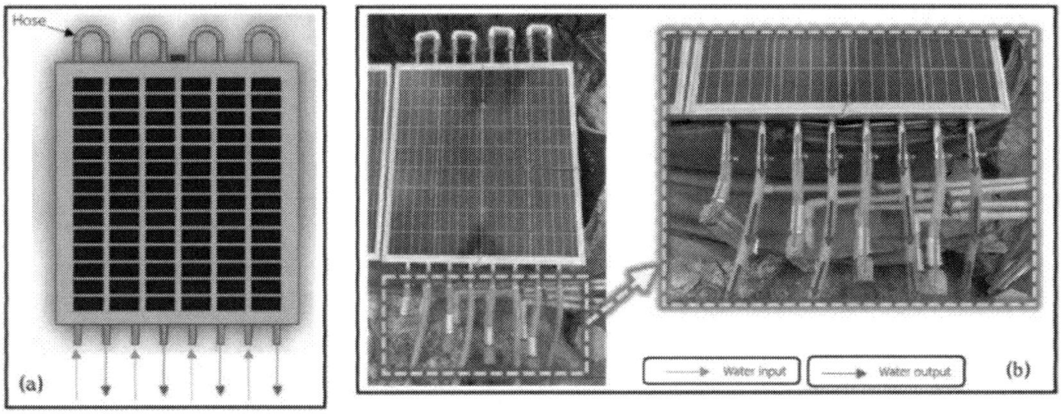

Fig. 3. (a) Model of U-type connection and (b) Real connection of U-type

In this experimental setup, a U-type connection is employed, as depicted in Fig. 3 (a) - (b). The Reynolds numbers of 1100, 2200, 3300, 4400, and 5500 were employed in the experimentation conducted on the water system. Each the experiment was repeated three times. Consequently, the values of parameters are average data.

Fig. 4. Equipment utilized for the measurement of parameters

3. Results and Discussions

In order to investigate the impact of alternative cooling methods on the performance of PV modules, the experiments were conducted in December 2023, and the results were averaged. Experimental investigations were carried out at the engineering faculty of Khon Kaen University, located in Thailand.

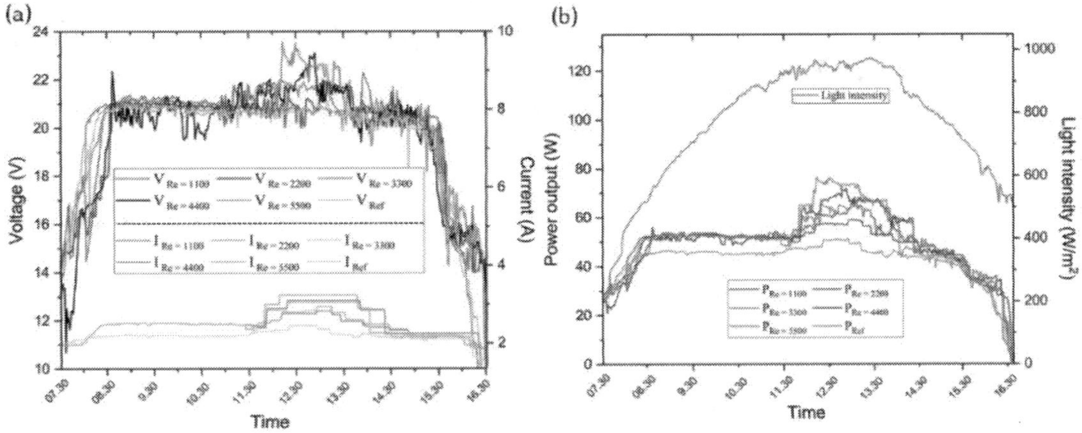

Fig. 5. (a) Voltage and current generation and (b) Power output of the PV modules

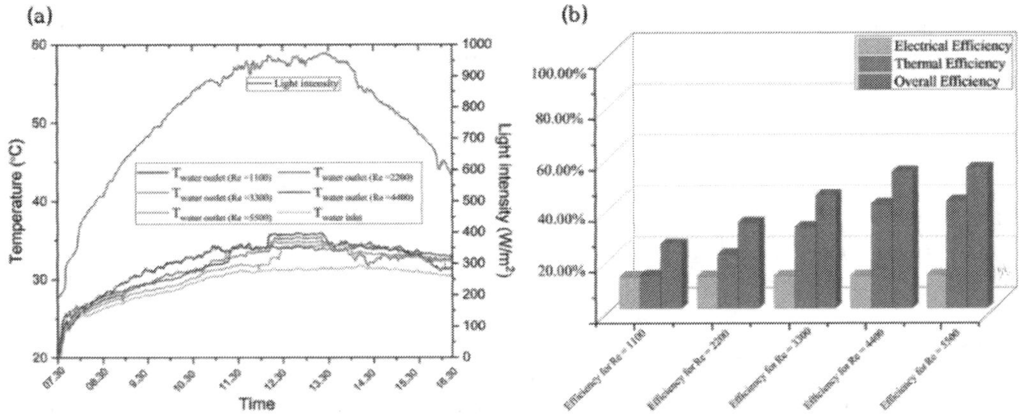

Fig. 6. (a) The temperature of water outlet each for Re and the light intensity (b) The comparisons for average overall efficiency.

Fig. 5 (a)-(b) depicts the voltage, current generation, and power output during the period 7.30–16.30. The average voltages of PV modules without cooling ($V_{avg\text{-}ref}$ = 19.32 V) and PV modules with nc-PCM (LA: MgO) exhibited the following average voltages: 19.60 V (Re = 1100), 19.73 V (Re = 2200), 19.63 V (Re = 3300), 19.45 V (Re = 4400), and 19.86 V (Re = 5500). In addition, it is found that the maximum voltage outputs are 21.21 V, 21.91 V, 22.75 V, 23.1 V, and 23.54 V, respectively, while the maximum voltage output of PV modules without cooling is 21.08 V. Moreover, the increase in light intensity, causes the temperature of the PV modules to rise. It is evident that photovoltaic systems utilizing PCM micro-channel containers have superior thermal management capabilities, as nc-PCM and micro-channel increase thermal conductivity and water at different Re is utilized for cooling purposes. The data presented in Fig. 5 (b) illustrates that the average power output (P_{avg}) at Re = 5500 was roughly 50.60 W. In comparison, the P_{avg} for the water-cooling inlet of PV at various Re values (1100, 2200, 3300, and 4400, respectively) were 47.64 W, 48.66 W, 49.42

W, and 49.82 W, while the P_{avg} for the PV reference case was 42.42 W. During the peak time of 12:00 to 13:00, the power output increased because the solar irradiance was at its highest point of the day. As a result, the power outputs at the peak time are higher than in other periods.

Table 1. Average difference of four parameters is as follows: the temperature of the PV's front, the temperature of the water system, the power output, and the electrical efficiency for each Re.

PV module	$\Delta T_{PV\text{-}avg}$ (°C) [$\Delta T_{PV\text{-}avg}$ (%)]	ΔP_{avg} (%)	$\Delta \eta_{electrical\text{-}avg}$ (%)	$\Delta T_{water\text{-}avg}$ (°C) [$\Delta T_{water\text{-}avg}$ (%)]
At Re = 1100	1.91 [4.17]	12.29	10.58	2.71 [9.04]
At Re = 2200	2.12 [4.92]	14.69	12.49	2.09 [6.93]
At Re = 3300	2.77 [6.25]	16.48	13.30	1.99 [6.63]
At Re = 4400	3.13 [6.50]	17.44	13.57	1.94 [6.49]
At Re = 5500	3.49 [7.11]	19.17	16.01	1.57 [5.17]

Table 1. depicts the average temperature difference between the front surface of PV with cooling and the surface of PV reference, as well as the average temperature difference between the water outlet and inlet. The average temperature reductions on PV's surface were as follows: 1.91 °C (4.17%), 2.12 °C (4.93%), 2.77 °C (6.24%), 3.13 °C (6.50%), and 3.49 °C (7.11%) when Re = 1100, 2200, 3300, 4400, and 5500, respectively. Furthermore, the cooling technique implemented in the PV module reduced the maximum temperature of the PV's surface by 3.50 °C (8.03%), 4.47 °C (12.08%), 4.67 °C (12.64%), 5.43 °C (13.82%), and 6.21 °C (14.25%), respectively, compared to the PV reference. Because the direction of the relationship between Re and flow rate is identical. Consequently, an increase in the Re or flow rate results in a rise in convection heat transfer, leading to enhanced PV cooling performance. Meanwhile, the temperature of the water outlet is reducing. Therefore, the difference in water average temperature (ΔT) was 2.71 °C (9.04%), 2.09 °C (6.93%), 1.99 °C (6.63%), 1.94 °C (6.49%), and 1.57 °C (5.17%), respectively. For the aforementioned reasons, at Re = 1100, we observed that the water outlet temperature reached its maximum. This is attributed to the long time it takes for water to pass through the tube at low flow rates, which causes heat to accumulate within the water. Conversely, the water outlet temperature was lowest at Re = 5500. Because the water has a high flow rate, there is less heat accumulation per unit time. Thus, the water outlet temperature will increase or decrease as a result, depending on Re value.

https://doi.org/10.1051/e3sconf/202453005003

Table 2. Average efficiencies at various Reynolds numbers.

PV module	Electrical efficiency (%)	Thermal efficiency (%)	Overall efficiency (%)
At Re = 1100	12.28	12.81	25.09
At Re = 2200	12.49	21.24	33.73
At Re = 3300	12.58	31.84	44.42
At Re = 4400	12.61	40.89	53.50
At Re = 5500	12.88	42.03	54.91
PV reference	11.10	-	11.10

Fig. 6 (a) depicts the fluctuation of light intensity and the temperature of the water inlet and outlet recorded during the experiment. The results indicate that the intensity of light radiation fluctuated over the course of the experiment; the average and maximum values of light radiation intensity were roughly 761.85 W/m2 and 974 W/m2, respectively. In this case, electrical efficiency depends on the average power output of PV. The incorporation of enhanced PV modules resulted in a notable improvement in the average electrical efficiency of 10.58%, 12.49%, 13.30%, 13.57%, and 16.01% for Re = 1100, 2200, 3300, 4400, and 5500, respectively, when compared to PV reference as shown in Table 1. However, thermal efficiency is contingent upon two factors in this particular instance: the difference in temperature between the water inlet/outlet and the mass flowrate of the water, as shown in equation (4). According to Fig. 6 (b), although the ΔT of Re = 1100 has the highest compared to other Re values, the mass flowrate ($\dot{m}$) of water is influenced by the Re. A decrease in Re leads to a decrease in the mass flow rate. Consequently, in this paper, thermal efficiency at Re = 5500 is the highest value (42.03%) due to the mass flowrate at Re = 5500 being higher than other Re. The overall efficiency is determined by combining the electrical efficiency and the thermal efficiency, as shown in Equation (7). The results indicate that there is no substantial variation in the electrical efficiency for each Re value. On the contrary, thermal efficiency experiences a substantial surge. Thus, the overall efficiency is contingent mainly upon the thermal efficiency. Therefore, at Re = 5500, the best overall efficiency is achieved. In addition, PV modules equipped with micro-channel PCM containers have a considerably greater overall efficiency compared to the PV reference. Fig. 6 (b) and Table 2 present average efficiency comparisons.

4. Conclusion

The experiments occurred at the Faculty of Engineering, Khon Kaen University, Thailand, to determine the impact of cooling techniques on the thermal and electrical performance of PV modules installed at a 15-degree inclination angle with their surfaces oriented towards the south. The purpose of the research was to determine how the various cooling techniques and Reynolds numbers (Re) of the water inlet influence the modules' behaviour and to select the best-suited Re. The experimental setup utilized Re of 1100, 2200, 3300, 4400, and 5500 in relation to the water inlet. The electrical efficiency of the enhanced photovoltaic modules is impacted by cooling and the increase in thermal conductivity of the nc-PCM and micro-channel containers, as shown by the results. Consequently, the conclusion of the present research outlines the subsequent essential aspects:

- The average surface temperature reductions of PV were as follows: 1.91 °C (4.17%), 2.12 °C (4.93%), 2.77 °C (6.24%), 3.13 °C (6.50%), and 3.49 °C (7.11%) at Re of 1100, 2200, 3300, 4400, and 5500, respectively.
- The variation in the average temperature of water (ΔT) was 2.71 °C (9.04%), 2.09 °C (6.93%), 1.99 °C (6.63%), 1.94 °C (6.49%), and 1.57 °C (5.17%) at Re of 1100, 2200, 3300, 4400, and 5500, respectively.
- The average electrical efficiency of the enhanced PV modules increased by 10.58% (Re = 1100), 12.49% (Re = 2200), 13.30% (Re = 3300), 13.57% (Re = 4400), and 16.01% (Re = 5500).
- The increased Re influenced the thermal efficacy of the enhanced photovoltaic modules; thus, the thermal efficacy of the average PV module was enhanced by 12.81% (Re = 1100), 21.24% (Re = 2200), 31.84% (Re = 3300), 40.89% (Re = 4400), and 42.03% (Re = 5500).

Based on the results, we recommend utilizing the water inlet at Re = 5500 due to its superior overall efficacy (54.91%). However, at this Re, the overall efficiency is slightly higher than that of Re=4400, therefore further investigation will be carried out the find the optimal Re.

The research team would like to thank the Center of Alternative Energy Research and Development, Khon Kaen University, for supporting our work.

References

1. H.S. Min, S.D. Fertahi, T. Bouhal, N.S. Naa and M.A.C. Munaaim: Solar Energy development:Case study in Malaysia and Morocco. International Journal on Emerging Technologies. **10**, 106-113 (2019).
2. S. Sreenath, K. Sudhakar and A.F. Yusop: SWOT Analysis of Solar PV Systems in Airport Environment. International Journal on Emerging Technologies. **10(2)** , 1-7 (2019).
3. M.W. Rahman, et al.: Comparative Analysis of Developed Incremental Conductance (IC) and Perturb & Observe (P&O) MPPT Algorithm for Photovoltaic Applications. In: 2016 10th International Conference on Intelligent Systems and Control (ISCO). IEEE, 1-6 (2016).
4. K. Velmurugan and S. Sukchai: Thermal Investigation of Paraffin Wax for Low-Temperature Application. Journal of Advanced Research in Dynamical and Control Systems. **11**, 1437-1443 (2019).
5. V. Karthikeyan , C. Sirisamphanwong and S. Sukchai: Investigation on Thermal Absorptivity of PCM Matrix Material for Photovoltaic Module Temperature Reduction. Key Engineering Materials. **777**, 97-101 (2018).
6. K. Velmurugan, et al.: Selection and preparation of suitable composite phase change. Material for PV Module Cooling. **10**, 385-394 (2019).
7. Li, Y., Tie, W.C., Zhu, Q.Z. et al.: A Study of $LiNO_3$–NaCl/EG Composite PCM for Latent Heat Storage. Int J Thermophys. **42**, 155 (2021).
8. K. Stejskalová, D. Bujdoš, L. Procházka, B. Smetana, S. Zlá, J. Teslík: Mechanical, Thermal, and Fire Properties of Composite Materials Based on Gypsum and PCM. Materials. **15**, 1253 (2022).

9. T. Wongwuttanasatian, T. Sarikarin, A. Suksri, Performance enhancement of a photovoltaic module by passive cooling using phase change material in a finned container heat sink, Sol. Energy. **195**, 47–53 (2020).
10. S. Homlakorn, K. Velmurugan, A. Suksri, T. Wongwuttanasatian: Comparative study for photovoltaic cooling using metal mesh inserted eutectic phase change material enclosure. Case Studies in Thermal Engineering. **48**, 103024 (2023).
11. Govind S. Menon, et al.: Experimental investigations on unglazed photovoltaic-thermal (PVT) system using water and nanofluid cooling medium. Renewable Energy. **188**, 986-996 (2022).
12. P. Seepana, et al.: Thermo-enviro-economic analysis of solar photovoltaic/thermal systemincorporated with u-shaped grid copper pipe, thermal electric generators and nanofluids: An experimental investigation. Journal of Energy Storage. **60**, 106611 (2023).
13. W. Phukaokaew, A. Suksri, T. Wongwuttanasatian. The Modification of Thermal Conductivity of Phase Change Material Using Nano Metal-Oxide Particles. Materials Science Forum. **1113**, 69-74 (2024).

Fabrication and Performance Analysis of an S1046 H-Darrieus VAWT using Mahogany as Blades

Michelle Angeli Viktoria Aquino[1], *Mikhail Alexander* Calibjo[1], *Christine Marie* Nulada[1], *Aira Crishel* Rodriguez[1], *Reylina* Tayactac[1,2*], *Jaime* Honra[1,2], *Ricky* Umali[1,2], and *Jennifer* Buenconcejo[1,2]

[1]School of Mechanical, Manufacturing, and Energy Engineering, Mapúa University, 658 Muralla St., Intramuros Manila, 1002 Philippines
[2]School of Graduate Studies, Mapúa University, 658 Muralla St., Intramuros Manila, 1002 Philippines

Abstract. Wind turbines are systems that generate electricity from the wind. These turbines vary in design and performance. For this study, the goal is to design an H-Darrieus type Vertical Axis Turbine that utilizes mahogany for its blade material. Mahogany is a natural composite type of material which addresses pollution concerns when using nanocomposites and is known for its high strength, durability, and workability. The airfoil to be made will follow an S1046 profile which yields the best performance when it comes to Tip Speed Ratio (TSR). The wind tunnel is utilized for the testing process wherein it controls the wind speed from 4m/s to 10m/s. Data showed that the maximum rotational speed in rpm achieved by the designed turbine is 53.18 at 10 m/s wind speed. Based on the results, the performance would not surpass the original material used for the design which is fiberglass polyester. However, the maximum TSR generated is 0.1879, that is comparatively lower than the maximum value from the specified theoretical considerations, specifically TSR of four.

1 Introduction

In the Philippines, energy consumption is predicted to increase by an average rate of 4.2% every year starting in 2015 to 54.9% in 2030 based on the data obtained by the Department of Energy [1]. Thus, the Philippine government aims to push the use of renewable energy to prevent the diminishing of non-renewable supplies. According to the National Renewable Energy Plan, the government envisions harvesting 1548 MW of renewable energy by 2030, and almost 38% is from wind power. Consequently, the country is known for areas that have strong wind velocity. This energy can be harnessed at a small scale by households to supply energy to some utilities at home.

Existing studies basically explored two different types of wind turbines to improve their current performance and efficiency, particularly horizontal and vertical wind turbines. One of the vertical axis wind turbine types is the Darrieus wind turbine. This type of wind turbine

* Corresponding author: rgtayactac@mapua.edu.ph

is made up of several curved aerofoil blades mounted on a rotating shaft or framework. It is a high-speed, low-torque device that produces alternating currents. Its low torque property is the basis of the physical momentum requirement as the system acts [2].

Common materials used in wind turbines are steel, fiberglass, resin, plastic, and cast iron. As for the turbine blades, the materials are categorized by different types such as the natural composites, hybrid composites, thermoplastic composites, and nano composites. An existing study by Pradeep et al. [3] was able to conduct a study which concluded that carbon nanotubes, as part of the nanocomposites, are recommended for the material structure of the wind turbine blade. However, it was rebutted by the study by Temesgen Batu et. al [4] that the challenge on this type of material is its contribution to pollution in the environment; thus, focusing on the natural composites instead. Natural fiber composites are materials made up of at least one reinforcing fiber obtained from renewable and carbon-neutral supplies such as wood. In the Philippine context, mahogany is a natural composite that is rarely used for the turbine blade material, yet it is one of the most abundant renewable resources in the country.

Due to the pollution-related concerns of nanocomposites, natural composite is encouraged to be used as the turbine blade material. Mahogany, as a natural composite, is a material that is worth evaluating for an airfoil material as it is utilized in existing studies in foreign countries, and it is abundant in the researcher's locality. As for the performance analysis, the Tip Speed Ratio (TSR) of the wind turbine is a significant parameter in assessing a wind turbine's performance. However, there is limited data on the turbine performance using mahogany as blades, focusing on the TSR. Additionally, the researchers intend to study the fabrication of an H-Darrieus vertical axis wind turbine with mahogany blades due to the rising demand for renewable energy and the available wind resources in the country.

1.1 Natural Composite Materials as Turbine Blade

A study conducted by Thomas [5] indicated that several properties such as high strength, high fatigue resistance, high stiffness, and low weight are needed when designing wind turbine blades. With that, natural composites such as wood are proposed to be a better option compared to synthetic composites as it includes similar properties like high specific strength, light weight, recyclable, low cost, and easy workability. In addition, the study specified that synthetic composites used in wind turbines have good mechanical properties but lacks biodegradable properties which makes it harmful for the environment.

Having a similar concern, the study of Batu et al. [4] acknowledged the environmental concern of using nanocomposite materials as wind turbine blades. As such, the researchers focused on the use of natural composites and their potential in the application of small-scale wind turbine blades. The results show that there is a significant improvement in the performance of the wind turbine simply by reducing the weight of the turbine blade for the specified natural composites. With that, the study indicated that composite materials are a good consideration for reducing the weight when designing a wind turbine and is therefore, beneficial in applications in wind energy consumption. Both studies indicated that the use of synthetic and nanocomposites poses an environmental concern in which natural composites can solve.

1.2 Performance of H-Darrieus Wind Turbines

The study of Kumar et al. [6] presented a comprehensive literature review to help in determining the most feasible solution in enabling a Darrieus turbine rotor to generate power instead of self-starting. This is due to the issue regarding the starting characteristics and poor performance in low wind speeds of H-Darrieus vertical axis wind turbines despite being highly efficient. Although the researchers presented several solutions for enhancing the performance of Darrieus turbines, it is still concluded that further research is required to determine the optimum solution to make a Darrieus turbine a low-speed device and improve its overall performance.

Similarly, a study conducted by Hashem & Mohamed [7] focused on increasing the performance of an H-Darrieus wind turbine specifically, the power coefficient, by choosing the parts that contribute to the increase of the turbine's efficiency. The researchers conducted a 2D simulation and determined that that the symmetric S1046-type is the best performing airfoil. Additionally, the researchers found that a three-bladed Darrieus wind turbine with an S1046-type airfoil would obtain the maximum power coefficient of 0.3463. The two studies solidified that there is a need to conduct further research to improve the performance of H-Darrieus turbines and increase its efficiency.

1.3 Wind Turbine Parameters

The study of Petković & Shamshirband [8] utilized the process of variable selection to identify the relevant and influential subset of parameters in optimizing the power produced in a wind turbine. The researchers utilized the method of ANFIS (Adaptive Neuro Fuzzy Inference System) to determine how the four most influential parameters which are blade pitch angle, rotor speed, wind speed and rotor radius affect the wind turbine power coefficient which in turn affects the wind turbine output energy as well. The results show that that the blade pitch angle is the parameter that is most influential in the prediction of wind turbine power coefficient and is also the best predictor of accuracy.

In align with this, Tahzib et al. [9] conducted a performance analysis comparing two airfoil profiles, NACA 0018 and S1046 for a range of tip speed ratios and various blade pitch angles for a 3-bladed H-Darrieus vertical axis wind turbine. The result of the study shows that the airfoil profile design, S1046, has executed a better performance over the NACA 0018 in almost all blade pitch angles and TSRs. Specifically, the S1046 VAWT achieved a maximum Cp of 0.537 when it is run at TSR 4 and at a blade pitch angle of -2 degrees. As such, the researchers concluded that an optimum vertical axis wind turbine should consist of an S1046 airfoil profile with a -2-degree blade pitch which is needed to be run at TSR 4. With this, both studies indicated that several parameters specifically, blade profile blade pitch angle, and tip speed ratio are needed to be considered in designing VAWT's to maximize its power coefficient and improve its performance.

2 Methodology

The flowchart in Figure 1 illustrates the process of fabricating the turbine and the testing of parameters. This consists of three phases which will be elaborated one step after the other. For the first phase, it focuses on data gathering. Existing studies on the performance of H-Darrieus turbine were collated to determine what parameters will be focused on to test the performance of the designed turbine. Furthermore, the mechanical properties of mahogany were determined using a universal testing machine. Lastly, using existing studies, the researchers then determined the dimensions of the turbine and its blades.

Phase 2 then focused on the fabrication and the set-up of the wind turbine. Before fabrication, a 3D model of the S1046 blade profile and the body of the turbine was first made as reference. The material needed in fabrication was then gathered. The dimensions from the 3D model were then followed by the fabricators in making each component. These components were then attached to each other, and the overall wind turbine was brought to the location of the wind tunnel in preparation for phase 3.

Then, the last phase focused on the evaluation and the fulfillment of the study's' objectives. The wind turbine was placed inside a wind tunnel and its rotations per minute were recorded. Other parameters such as wind speed and wind density of the wind tunnel were also collected. The data was then analyzed, and conclusions were made from the experiment.

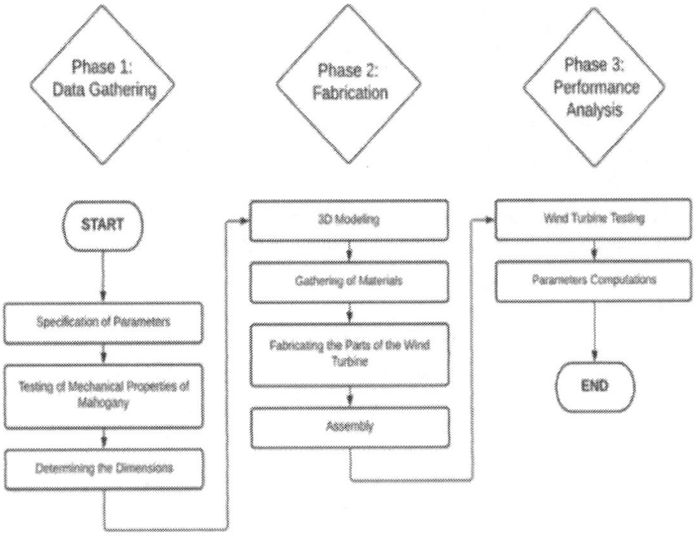

Fig. 1. Process Flowchart

2.1 Determining the Dimensions and Designing using Fusion 360

The airfoil named S1046 17% (Danny Howell) was taken from an online airfoil plotter. Coordinates were downloaded from the online airfoil plotter and were transferred to Fusion360. The dimensions of the airfoil were then achieved using the inspect tool.

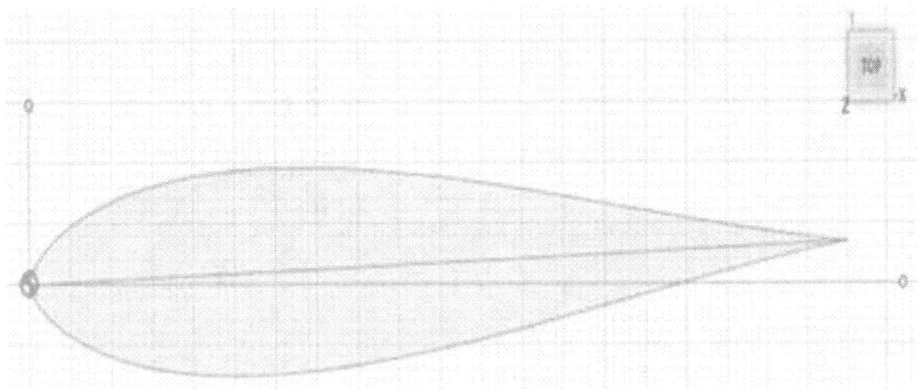

Fig. 2. S1046 Blade Profile

Figure 2 illustrates the blade profile of the wind turbine where this is the airfoil design that the researchers used for shaping the mahogany slabs while the table 1 shows the dimensions of the airfoil.

 https://doi.org/10.1051/e3sconf/202453005004

Table 1. Dimensions of the Airfoil

Part	Size
Thickness of Airfoil	0.75 in
Camber Length	3.5 in
Blade Height	17 in
Wall Thickness	0.01 in

To visualize the design of the wind turbine set-up, the researchers will use Fusion 360 as the software for 3D modeling. The airfoil design of wind turbine blades was based on the standard design of the S1046 made by Cubas 2022 while the strut, shaft, and body of the wind turbine is proportioned to the said design and to the size of the wind tunnel to guarantee that it will fit inside it. The coordinates of points for the outline of the airfoil design are tabulated in an excel file, which will then be imported in the said software; thus, ensuring the right specifications of S1046. Figure 3 illustrates the appearance of the whole wind turbine.

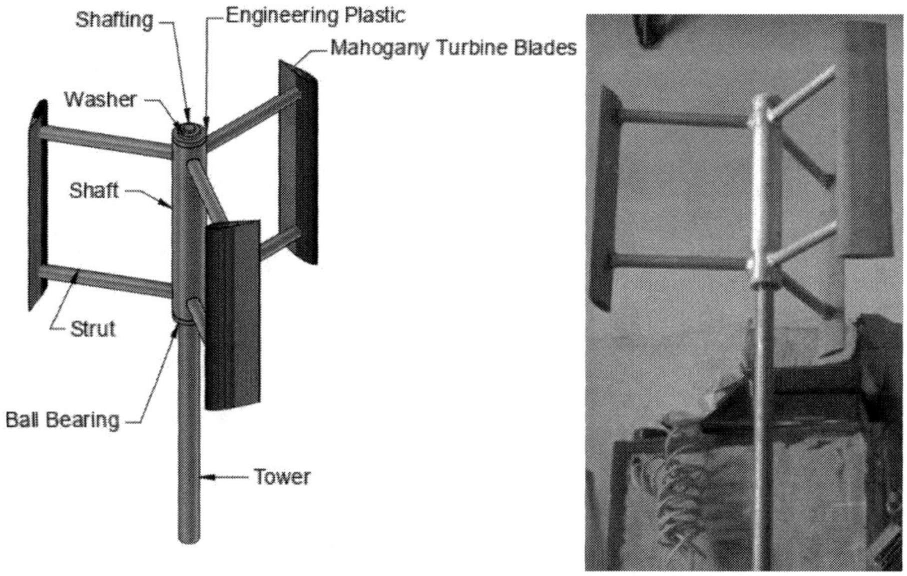

Fig. 3. Isometric View of the Wind Turbine (a) Fabricated Wind Turbine (b)

2.2 Fabrication and Assembly

The H-Darrieus VAWT has different parts in which each are individually made. Each blade is made by drawing the S1046 blade profile into the mahogany slab which is then cut and filed to ensure smooth surface. It is made hollow by cutting first big chunks of wood and then sanding it as well to make the thickness accurate and the surface smooth. Meanwhile, the shaft, strut, and tower are all made by cutting an aluminum round tube into its respective sizes. The shaft has 2 ½ inches of diameter and 15 inches diameter. The strut has 1 ½ inch diameter and 10 inches length. The tower has 0.75 inch diameter and 20 inches length. Finally, the bearing is made from a metal plate that has 1.34 inch outside diameter and 0.5 inch inside diameter. A groove is done on one surface of the metal plate to serve as a guide for the 4 mm ball bearings to rotate. The dimensions of the turbine shows in Table 2.

Table 2. Dimensions of the Turbine

Part	Size
Shaft (diameter)	2.5 in
Shaft (height)	15.25 in
Strut (length)	10 in
Strut (thickness)	1.5 in
Tower (diameter)	0.75 in
Tower (height)	20 in

Upon completing the fabrication of each component, the engineering plastic is attached at both ends of the shaft. Then, the struts are welded on equal distances in the outer part of the shaft. Meanwhile, two U-shaped metals are screwed to the upper portion and lower portion of each blade. The holes on the U-shaped metal are aligned with the holes on the other end of the strut and a screw is inserted through those holes to combine the blade and the strut. After completing the upper portion of the wind turbine, a shafting is inserted to the tower wherein it will hold all the upper parts altogether. This will then be locked using a screw. This is followed by an insertion of the bearing and the whole upper part of the wind turbine into the shafting. To ensure that the upper part won't be removed once there is a rotation, a washer is inserted on the top part of the shafting. The figure above shows the finalized designed wind turbine.

3 Results and Discussion

The findings are reviewed by comparing data in terms of TSR with related existing studies done through blade design and material modifications. There are several tables and graphs that will visually represent the data with descriptions as well as explanations to support them. It presents tabular data analysis as well as a thorough evaluation of the structural properties of mahogany, wind speed, power available from the wind, shaft power (theoretical), and TSR. The first three tables and figures are the data that were obtained when using the Universal Testing Machine (UTM) in Mapua University. This includes the mechanical properties of the mahogany as the turbine material shown in Table 3.

Table 3. Mahogany Mechanical Data

	Tensile	Compressive	Flexural
Maximum Force, kN	20.5906	233.750	2.79063
Maximum Stress, MPa	32.9450	374.000	53.5800
Break Stress, MPa	28.9450	-	-

The maximum stress is 32.9450 MPa at the maximum force of 20.5906 kN. This is the amount of force when the rupture of the mahogany plate happened. It is evident on the figure below that the break stress is lesser than the maximum stress. This is true as to the stress-strain curve because mahogany plate experiences the stress at maximum before the necking is encountered. When the necking happens, the stress decreases at a constant percentage of strain until the plate breaks.

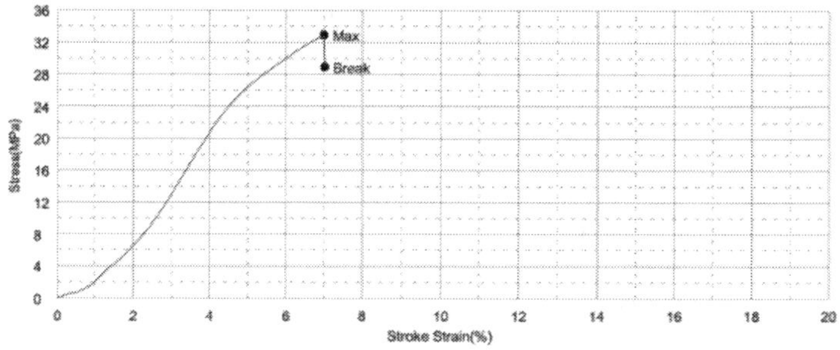

Fig. 4. Tensile Stress (MPa) vs. Stroke Strain (%)

This is the amount of force when the rupture of the mahogany plate happened. It is evident on the figure below that the break stress is lesser than the maximum stress. This is true as to the stress-strain curve because mahogany plate experiences the stress at maximum before the necking is encountered. When the necking happens, the stress decreases at a constant percentage of strain until the plate breaks. The stress-strain curve shows where the maximum stress happens upon applying tension to mahogany, as shown in Figure 4. As the stress increases the stroke strain also increases, which follows the proportional limit of Hooke's Law.

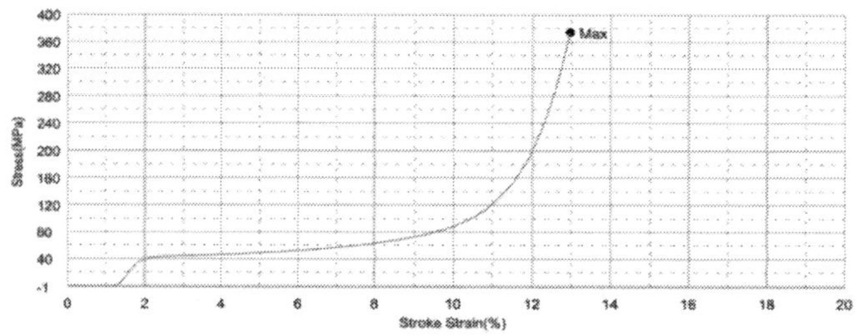

Fig. 5. Compressive Stress (MPa) vs. Stroke Strain (%)

Figure 5 above shows the stress-strain curve of mahogany when it was subjected to compressive stress. The y-axis represents the stress in MegaPascals (MPa) on the mahogany while the x-axis represents the stroke strain in percentage. It can be seen that the graph shows an exponential growth relationship between stress and strain. As strain increases, the stress also increases.

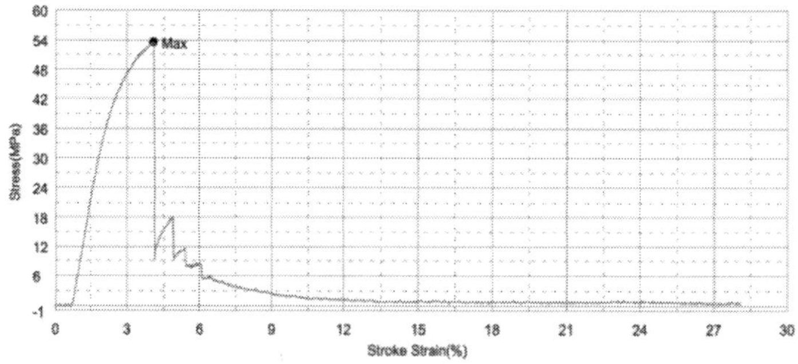

Fig. 6. Flexural Stress (MPa) vs. Stroke Strain (%)

Figure 6 depicts the flexural stress (MPa). For this type of testing, a three-point bending test was performed. As seen on the graph, the mahogany is resisting the bending force applied to it until it reaches the maximum stress it can attain and reaches its breaking point. Additionally, before the mahogany reaches its maximum stress, the flexural stress and stroke strain have a direct proportionality relationship. However, after the maximum stress, the mahogany has three visible fluctuating data which means that it is still inducing stress but significantly lowered compared to the maximum stress. After the fluctuation, the mahogany can no longer experience stress as it reached its final breaking point. All throughout the testing, even though the flexural stress has a non-linear pattern, the strain (or amount of deformation) experienced by the mahogany is continuously increasing.

The following results are gathered from the wind tunnel testing in DOST PAGASA facility. As shown in Table 4, the density varied through some of the trials but at little discrepancy. The density was recorded for the power computations. These densities are relatively lower than the theoretical value of 1.2 $\frac{kg}{m^3}$ because the wind tunnel facility reads the actual data for this parameter. Hence, these values are the actual values for the densities of air which will be used specifically for shaft power (theoretical) and wind power.

Table 4. Wind Speed, Density and Rotational Speed

Trial	Wind Speed $\left(\frac{m}{s}\right)$	Density $\left(\frac{kg}{m^3}\right)$	RPM
1	4	1.145	16.28
2	4	1.145	18.2
3	5	1.146	25.77
4	5	1.145	25.99
5	6	1.145	33.99
6	6	1.146	34.52
7	7	1.145	38.04
8	7	1.146	39.27
9	8	1.145	43.44
10	8	1.145	44.12
11	9	1.145	43.28
12	9	1.146	44.21
13	10	1.146	51.02
14	10	1.146	53.18

The revolutions per minute of the designed wind turbine was also recorded using a laser tachometer in the 14 trials inside the wind tunnel. The highest value achieved was 53.18 rpm at 10 m/s and lowest value was 16.28 rpm at 4 m/s.

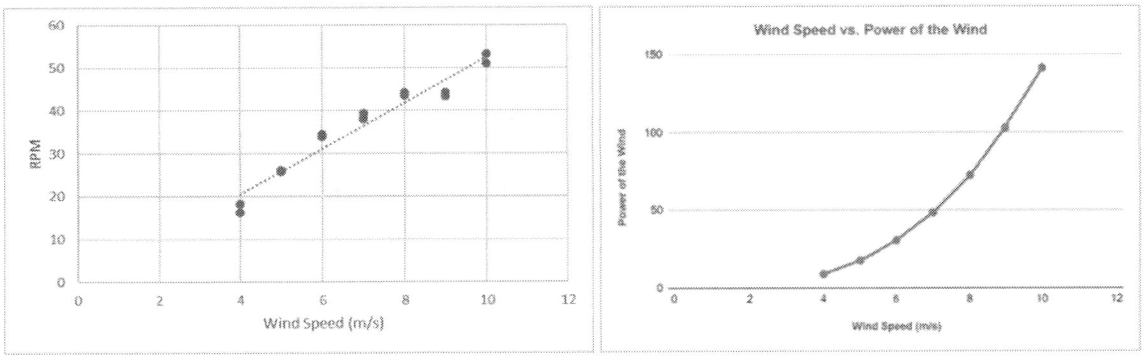

Fig. 7. Wind Speed vs RPM (a) Wind Speed vs Power of the Wind (b)

Figure 7a is a graph with a reference line that shows the relationship between wind speed and the revolutions per minute of the wind turbine. The two parameters have a directly proportional relationship whereas the wind speed increases, the rpm of the turbine also increases. This shows that the turbine blades take in most of the air flow which causes it to turn as much as the wind speed increases in each trial. This is also indicating that the ball bearing that is made from scratch is effectively functioning as it should in the turbine design. Figure 7b shows the generated line graph that depicts the relationship between wind speed and power of the wind. It can be observed from the graph that that the obtained results for the power of the wind is directly proportional to the value of the wind speed. As such, when the wind speed increases, the value for the power of the wind also increases.

The power computation of the wind turbine is vital to the study as this is one of key components in the analysis of turbine performance. With that, the first power parameter, power of the wind, is obtained through using different variables such as the density of air, turbine swept area, and the wind speed. This parameter would determine which wind speed the turbine would produce maximum power generated from the wind. The results obtained can be seen in table 5 shown below wherein it contains the values of the power of the wind for each trial and its corresponding wind speed. From the tabulated data it can be observed that the highest obtained power of the wind is 141.40 W at 10 m/s.

Table 5. Wind Speed and Power of the Wind

Trial	Wind Speed $\left(\frac{m}{s}\right)$	Power of the Wind (W)
1	4	9.0418
2	4	9.0418
3	5	17.6752
4	5	17.6598
5	6	30.5162
6	6	30.5428
7	7	48.4585
8	7	48.5009
9	8	72.33
10	8	72.3346
11	9	102.9920
12	9	103.0820
13	10	141.4019
14	10	141.4019

The second power parameter is the shaft power which could be determined through using the variables: maximum power coefficient, area, and wind speed. This parameter determines the capability of the wind turbine in its conversion of mechanical energy from blades into rotational energy which in turn, will be converted into electrical work. With this, the results obtained is shown in table 6 below. From the data gathered, it can be observed that the highest shaft power obtained is 51.27 W at 10 m/s.

 https://doi.org/10.1051/e3sconf/202453005004

Table 6. Wind Speed and Shaft Power

Trial	**Wind Speed** $\left(\frac{m}{s}\right)$	**Shaft Power (W)**
1	4	3.2816
2	5	6.4094
3	6	11.0754
4	7	17.5873
5	8	26.2527
6	9	37.3794
7	10	51.2748

As observed in Figure 8a, the experiment yielded results where the value of the shaft power is directly proportional to the value of the wind speed. This is mainly due to the dependency of the shaft power to the wind speed, where it is the product of maximum power coefficient, area (which is the product of diameter and height), and cube of wind speed. Hence, when the wind speed was increased, there was also an increase in the number of rotations of the turbine thus increasing the value of shaft power because the other variables (maximum power coefficient, diameter, and height) are held constant.

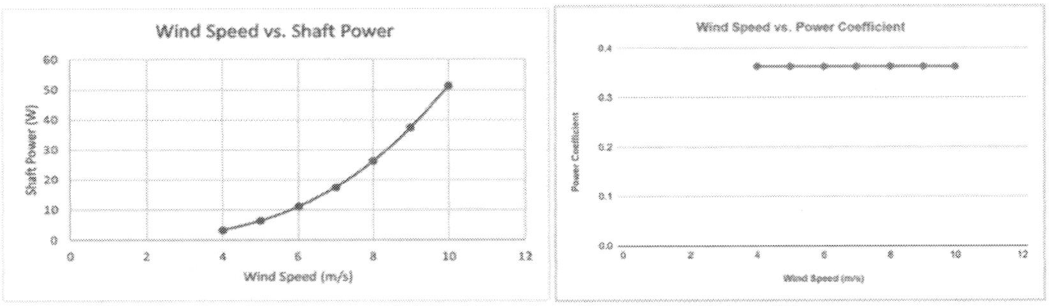

Fig. 8. Wind Speed vs Shaft Power (a) Wind Speed vs. Power Coefficient (b)

Figure 8b is a line graph that shows the relationship of wind speed and power coefficient where it can be observed that the power coefficient throughout the trials has not changed significantly. However as seen from the table, the values for the power coefficient varies for each of the trials as the wind speed increases. Among all the trials, the maximum power coefficient obtained is 0.362934. The behavior of the graph shows that the results obtained for power coefficient have minimum discrepancies.

The analysis of the turbine performance is a substantial part of the study where the researchers could assess if the S1046 H-Darrieus VAWT with mahogany as its blades is well fabricated and could achieve the specific objectives. The first performance parameter, power coefficient, is computed as the ratio of shaft power (theoretical) to the power available from the wind, using their corresponding wind speed for each trial and wind speed. This would determine which wind speed would produce maximum power.

The obtained results can be seen in Table 7 shows the computed values of power of coefficient and TSR with corresponding wind speed. The highest power coefficient was obtained is 0.362934 and the lowest value is 0.362618. Based on these values, it can be observed that the discrepancies are low hence, it could maintain a certain value for Cp as the wind speed increases. The second performance parameter, tip speed ratio, is defined as the ratio of the angular velocity multiplied by the radius of rotation to the wind speed. This parameter is significant factor in designing a wind turbine as it is related to its efficiency.

Table 7. Wind Speed, Power Coefficient and Tip Speed Ration

Trial	Wind Speed $\left(\frac{m}{s}\right)$	Power Coefficient	Tip Speed Ratio
1	4	0.362934	0.1218
2	4	0.362934	0.1362
3	5	0.362618	0.1542
4	5	0.362934	0.1555
5	6	0.362934	0.1695
6	6	0.362618	0.1722
7	7	0.362934	0.1897
8	7	0.362618	0.1679
9	8	0.362934	0.1625
10	8	0.362934	0.1650
11	9	0.362934	0.1439
12	9	0.362618	0.1470
13	10	0.362618	0.1527
14	10	0.362618	0.1591

As shown in Figure 9a, the TSR is increasing along with an increase in wind speed between the range of 4m/s to 7m/s. However, at the range of 7m/s to 10m/s, the TSR has a downward trend. Overall, the average behavior of these points on the line graph is facing downward as represented by the broken line. Moreover, the highest TSR value is 0.1897 at 7m/s wind speed. As compared to the maximum TSR that the turbine can obtain with 3 blades, the highest TSR value is comparatively low. The maximum TSR is approximately 4 based on the theoretical considerations of this study.

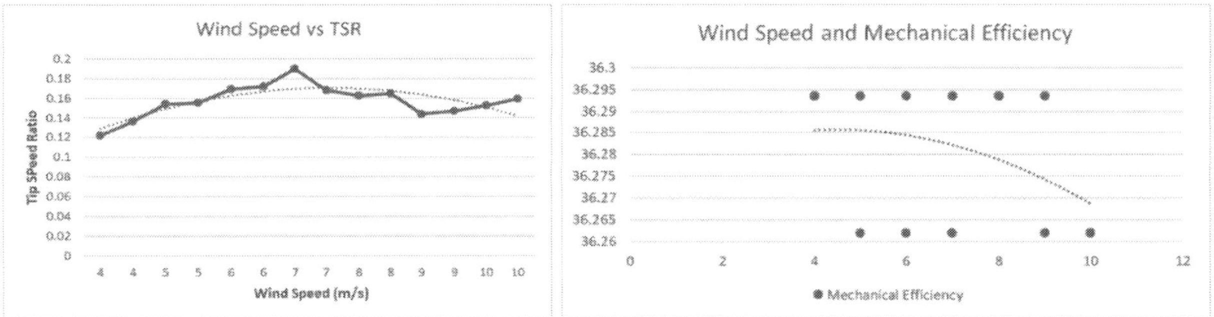

Fig. 9. Wind Speed vs. TSR (a) Wind Speed vs. Mechanical Efficiency (b)

Based on the theoretical consideration for the power coefficient, it is equivalent to the efficiency of the turbine but in terms of percentage. Hence, they have the same behavior when plotted in the graph. To know the exponential trend of the mechanical efficiency versus the wind speed, it is represented by the broken line in Figure 9b, which shows a downward behavior that is similar to the result of the tip speed ratio. The third and final performance parameter is mechanical efficiency. This is generally computed as the ratio of the power output to its power input. As for this study, the output is the shaft power (theoretical) while the input is the power of the wind. From Table 8, the highest efficiency that the fabricated wind turbine has reached is 36.29% which is acceptable in the standard of the H-Darrieus wind turbine.

Table 8. Wind Speed and Mechanical Efficiency

Trial	Wind Speed $\left(\frac{m}{s}\right)$	Mechanical Efficiency (%)
1	4	36.29345
2	4	36.29345
3	5	36.26178
4	5	36.29345
5	6	36.29345
6	6	36.26178
7	7	36.29345
8	7	36.26178
9	8	36.29345
10	8	36.29345
11	9	36.29345
12	9	36.26178
13	10	36.26178
14	10	36.26178

As the wind speed increases, efficiency decreases, which indicates a need to further improve the design of the rotor blades. Although there is a slight decrease in mechanical efficiency, the data show that the efficiencies are still precise wherein it fluctuates at a minimal discrepancy. This indicates that the fabricated wind turbine could maintain its efficiency at specific wind speeds to produce mechanical power.

4 Conclusion

Regarding the crucial component of this study, the researchers were successful in proving the viability of the construction employing mahogany as blades. It may be inferred that the wind turbine's blade design is capable of efficiently absorbing the entire airflow. Additionally, we observed that the highest TSR generated, at 0.1879, is considerably lower than the highest TSR of 4. This experimental data achieved can be used to further improve the design and fabrication of mahogany turbine blades. A low TSR results in an increase in the angle of attack which causes dynamic stalls and low-lift generation. Dynamic stalls in turbines can cause unsteady aerodynamic loads on the airfoils which can affect its rotation. Uneven and rough surfaces can cause stalls. The low TSR achieved in the experimentation may indicate that the surface finish of the blade was not smooth enough or that the S1046 blade profile was not accurately followed. As for the performance of the wind turbine, the designed wind turbine has not achieved the desired optimum performance. Hence, the performance would not be superior to the fiberglass polyester original material utilized for the design. However, mahogany can still be used as an alternative material for the blades of the manufactured wind turbine.

https://doi.org/10.1051/e3sconf/202453005004

References

1. djsolis, PHILIPPINE ENERGY PLAN 2012-2030 Update (2015)
2. A. Mahajan, P. Verma, S. Sharma, and P. Saini, Exploration of Materials Used to Design Airfoil the Vertical Axis Wind Turbine, **3**, 8 (2020)
3. A. V. Pradeep, S. V. Satya Prasad, L. V. Suryam, and P. Prasanna Kumari, A comprehensive review on contemporary materials used for blades of wind turbine, in *Materials Today: Proceedings*, Elsevier Ltd (2019)
4. T. Batu, H. Lemu, and B. Sirhabizuh, Study of the Performance of Natural Fiber Reinforced Composites for Wind Turbine Blade Applications, *Advances in Science and Technology Research Journal*, **14**, 2 (2020)
5. L. Thomas, Advanced materials for wind turbine blade-A Review (2018)
6. P. M. Kumar, K. Sivalingam, T. C. Lim, S. Ramakrishna, and H. Wei, Strategies for Enhancing the Low Wind Speed Performance of H-Darrieus Wind Turbine—Part 1, *Clean Technologies*, **1**, 1 (2019)
7. I. Hashem and M. H. Mohamed, Aerodynamic performance enhancements of H-rotor Darrieus wind turbine, *Energy*, **142** (2018)
8. D. Petković and S. Shamshirband, Soft methodology selection of wind turbine parameters to large affect wind energy conversion, *International Journal of Electrical Power & Energy Systems*, **69** (2015)
9. T. Tahzib, M. A. Hannan, Y. A. Ahmed, and I. Z. M. Kamal, Performance Analysis of H-Darrieus Wind Turbine with NACA0018 and S1046 Aerofoils: Impact of Blade Angle and TSR, *CFD Letters*, **14**, 2 (2022)

Efficient Risk Management for Distributed Clean Energy: Principal Component based Weather Derivatives

Takuji Matsumoto[1*], and *Yuji* Yamada[2]

[1] Faculty of Transdisciplinary Sciences for Innovation, Kanazawa University, Ishikawa 920-1192, Japan
[2] Faculty of Business Sciences, University of Tsukuba, Tokyo 112-0012, Japan

Abstract. As global efforts to achieve net-zero emissions intensify, the integration of renewable energy has brought to the critical need for effective volumetric risk hedging strategies, particularly at the local level. However, existing financial instruments based on total power output, such as wind power futures, fall short in local hedging. This study introduces Principal Component (PC) derivatives designed for the solar power sector, using multi-regional solar radiation as the underlying to overcome data handling complexities. In particular, by incorporating our previous concept of prediction error derivatives, we provide a unique solution to complex pricing to help manage cash flow volatility risks. In addition, we propose PC derivatives based on solar radiation residuals to hedge volumetric risks. Empirical analysis shows that our PC derivatives outperform existing wide-area derivatives in terms of hedge effectiveness, with a 20% increase over area-specific derivatives. Using as few as three or four PC derivatives can provide comprehensive coverage across different areas, enhancing market liquidity and creating an efficient transaction framework. Our results highlight the practical benefits of this approach, including the potential to reduce transaction costs by countertrading in different regions.

1 Introduction

In recent years, with the global imperative to achieve net-zero emissions, renewable energy sources such as solar power have seen rapid deployment [1][1]. This surge underscores the critical need for clean, sustainable energy solutions to combat climate change. At the same time, overcoming the risk management challenges associated with the intermittent nature of renewable energy generation, which is highly dependent on varying weather conditions, has emerged as a critical issue to ensure the reliability and effectiveness of these clean energy systems. However, few financial instruments have been traded to hedge the volumetric risk

* Corresponding author: mtakuji@staff.kanazawa-u.ac.jp

[1] According to the International Energy Agency [1], under current energy policies, renewable energy will account for 80% of new power generation capacity by 2030, with half of this (40% of the total) coming from "clear frontrunner" solar power.

of renewable energy. One of the few examples is the Wind Power Futures (WPF) listed on the European Energy Exchange (EEX) and Nasdaq, but WPF is limited in its ability to effectively hedge local areas because its underlying assets are power generation over a wide area. Recognizing this problem, Thomaidis et al. [2] performed factor decomposition based on principal component analysis (PCA [3]) on wind power generation in different areas. They then proposed a hedging contract concept where the principal risk components obtained from PCA are used as underlying assets. However, it is still an open question as to a realistic hedging scheme and pricing for PCA-based derivatives. In other words, there are still issues that need to be examined in terms of their practical applicability.

In general, PCA has been employed in the analysis of distributed renewable generation due to its effectiveness in compressing dimensionality and enhancing the understanding of spatial data features [4]. For example, Burke et al. [5] found that PCA is effective in limiting the number of random statistical variables needed to represent distributed wind power; Fonseca Junior et al. [6] proposed a highly accurate model using PCA to predict solar power generation at local sites in Japan; Christodoulou et al. [7] conducted a PC risk analysis of wind power in the context of site selection for wind power investment. As these studies show, PCA exhibits a natural affinity for the analysis of regionally distributed renewable energy generation and holds significant potential for modeling geographic time series data through dimensionality compression.

In this initial study, we extend the PCA-based wind power generation volume derivative concept of Thomaidis et al. [2] to the solar photovoltaic (PV) sector and explore its potential application in actual transactions that are both intuitive and easy to trade. To this end, we introduce PC derivatives by incorporating the design ideas of weather prediction error derivatives based on nonparametric trend estimation addressed in our previous works [8–9] to enhance transparency in pricing and product design. We propose prediction error-based PC derivatives on "solar radiation residuals" for multiple areas calculated by first removing annual periodic trends from time-series data of solar radiation in each of the multiple areas and then performing principal component analysis on the bundled data set, i.e., by making the underlying values in each area orthogonal to the time series direction in advance. This calculation allows us to estimate a loading matrix (transformation matrix to principal component series) that is commonly used throughout the year, resulting in an intuitive and robust commoditization and a practical hedging scheme.

An important point to note in proposing a hedge for solar generation risk is that we use a weather derivative [10] based on solar radiation, not solar generation, as the underlying. When power generation serves the underlying asset for derivatives, as in the existing WPF, it becomes challenging to employ it as an indicator for consensus building in the settlement process. This is due to issues such as emergency curtailment, which has seen a rapid increase both in the Kyushu region of Japan [11] and globally [12] in recent years, as well as changes in installed capacity, delayed information disclosure, and opaque measurement errors. In response to these challenges, we opt for a more commonplace and homogeneous observed data, i.e., solar radiation, as in recent related studies [13–16], as an alternative underlying asset.

The empirical analysis shows that, similar to Thomaidis [2], the use of PC derivatives significantly increases the hedging effectiveness compared to existing wide area derivatives such as WPFs. It is also shown that the use of multiple PC derivatives improves the hedging effectiveness by up to 20% compared to the use of multiple area-specific solar radiation derivatives. In particular, the use of a maximum of three to four PC derivatives can improve the overall area hedge effectiveness. Even better, the loading matrices of the four PC derivatives, which should be published by the product providers (exchanges), can also provide an intuitive interpretation (semantics of the instruments) by displaying them in a heat map on a map. PC derivatives allow for liquid and efficient trading as players in multiple regions can trade simultaneously. In addition, the design concept of prediction error derivatives makes it possible to have an expected payoff of approximately zero for the entire year, eliminating the need for premium payment procedures at the time of contracting and providing a solution to the pricing problem. In other words, only the risk of cash flow fluctuations can be effectively reduced at no additional cost, and this advantage may also be a factor encouraging the widespread use of this derivative. Furthermore, we show that PC derivatives have the additional advantage of allowing efficient countertrades between different areas, which could reduce potential transaction costs (i.e., risk premiums).

The following sections of this paper are organized as follows. First, in the next section, we introduce the specific composition method of PC derivatives and then construct the hedging models. In Section 3, we present the empirical analysis results of principal component analysis and hedging effects. Finally, in Section 4, we present our conclusions.

2 Methodology

Typical hedging problems using traditional derivatives usually entail applying the observed underlying indexes as they are. However, in our proposed PC-based weather derivatives, as we envision exchanges designing highly liquid instruments that can be traded among multiple hedgers, the process of designing effective underlying indexes is key. With this in mind, this section first presents the steps in designing PC derivatives, constructs a specific hedging model, and defines the metric for hedging effectiveness.

2.1 PC derivatives design procedure

First, we show how to design the solar radiation PC derivatives ("Rad PC derivatives") using only time series data of solar radiation.

2.1.1 Calculating the loading matrix using in-sample data

i. For the daily integrated solar radiation (N samples) observed for each area $i \in \{1, \dots, A\}$, the annual periodic trend is removed and the residual solar radiation $R^{(in)}_{[N\times A]}$ (corresponding to the deviation from the climatological normal) is calculated[2]. $R^{(in)}_{[N\times A]}$ corresponds to the in-sample payoff of the "rad area derivatives", and we also use this derivative as a benchmark for the PC derivatives in our empirical analysis.
ii. By scaling the area-specific radiation residual series by its standard deviation, the standardized (z-score normalized) radiation residual matrix $\bar{R}^{(in)}_{[N\times A]} \coloneqq R^{(in)}_{[N\times A]} D_{R\,[A\times A]}$ is calculated (note that D_R is a scale-transformed diagonal matrix with the inverse of the standard deviation σ_i of each column of $R^{(in)}_{[N\times A]}$ as the diagonal component)[3].

[2] In this study, seasonality is removed by estimating annual periodic trends for daily integrated solar radiation with a generalized additive model (GAM [17]) (see, e.g., [9, 18]), and the same can be done using published normal values.

[3] Before scale transformation, each column should be centered on the mean if there is a bias in $R^{(in)}_{[N\times A]}$. However, the solar radiation residuals used in this study are pre-centered because they are GAM residual series as mentioned in footnote 2. Although standardization is used in this study for ease of interpretation of the empirical analyses, the hedging effects shown in Section 3.3 would hardly change even without standardization.

iii. By performing PCA on $\bar{R}^{(in)}_{[N\times A]}$, a loading matrix $V_{R[A\times A]}$ and a principal component score $T^{(in)}_{R[N\times A]}$ (equivalent to the "Rad PC derivatives" payoffs in the in-sample) are obtained as follows[4].

$$T^{(in)}_{R[N\times A]} = \bar{R}^{(in)}_{[N\times A]} V_{R[A\times A]}. \tag{1}$$

2.1.2 Method for determining out-of-sample payoffs

iv. On each observation day (payment day), the annual periodic trend is subtracted from the observed solar radiation for each area to produce a column vector $R^{(out)}_{[1\times A]}$ ("Rad area derivatives" out-of-sample payoffs); a scale transformation ($\bar{R}^{(out)}_{[1\times A]} := R^{(out)}_{[1\times A]} D_{R[A\times A]}$) is performed on it by the same operation as ii to obtain the standardized series $\bar{R}^{(out)}_{[1\times A]}$.

v. The out-of-sample payoff of the Rad PC derivatives $T^{(out)}_{R[1\times A]}$ (the derivatives on the 1st - Ath principal components) is calculated by using the pre-calculated loading matrix $V_{R[A\times A]}$ as

$$T^{(out)}_{R[1\times A]} = \bar{R}^{(out)}_{[1\times A]} V_{R[A\times A]}. \tag{2}$$

2.2 Constructions of Hedging models

For the hedging model, we consider hedging the daily fluctuation risk, which is defined as the daily total electricity generation from the PV unit $V_{i,t,h}$ at time h on day t in area i: $V_{i,t} := \sum_{h=1}^{24} V_{i,t,h}$. We employ two strategies based on "area derivatives" and "PC derivatives". Assuming that $n \in \{1, \dots, A\}$ types of derivatives are tradable, the GAM-estimated hedging models are constructed as follows[5].

Model for Area derivatives:
$$V_{i,t} = Trend_{i,t} + \sum_{j=1}^{n} f_{i,R,a(j)}(Seasonal_t) Rad_{a(j),t} + \eta_{i,t}. \tag{3}$$

Model for PC derivatives:
$$V_{i,t} = Trend_{i,t} + \sum_{k=1}^{n} f_{i,PR,k}(Seasonal_t) PCrad_{k,t} + \eta_{i,t}. \tag{4}$$

where $Trend_{i,t} := \beta_i Period_t + f_i(Seasonal_t)$, $Period_t$ denotes number of elapsed days, $Seasonal_t$ $(= 1, \dots, 365(366))$ denotes annual periodic dummy variables. In other words, $Trend_{i,t}$ corresponds a discount bond payoff that depends only on deterministic date information. $Rad_{a(j),t}$ denotes the payoff of the Rad area derivative in area $a(j)$ (corresponding to the element $R_{t,a(j)}$ in the matrix introduced in Section 2.1). We assume that the Rad area derivatives are commoditized in the order of the areas with the highest installed PV capacity, and $a(j)$ denotes the area number with the j-th highest PV capacity. Also, $PCrad_{k,t}$ denotes the payoff of the Rad PC derivative in the k-th component (corresponding to the principal component score $T_{R_{t,k}}$). The $f_{i,R,a(j)}$ or $f_{i,PR,k}$, estimated as

[4] In this study, PCA is performed using the function prcomp() of the R package "stats".
[5] In this study, we use gam() in the R package "mgcv" 19 for estimation. When estimating all functions f, we apply periodic cubic splines 20 and incorporate a constraint that the starting and ending points are joined to ensure the robustness of the hedging model (as detailed in 18).

a smooth annual periodic trend, represents the contract volume of these derivatives[6]. Note that $\eta_{i,t}$ denotes the residual of a mean of 0 and provides hedging errors.

2.3 Measurement of hedge effect

The variance reduction rate (VRR) is defined as follows for hedging models, and we refer to $1 - VRR$ as hedge effect.

$$VRR_i := \mathrm{Var}[\eta_{i,t}]/\mathrm{Var}[V_{i,t} - Trend_{i,t}]. \quad (5)$$

Note that when calculating VRRs for the out-of-sample data, they are obtained by substituting the out-of-sample exogenous variables into the hedging models using the in-sample estimated β and f (the hedging error $\eta_{i,t}$ is obtained in the same way). This way, we can precisely calculate the variance reduction effect (or contribution ratio) of the introduced derivatives. In other words, our metric quantifies the additional contribution of the introduced solar radiation derivatives to the variance reduction, by assuming that the hedger has pre-hedged the volatility of the deterministic seasonal fluctuations by the discount bond $Trend_{i,t}$.

3 Result

3.1 Data description

The data and sources of information used in this study are described as follows[7]:

- Daily integrated solar radiation Rad [MJ/m^2]: Japan Meteorological Agency, https://www.data.jma.go.jp/gmd/risk/obsdl/
- Solar power generation V [MWh]: each general electric utility (e.g., Tokyo Electric Power Company: https://www.tepco.co.jp/forecast/html/area_data-j.html)[8]

For solar generation, data from the following nine areas are used for each: 1.Hokkaido; 2.Tohoku; 3.Tokyo; 4.Chubu; 5.Hokuriku; 6.Kansai; 7.Chugoku; 8.Shikoku; 9.Kyushu (see Figure 1). On the other hand, for solar radiation, data for one major city in each area is used (1.Sapporo; 2.Sendai; 3.Tokyo; 4.Nagoya; 5.Toyama; 6.Osaka; 7.Hiroshima; 8.Takamatsu; 9.Fukuoka). We estimate the PCA loading matrices and hedging models using data from the in-sample period of 2019-2021, and verify the hedging effects in the out-of-sample period of 2022.

---

[6] Whether using area derivatives or PC derivatives, if n different instruments are listed in the market, we assume that all instruments are used to hedge to maximize the hedging effect.

[7] Data from the following nine areas are used for each: 1.Hokkaido; 2.Tohoku; 3.Tokyo; 4.Chubu; 5.Hokuriku; 6.Kansai; 7.Chugoku; 8.Shikoku; 9.Kyushu. However, for solar radiation, data for one major city in each area is used.

[8] The amount of electricity generated for the entire area is divided by the generation capacity within the area (the amount of installed capacity at the end of each year downloaded from https://www.fit-portal.go.jp/PublicInfoSummary is converted to daily granularity using GAM) to calculate the amount of electricity generated by a 1 MW PV system, which is used as V.

 https://doi.org/10.1051/e3sconf/202453005005

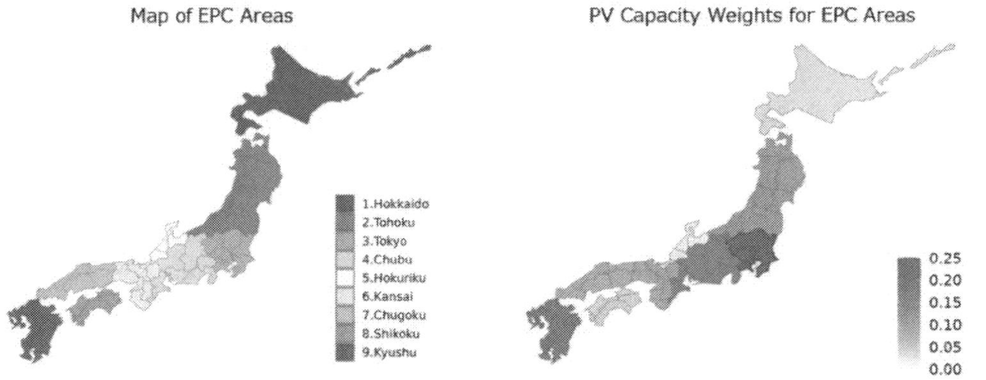

Fig. 1. Overview of Electric Power Company (EPC) areas and share of installed solar power capacity per area

3.2 Principal Component Analysis

Figure 2 shows the first four components (i.e., $V_{R_{:,(1:4)}}$ of PC1–4) of the estimated loading matrix for the Rad PC derivatives as a heat map in Japan. The principal component (PC1) has the same sign in all areas, indicating the general direction of the nationwide solar radiation. PC2 through PC4 exhibit a gradation from north to south, a mountain-like shape with the Tokyo area at the top, and stripe pattern, respectively. In such cases, it would be useful to label the loading matrices with interpretable instrument labels. That is, for example, PC1 could be referred to as a "whole area" type product, PC2 as a "north-south" type product, PC3 as a "radial" type product, PC4 as a "wave" type product, etc., to facilitate trading decisions.

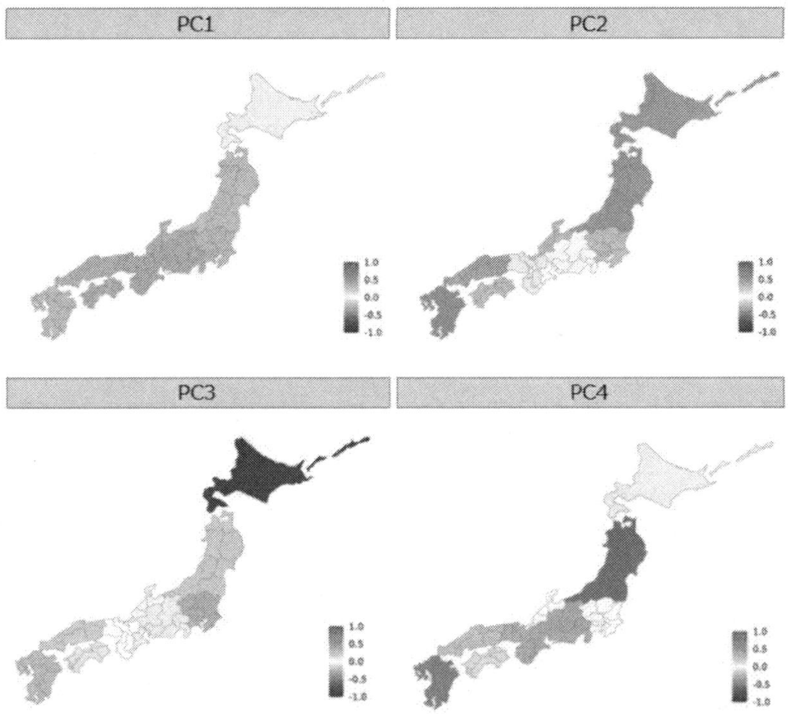

Fig. 2. Mapping of loading vectors for four principal components

If PC derivatives are allowed to be traded continuously over a long period of time, the loading matrix could be reviewed and updated, for example annually, to maintain the trader's hedging effectiveness. In such cases, it is desirable for the loading matrix not to fluctuate with the usage sample (estimation period). In other words, the robustness of the loading matrix, as well as the interpretability mentioned earlier, may influence the exchanges' decision to commoditize. Therefore, Figure 3 shows the loading matrices for all PC derivatives, including the top four components shown in Figure 2, which also represents the estimation results when the estimation period is shifted by one year along with their numerical values. Overall, the changes with different estimation periods are small from the first to the fourth components. This suggests that the estimation results of the loading matrices, i.e. the instrument specifications of the PC derivatives of the prediction errors, are robust for the upper principal components.

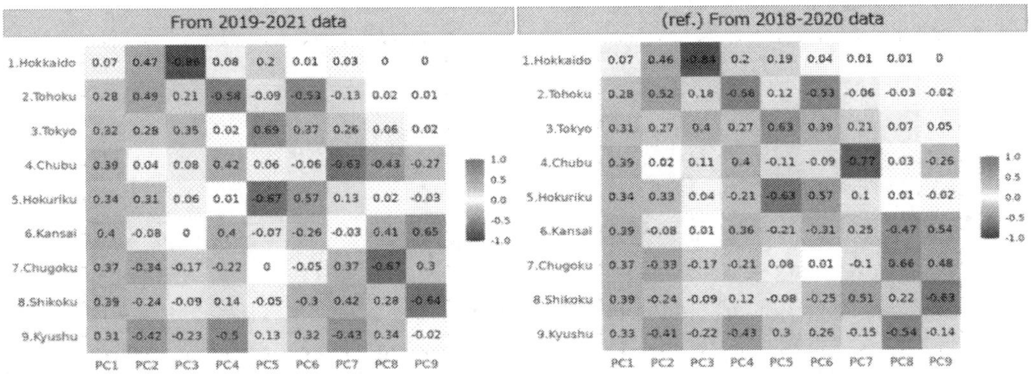

Fig. 3. Estimated loading matrices

3.3 Hedge effect

In terms of the hedging effect, Figure 4 illustrates the nine-area average (i.e. $1 - (1/A)\sum_{i=1}^{A} VRR_i$). It is worth noting a couple of remarks. First, PC derivatives are generally more effective as hedges than conventional "power futures" that use total power generation as the underlying asset (created in the same way as the existing WPF). In addition, PC derivatives using the first four components show, on average, a 20% higher effectiveness than area derivatives using the same number of instruments. Second, the out-of-sample hedging effects are generally high, suggesting that a robust hedging model can be constructed. Note that, considering the practical incentives of exchanges to commoditize area-based derivatives, we assume that instruments are listed in the order of areas with the largest amount of electricity generation (installed capacity). The order of instruments in hedging models (3) is $a = [3,9,4,2,6,7,8,1,5]$ (i.e., Tokyo, Kyushu, ..., Hokuriku, as shown in Figure 1).

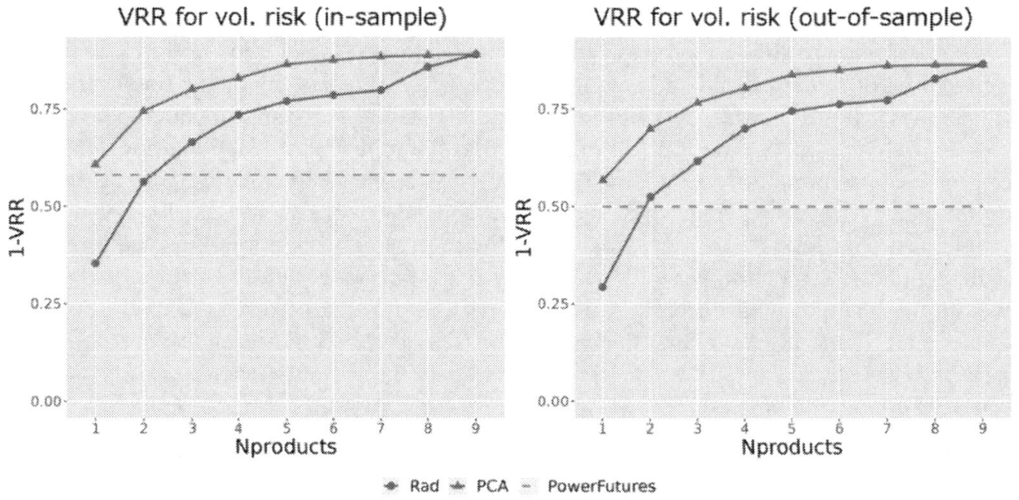

Fig. 4. Cumulative hedge effects

3.4 Contract volume trend

Figure 5 shows the seasonal trend of the contract volume of Rad PC derivatives ($f_{i,PR,k}(Seasonal_t)$ in the hedging model (4)). Despite seasonal changes, it can be observed that the magnitudes of the trading volumes and their signs are generally consistent with those of the PC loading matrices. In PC1, the seasonal changes by region are well reflected in the trading volume. For instance, the trading volume is consistently lower throughout the year in Hokkaido, where there is less solar radiation, and the trading volume is lower in winter in Tohoku and Hokuriku, where snow cover is more prevalent. The higher the order of PC, the wider the confidence interval (gray band) tends to be, and the intensity of seasonality tends to be mitigated. Notably, regions with different signs mean that efficient trades can be made between them. For example, PC2 derivatives can be effectively traded between eastern Japan (Hokkaido, Tohoku and Tokyo) and western Japan (Chugoku, Shikoku and Kyushu) to reduce each other's risk. Similarly, PC4 derivatives can be effectively traded between "Chubu or Kansai" and "Tohoku or Kyushu".

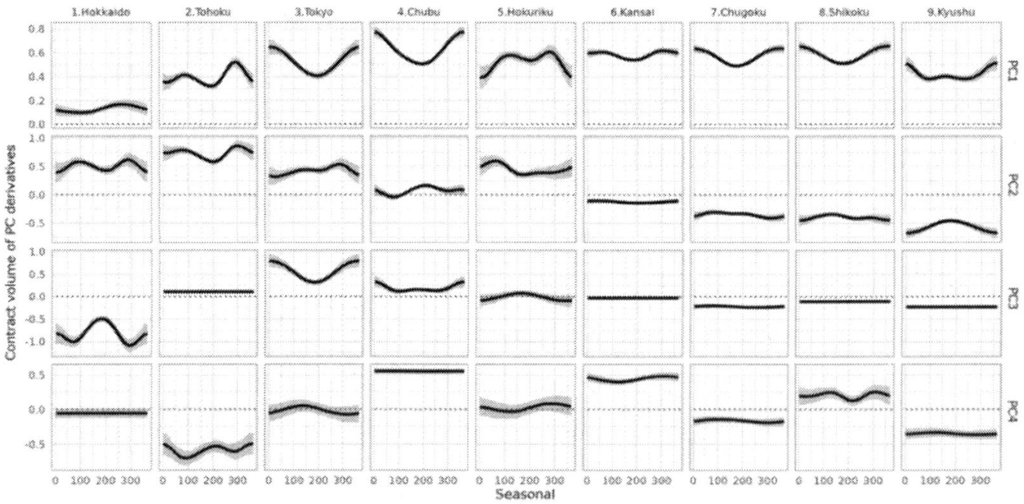

Fig. 5. Seasonal contract volume of Rad PC derivatives for each area

4 Conclusion

In this study, we extended the concept of PCA-based derivative contracts, as initially proposed by Thomaidis et al. [2], to the PV sector and investigated its applicability to actual market transactions from various perspectives. Leveraging the idea of weather prediction error derivatives introduced in our previous studies for PC derivatives, we developed an intuitive and easy-to-trade methodology for PC derivatives and improved the transparency of pricing and product design.

Our empirical results demonstrate, similar to [2], that the utilization of PC derivatives significantly enhances hedging effectiveness compared to existing wide area derivatives. It is worth noting that hedging effectiveness improved with three to four PC derivatives across the entire range by approximately 20%, relative to multiple area-specific derivatives. While PC derivatives come with the disadvantage of increased computational complexity, the benefits of commoditization are deemed to be significant, even considering this factor.

If PC derivatives are to be listed in the market, exchanges would need to specify loading matrices. Regarding this matter, we have confirmed that the loading matrix remains relatively unchanged with the estimation window for the first 3–4 components, indicating the feasibility of trading. In addition, when the loading matrices are visualized on a map, it is evident that semantic labeling of each PC derivative product up to about the fourth component (in the Japanese case, "whole area" type, "north-south" type, "radial" type, and "wave" type) is possible, helping avoid the issue of black boxing of instruments.

PC derivatives can be traded simultaneously by players in multiple areas, which facilitates liquidity. In addition, by applying the seasonal trend elimination idea of prediction error derivatives in previous studies, the loading matrix required for the derivative contract can be computed robustly and used throughout the year. Moreover, the expected payoffs are generally zero throughout the year, which also facilitates trading. It has also been shown that PC derivatives other than the first principal component can be traded by hedgers across regions in a way that they hedge each other's risk. Such trading would reduce potential transaction costs, because no additional risk premiums may be required by the sellers of derivatives. While this study has demonstrated the feasibility and utility of PC derivatives for practical PV risk management in Japan, further validation and application using data from other countries or areas is expected in the future.

This work was funded by Grants-in-Aid 20H00285, 19K22024, and 21K14374 from the Japan Society for the Promotion of Science (JSPS).

References

1. International Energy Agency, World Energy Outlook 2023 https://www.iea.org/reports/world-energy-outlook-2023/ (2023)
2. N. S. Thomaidis, T. Christodoulou, F. J. Santos-Alamillos, Appl. Energy **339**, 120925 (2023)
3. S. Wold, K. Esbensen, P. Geladi, *Chemom. Intell. Lab.* **2**, 37–52 (1987)
4. U. Demšar, P. Harris, C. Brunsdon, A. S. Fotheringham, S. McLoone, *Ann. Am. Assoc. Geogr.* **103**, 106–128 (2013)
5. D. J. Burke, M. J. O'Malley, *IEEE Trans. Power Syst.* **26**, 2084–2092 (2011)
6. J. G. D. S. Fonseca Junior, T. Oozeki, H. Ohtake, K. I. Shimose, T. Takashima, K. Ogimoto, *Renew. Energy* **68**, 403–413 (2014)
7. T. Christodoulou, N. S. Thomaidis, I. Pytharoulis, S. Kartsios, Preprints, 2023120108. https://doi.org/10.20944/preprints202312.0108.v1 (2023)

8. Y. Yamada, T. Matsumoto, *Energies* **14**, 7311 (2021)
9. T. Matsumoto, Y. Yamada, *Energy Econ.* **95**, 105101 (2021)
10. A. Alexandridis, A. D. Zapranis, *Weather derivatives: modeling and pricing weather-related risk* (2012)
11. A. Bunodiere, H. S. Lee, *Energies* **13**, 4703 (2020)
12. Y. Yasuda, L. Bird, E. M. Carlini, P. B. Eriksen, A. Estanqueiro, D. Flynn, ..., T. K. Vrana, *Renew. Sust. Energ. Rev.* **160**, 112212 (2022)
13. C. F. Boyle, J. Haas, J. D. Kern, *Renew. Energy* **164**, 1230–43 (2021)
14. S. Mosquera-López, J. M. Uribe, *Appl. Energy* **322**, 119476 (2022)
15. O. Alao, P. Cuffe, *IEEE Trans Smart Grid* **13**, 4730–46 (2022)
16. T. Matsumoto, Y. Yamada, *Asia-Pac. Financ. Mark.* **26**, 211–27 (2019)
17. T. Hastie, R. Tibshirani, *Generalized Additive Models* (1990)
18. T. Matsumoto, Y. Yamada, *Energies* **14**, 3351 (2021)
19. S. N. Wood, Package 'mgcv' https://cran.r-project.org/web/packages/mgcv/mgcv.pdf (2023)
20. S. N. Wood, *Generalized Additive Models: An Introduction with R, 2nd ed.*(2017)

https://doi.org/10.1051/e3sconf/202453005006

Hydrogen Integration for Achieving Net-Zero Emissions with Plug-in Fuel Cell Hybrid Electric Vehicles

Luis D Cortez[1*], *C V* Tapia-Bastidas[1,2], *Eduardo* Alava[1], *José* Hidalgo'Crespo[3] and *Jorge L* Amaya Rivas[1]

[1]ESPOL Polytechnic University, Campus Gustavo Galindo km 30.5 Vía Perimetral, 90211, Ecuador
[2]The University of Queensland Sciences, Division of Materials, St.Lucía, 4072, Australia
[3] Univ. Grenoble Alpes, CNRS, G-Scop, 46 Avenue Felix Viallet, Grenoble, 38000, France

Abstract. Currently, to achieve net-zero emissions, the transport sector is going through a decarbonization process, with Battery Electric Vehicles (BEVs) leading the race. However, these vehicles present a limited range and high charging time as barriers to replacing 100% of the transport sector. Also, BEVs cannot achieve net-zero emissions given that the electric rechargeable stations are powered by the local grid electricity. Thus, even though electric vehicles do not produce greenhouse gas (GHG) emissions directly; there are indirect emissions linked to the electricity used, relying on the balance between renewable and non-renewable energy sources in the local network. In this study, green Hydrogen is assessed as a possible solution to reach net-zero emissions with Fuel Cell Plug-in Hybrid Electric Vehicles (FC-PHEVs) that count with a greater range. A specific model of a cradle-to-grave life cycle is developed, allowing the assessment of its environmental impacts. Consequently, 3 scenarios are implemented to be assessed and compared using the model. The analysis shows that a considerable part of GHG emissions of Hydrogen implementation is found in its transportation, although research is being carried out on alternative solutions to mitigate this drawback.

1 Introduction

Climate change is considered one of the most critical challenges for this century. At the United Nations Climate Change Conference in 2015, an agreement was established to keep global warming below 2 °C [1]. To reach this goal, the world should decarbonize in different sectors, including energy production and transportation. Concerning the energy sector, the implementation of renewable sources plants continuously increases, aiming to reduce fossil fuel dependence [2]. Likewise, Green Hydrogen plays a key role in the energy transition due to its potential applications like electricity generation where no GHG emissions are produced [3]. On the other side, concerning the transportation sector, specifically road transportation, the scientific community, governments, and private companies have put their efforts into

* Corresponding author: ldcortez@espol.edu.ec

developing and supporting alternative fuel vehicles (like electricity, biodiesel, etc.) to reduce GHG emissions [4].
Battery Electric Vehicles (BEVs) are leading the race to decarbonize the transportation sector with the increase in their implementation over the world in the last years. However, BEVs have market disadvantages due to their relatively low autonomy range and high battery refilling time [5]. For this reason, Hydrogen has been assessed as an alternative to decarbonize the transportation sector being used as fuel for vehicles using Fuel Cells. These electrochemical devices are used to generate electricity from the chemical reaction of Hydrogen with air and only emitting water vapor as a byproduct. The vehicles that use them are called Fuel Cell Electric Vehicles (FCEVs) [6]. However, considering the current limitations in the production of Hydrogen to supply the fuel demand, it was introduced the idea of hybrid configurations for vehicles based on Hydrogen, giving as a result, the Plug-in Fuel Cell Hybrid Electric Vehicles (FC-PHEVs) [5]. In this way, Hydrogen can deliver that extended range needed for BEVs and decrease the refill time. Nevertheless, despite all the advantages of using Hydrogen as a fuel in transportation, the main barriers to its use are the currently limited Hydrogen production and the lack of refueling infrastructure [5].

2 State of the art

For the implementation of Hydrogen to achieve the goal of decarbonization, first, it is necessary to consider the Hydrogen production pathways and evaluate which one is the cleanest option. Currently, the most common way to produce Hydrogen is Steam Methane Reforming, a thermochemical process where natural gas reacts with steam at high temperatures. However, a significant amount of greenhouse gases is emitted from this process, although it could be captured up to 90% using carbon capture and storage systems, meaning that this is a low-carbon emissions production pathway. This Hydrogen has been denominated as Blue Hydrogen [3]. Another method for Hydrogen production is based on electrolysis, which is considered the cleanest Hydrogen production pathway. This process requires deionized water as feedstock and electricity. The GHG emissions linked to Hydrogen production could be avoided entirely if the electricity used is generated from renewable sources, obtaining a zero emissions Hydrogen known as Green Hydrogen [7].
Regarding renewable energy and its use to produce Hydrogen, in Europe, there are renewable energy plants (REPs) already working on generating electricity for the local grid and producing Hydrogen, as is the case of Iberdrola in Spain, which has developed a 20 MW Electrolysis plant powered by a 100 MW solar energy park with a production nominal of 3000 ton per year [8]. Also, there are cases like the Energy Park Haringvliet Zuid in the Netherlands, where a Hydrogen production unit is planned. This last case presents an interesting project of hybridizing two common renewable sources, solar and wind, clarifying the advantages that a Hybrid Power Plant (HPP) offers. The HPPs can prolong the electricity generation timing, in comparison with a standalone solar plant or a wind farm, and, at the same time, control the widespread intermittency production issue, complementing one another because there is more solar energy production during the day and in the summertime, and more wind energy production at night and in wintertime [9].
In this way, several studies have developed possible scenarios and requirements for Hydrogen implementation in the transport sector. The most common scenario is that Hydrogen is produced in a central production plant (off-site), which means it may be transported to the refueling stations [10]. This scenario needs to consider the Hydrogen transportation pathway and the storage technology which depend on the Hydrogen's physical state. For gaseous Hydrogen (GH2) transportation, it is necessary to compress it into tube trailers at 25-50 MPa in a distribution gas terminal suited next to the electrolysis plant preferably; these vessels must be mounted on trailers and transported by trucks. In the

refueling station, the GH2 must be further compressed at 70 – 90 MPa and stored in high-pressure buffer storage before dispensing [11]. Another GH2 transportation pathway is through pipeline systems, which need compressor stations and storage facilities; that makes this alternative a highly cost-effective option for high-scale hydrogen production plants [12]. One of the main barriers to the use of pipelines as a transport medium for Hydrogen is the phenomenon known as Hydrogen Embrittlement [12,13], which affects metallic materials commonly used for long-distance gas transportation, reducing their mechanical properties and degrading their load-bearing capacity when Hydrogen is present [13].

On the other hand, Liquid Hydrogen (LH2) needs to be cooled down at cryogenic temperatures, stored in cryogenic tanks at 0.4 MPa, and transported by trucks. In the refueling station, LH2 must be evaporated and stored in its gaseous state at 70 – 90 MPa. LH2 has an advantage in refueling energy usage but is inefficient in energy consumption due to the liquefaction and evaporation process. GH2 has the advantage of the current infrastructure, but in comparison with LH2, it can only be transported in low amounts [10]. Finally, vehicles based on Hydrogen, FCEVs, and FC-PHEVs have already been developed by several automotive companies. The difference between FCEVs and FC-PHEVs is the powertrain configuration. On one side, FCEVs can only be refueled by Hydrogen, which means it has a higher storage level of Hydrogen than FC-PHEVs. However, besides the Hydrogen refueling, FC-PHEVs can also be plugged-in to be charged [5]. Finally, these vehicles can be refueled in less than 5 minutes, representing a significant advantage against the BEVs (hours of charging time) [6].

3 Model of green Hydrogen implementation in the transport sector with Plug-in Fuel Cell Hybrid Electric Vehicles

This study is focused on developing a model to assess the environmental impacts of a cradle-to-grave scenario using green Hydrogen as a fuel to power Plug-in Fuel Cell Hybrid Electric Vehicles (FC-PHEVs). A life cycle assessment is suited as the best option for this type of assessment [14]. The model proposed covers the production, distribution, end of use, and end of life of green Hydrogen as an alternative to reach net-zero emissions in the transport sector. The cradle-to-grave model starts with electricity generation from renewable sources corresponding to feedstock and energy inputs in the system boundaries [15]. Depending on the circumstances, solar power plants and wind farms are considered two of the best options for obtaining the clean energy needed. These are the most common renewable energy plants due to their continuously decreasing energy technology prices.

Nevertheless, considering the advantages of Hybrid Power Plants (HPPs) presented in Section 2, and the excellent complementation between solar and wind energy across the time of the day and seasons of the year, a Wind-Solar HPP is assumed for this study, having the wind and sunlight as the feedstock to generate the electricity. Consequently, the clean electricity generated is used to supply the model's local grid demands and other processes, as shown in figure 1. The next stage is Hydrogen production, in this case via electrolysis powered by the HPP from the first stage. In this way green Hydrogen is obtained; therefore, there are zero direct emissions during its production process. The electrolysis plant to supply enough Hydrogen for the scenario requires deionized water, high-voltage electricity, and a high-scale electrolyzer. Moreover, the electrolysis plant is assumed to be located near the HPP. In addition, the Hydrogen produced in the electrolysis plant can also be used to store energy for the cases of peak electricity demands in the local grid, as is shown in figure 1. In this way, Hydrogen could help to supply that demand using Fuel Cells to produce electricity [3], although this is not part of this assessment. Before the transportation of Hydrogen to the refueling stations, Hydrogen (in its gaseous state for this study) needs to be compressed into tube-trailer vessels in the compression station next to the Electrolysis plant. These vessels are mounted on trailers and then transported by diesel trucks to the refueling stations [11,16]. The next stage introduces the idea of Refueling and Charging Hybrid Stations (RCHS), where Hydrogen would be transported, further compressed, and stored in a high-pressure storage system before being dispensed to vehicles [16]. These stations offer the service of refueling Hydrogen to vehicles and the service of charging vehicles. The operations of the RCHSs are based on a compressor to refuel Hydrogen and points of charging to supply electricity to the vehicles. The energy used in these stations for the compressor and the charging points is also taken directly from the HPP, as shown in figure 1. Finally, the vehicle used in this model is an FC-PHEV that is supposed to be charged and refueled in the RCHS. This vehicle has a smaller battery than BEVs but uses Hydrogen Fuel Cell technology to generate electricity by recombining the Hydrogen with oxygen from the air for the electric engine to obtain an extended range. The only result of this reaction is water vapor that goes through an exhaust pipe and is emitted into the atmosphere [5].

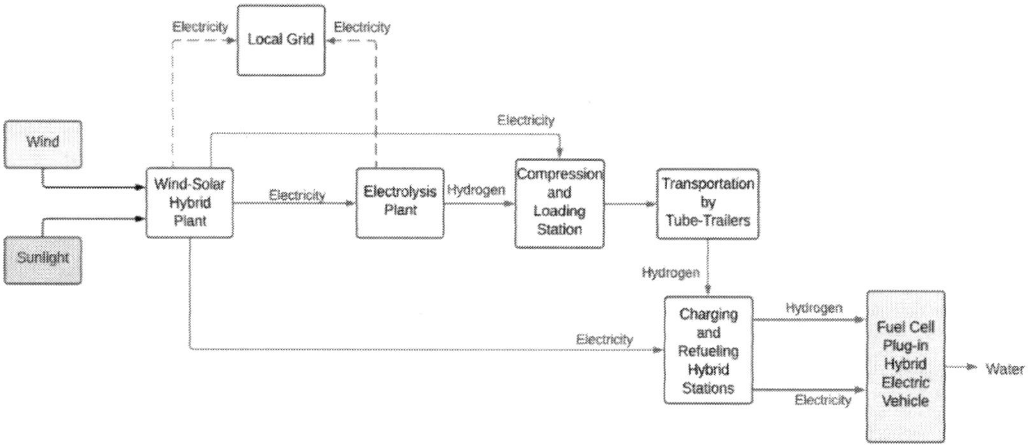

Fig 1. Model to integrate green Hydrogen to power Fuel Cell Plug-in Hybrid Vehicles

4 Case study – cradle-to-grave scenarios of using green Hydrogen to fuel Plug-in Fuel Cell Hybrid Electric Vehicles

For this study, three scenarios are created from the model shown in figure 1 changing only the renewable sources used to generate the electricity that powers the whole scenario. These scenarios assess the environmental impacts of the production and use of electricity and green Hydrogen to charge and refuel 1,000 FC-PHEVs daily. Thus, 10 “Refueling and Charging hybrid stations” with a capacity to refuel and charge 100 FCPHEVs each. For the specifications of the vehicle used for the assessment, there is a “Mid-power Plug-in Hydrogen Fuel Cell Electric” system that counts with a 10.5 kWh high-voltage lithium-ion battery and 3 Hydrogen storage tanks with a capacity of 4.4 kg at 70 MPa available on current market. Regarding electricity generation, the 60MW Energy Park Haringvliet Zuid located in the Netherlands mentioned in section 2 was chosen as a reference for the scenarios; but only due to its location and function of supplying electricity to the grid using 2 renewable sources; thus, the data for simulating the electricity generation from wind and solar energy used for the LCA is taken from the Ecoinvent database. For the stage of Hydrogen production, it is assumed a 10MW electrolysis plant with a nominal Hydrogen production of 4400 kg/d at 3 MPa of pressure; these specifications are found in commercial PEM electrolyzers. Based on the literature, 10 kg of deionized water and an electrolyzer consumption of 64.5 kWh is assumed per kg of Hydrogen produced [17]. Likewise, the Hydrogen produced would be compressed into tube-trailer vessels at 25 MPa to be transported by diesel trucks; then, 3 compressors would be needed, due to their inherent-flow capacity and the total energy consumption, assumed as 6440 kWh, based on commercial compressors. For the transportation of Hydrogen, based on literature, each truck could transport a total amount of 440 kg in 3 tube-trailer vessels [11], but in this case, to simplify the calculation, is assumed that each truck transports only 440 kg of Hydrogen. It is also assumed a round-trip route for every station with an average of 65 km distance from the Hybrid Energy Park to the Rotterdam Center, which means 20 travels of 65km per day. Finally, for the RCHSs, it is assumed that every station needs a compressor, and the calculated consumption for around 100 FC-PHEVs refueled by the compressor per day is 1104 kWh, then the total consumption for 10 compressors is 11040 kWh. The electric part has 4 charging points, and the total charging consumption for 1000 FC-PHEVs is 10500 kWh due to the battery size chosen. As mentioned, three scenarios are created and differentiated by the renewable sources used; thus, the amount of GHG emissions is expected to change in every scenario. The three scenarios developed are:

1. Wind Scenario: FC-PHEVs are refueled with Hydrogen produced using electricity from only wind energy; they are also charged with the same electricity (100% wind energy).

2. Hybrid Scenario: FC-PHEVs are refueled with Hydrogen produced using electricity from wind and solar energy and charged with the same electricity (50% wind and 50% solar energy).

3. Solar Scenario: FC-PHEVs are refueled with Hydrogen produced using electricity from only solar energy; they are also charged with the same electricity (100% solar energy).

5 Life Cycle Assement

The Life Cycle Assessment (LCA) is a method to quantify and assess environmental impacts linked to a product's life cycle from raw material extraction to the final life of the product. This method has 4 phases: Definition of goal and scope, Life cycle Inventory, Life Cycle Impact Assessment, and Interpretation [3].

5.1 Goal and Scope

This study aims to assess the feasibility of using green Hydrogen as an alternative to reach absolute net-zero emissions in the transportation sector from a GHG emissions point of view. The scenario presented in section 3, where most processes have no direct GHG emissions, tries to obtain the closest result to net-zero emissions. As in several studies of LCA about transportation, the functional unit is defined in terms of distances [17]. This study defines the functional unit as "1 km travelled by a Plug-in Fuel Cell Hybrid Electric Vehicle (FC-PHEV)". The system boundary used for this LCA is the "cradle-to-grave"; it comprises the entire life cycle of the Hydrogen (feedstock and energy inputs, hydrogen production and transportation, and end of use and end of life of Hydrogen) [15]. The scenario is developed in the Netherlands context, where is assumed that the electricity used is from the Energy Hybrid Park Haringvliet Zuid located 20 km outside Rotterdam, and the electrolysis plant is assumed close to that energy park [9]. Thus, it is also assumed that the proposed hybrid stations are in Rotterdam. This study excludes infrastructure construction, manufacturing, and decommissioning of capital goods, including the energy hybrid park, hydrogen production plant, hydrogen production devices, and refueling hybrid stations. Finally, the vehicle cycle is also excluded.

5.2 Life Cycle Inventory

Table 1 describes the specifications of each process stage used in the scenario developed with its respective sources. It is also added the data used to compare the three FC-PHEV scenarios with a BEV scenario charged with the electricity grid.

Table 1. Life Cycle Inventory.

Process stage	Specifications	Source
Solar Electricity	Electricity, high voltage {RoW}| electricity production, solar thermal parabolic trough, 50 MW	
Wind Electricity	Electricity, high voltage {NL}| electricity production, wind, >3MW turbine, onshore	
Electrolysis Plant	10 kg of deionized water and an electrolyzer consumption of 64.5 kWh per kg of Hydrogen produced. A total amount of 4400 kg of Hydrogen	[17]
Compression into tube-trailers	Commercial Hydrogen compressor at 25 MPa; 3 compressors necessary due to the flow capacity of each one; total energy consumption is 6440 kWh	[11] Adapted
Transportation by trucks	Transport, freight, lorry 3.5-7.5 metric ton, EURO3 {RER}, 4.4 tons of Hydrogen, 20 travels of 65 km	[11]
Refueling and Charging Hybrid station	*Hydrogen Compressor:* Power capacity of 110kW. Consumption for 100 FC-PHEVs refueled per day is 1104 kWh. The total consumption for 10 compressors is 11040 kWh *Electric Chargers:* The total charging consumption for 1000 FC-PHEVs is 10500 kWh	[11] From the available market FC-PHEVs
Fuel Cell Plug-in Hybrid Electric Vehicle (FC-PHEV)	10.5 kWh high-voltage lithium-ion battery and 3 hydrogen storage tanks with a capacity of 4.4 kg at 70 MPa Range of 400 km	From the available market FC-PHEVs
Battery Electric Vehicle (BEV)	BMWi with a 50kWh battery and 210 km of range	[4]
Local electricity grid	Electricity, high voltage {NL} production mix	

5.3 Life Cycle Impact Assessment

In the present study, the selected impact category to be assessed is the Global Warming Potential (GWP). The IPCC 2013 100a impact assessment method was used in a LCA software. This method can calculate and evaluate the GHG emissions in kg CO2-eq emitted by the entire scenario developed; the results are in gCO2-eq emitted per km travelled by the FC-PHEVs. For better understanding, the 3 scenarios developed in the case study were simulated. Finally, the charging of BEVs using the Netherlands grid electricity was also simulated to obtain actual results; thus, it could be compared with the FC-PHEV scenarios.

6 Results and discussion

The Carbon emissions in gCO2-eq / vehicle km of the three scenarios determined are presented in figure 2. The wind scenario has the least Life Cycle Carbon Footprint scenario with 7.93 g CO2/km, where almost all emissions are from transportation. On the other side, the Solar Scenario has the maximum contribution with 30.9 g CO2/km, mainly due to the electricity consumption of the electrolysis plant. Moreover, as expected, the hybrid scenario with 19 g CO2/km is in the middle of the range between wind and solar scenarios. Life Cycle Carbon Footprint of charging BEVs with the Netherlands grid electricity (147.24 g CO2/km) is also shown and compared with the FC-PHEV scenarios. The high difference in the results between the BEV scenario and the rest of the scenarios has a twofold explanation.

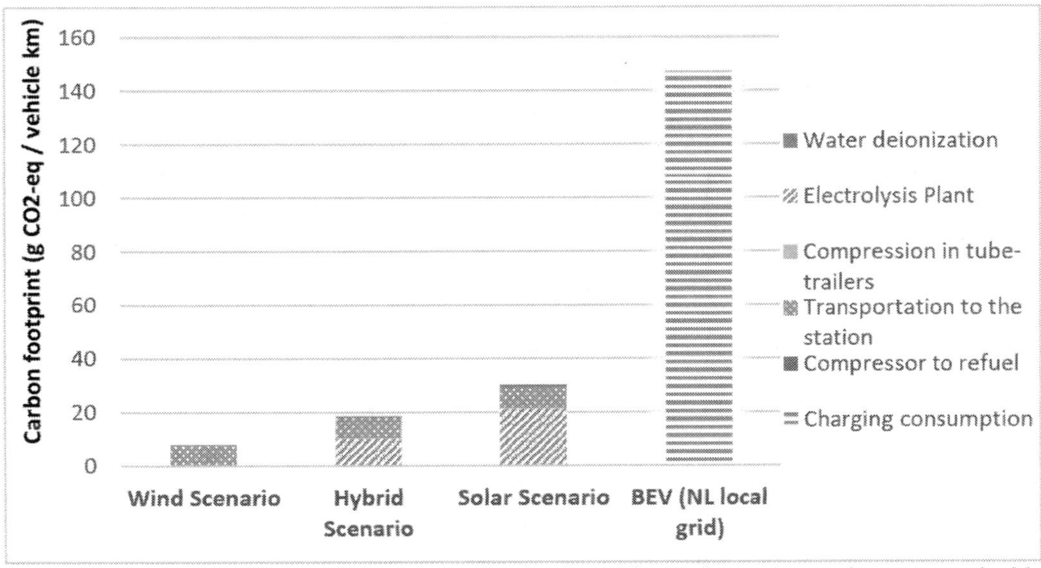

Figure 2. The carbon footprint of the 3 scenarios developed for Hydrogen integration compared with Battery Electric Vehicles charged with local grid scenario.

First, the BEVs are charged with a country grid electricity mix rather than using renewable energies like in the FC-PHEV scenarios, and as is known, the grid electricity generation of a country involves renewable and non-renewable energy sources. In this way and adding that no emissions are produced when BEVs are being driven, considerable BEV emissions are linked to the percentage of non-renewable energy in the local grid used to charge them, and this value changes in every country. However, it must be mentioned that whether BEVs were charged with electricity produced from renewable sources, their emissions would be lower than FC-PHEVs [17]. The second reason is the range of BEVs. The units of the Life Cycle Carbon Footprint for this assessment are "g CO2/km", which means that the total amount of CO2 emitted to charge the battery of the BEV has to be divided by the range of the vehicle, and based on the literature, one of the problems of using BEVs is the low range due to the battery size [4]; then, the low range of BEVs increases the magnitude of Carbon intensity per kilometer travelled. Moreover, BEVs would need to increase the battery size to obtain the same range of FC-PHEVs, which could worsen the charging time and vehicle mass [5]. In summary, having a lower range than FC-PHEVs (in this case almost half) and being charged by the selected grid electricity, BEVs produce 4.76 times the solar scenario emissions, 7.74 times the hybrid emissions, and 18.55 times the wind scenario, not considering the manufacturing of powertrains. Concerning the comparison between FC-PHEVs scenarios, Table 2 shows the specific values of carbon intensity per km travelled for each FC-PHEVs scenario detailed. Assessing these results and keeping in mind the advantages of HPPs presented in Section 2, the Hybrid scenario can be considered the best clean-efficiency option. Even though the wind scenario is a less polluting alternative (11.04 g CO2/km less than the Hybrid scenario), the HPP can increase the electricity generation hours energy compared with a solar plant or a wind farm. The advantage of this scenario is how solar and wind energy can complement one other in a HPP. There is more solar energy production during the day and summertime and more wind energy production at night and wintertime [9]. This characteristic of the HPPs ensures the production of enough energy to supply the local grid demand, hydrogen production, and the intermittency presented regularly in renewable energy plants.

Table 2. Global Warming Potential of the 3 scenarios developed for Hydrogen life cycle as fuel to FC-PHEVs.

	Wind Scenario ($g\ CO_2\ eq/km$)	**Hybrid Scenario ($g\ CO_2\ eq/km$)**	**Solar Scenario ($g\ CO_2\ eq/km$)**
Water deionization	0.110	0.110	0.110
Electrolysis Plant	0.085	10.200	21.130
Compression in tube-trailers	0.002	0.210	0.460
Transportation to the station	7.730	7.730	7.730
Compressor in the station	0.003	0.360	0.750
Charging points	0.003	0.360	0.680
GWP	**7.932**	**18.970**	**30.860**

On the other side, since in neither of the three scenarios, the methodology of transporting Hydrogen in the gaseous state to the refueling stations was changed, the emissions had to be the same for each, and it can be observed in figure 1. The transportation of Hydrogen represents an important amount of GHG emissions for each scenario, in the wind scenario is 97% of the total GWP, 41% for the hybrid scenario, and 25% in the solar scenario and that could be explained because it was assumed that Hydrogen is transported in the gaseous state by diesel trucks, which generate emissions due to the engine combustion. The generated emissions are directly related to the distance and the weight of Hydrogen transported, which is why the calculations show the carbon footprint impact of most of the transportation process expressed in "tkm" units (ton-kilometers). Additionally, as explained in Section 2, the two common ways to transport Hydrogen are defined by its state [11], which determines the amount of Hydrogen that can be transported for each truck. Liquid Hydrogen can be transported in a higher volumetric capacity than Gaseous Hydrogen [10], which can change the transportation methodology, the magnitude of tkm, and consequently, the GHG emissions. Liquid Hydrogen implementation for this scenario can be assessed in another study and compared with the results of this study. Several automotive companies are working on reducing Hydrogen transportation emissions by developing Fuel Cell drivetrains for trucks and trailers [17]. In this way, fuel cell trucks can be refueled in the electrolysis plant as well as in the refueling stations creating a natural synergy between the electrolysis plant, the refueling stations, and hydrogen transportation [17]. The other alternative to reduce hydrogen transportation emissions is implementing a pipeline system due to its advantages described in Section 2. However, first, it is necessary to solve the already-mentioned phenomenon known as Hydrogen Embrittlement.

7 Conclusions

FC-PHEV scenarios developed in this study were compared with BEVs charged by the Netherlands grid electricity. The results showed that FC-PHEV scenarios have lower Carbon emissions per km travelled because they are powered by renewable energy, and BEV scenarios with a mix of renewable and non-renewable energy. Moreover, even though BEVs could be less polluting if they were charged with electricity from renewable energy sources, the emissions also depend on the range of the vehicle, and FC-PHEVs have a higher range than BEVs; thus, FC-PHEVs could be considered a real good alternative to decarbonize transportation. Concerning the comparison between the FC-PHEV scenarios, despite the Wind scenario being the less polluting alternative, the Hybrid scenario presents a clean-efficiency option even though the 11.04 g CO2-eq/km of difference between them, and that is due to the advantages of the Hybrid Energy Park, they present a significant increase in energy production because of the solar and wind energy complementation in the times of the day or the seasons of the year, keeping a low life cycle carbon footprint, not too far from the wind scenario. However, it is necessary for a deep cost-effective assessment. Finally, in the process of reaching net-zero emissions in the transportation sector with the implementation of Hydrogen as a fuel, its distribution is a key point that can be improved concerning its environmental impacts.

References

1. United Nations Framework Convention on Climate Change. Paris Agreement. (2015)
2. Elisa Papadis, George Tsatsaronis, Challenges in the decarbonization of the energy sector, Energy, Volume 205, 118025, ISSN 0360-5442 (2020)
3. Osman, A.I., Mehta, N., Elgarahy, A.M. et al. Hydrogen production, storage, utilisation and environmental impacts: a review. Environ Chem Lett 20, 153–188 (2022)
4. Dario Pevec, Jurica Babic, Arthur Carvalho, Yashar Ghiassi-Farrokhfal, Wolfgang Ketter, Vedran Podobnik, A survey-based assessment of how existing and potential electric vehicle owners perceive range anxiety, Journal of Cleaner Production, Volume 276, 122779 (2020)
5. Roberto Álvarez Fernández, O. Pérez-Dávila, Fuel cell hybrid vehicles and their role in the decarbonisation of road transport, Journal of Cleaner Production, Volume 342, 130902, ISSN 0959-6526 (2022)
6. Oriana Perez-Dávila, Roberto Álvarez Fernández, Optimization algorithm applied to extended range fuel cell hybrid vehicles. Contribution to road transport decarbonization, Energy, Volume 267, 126519, ISSN 0360-5442 (2023)
7. Bhandari, Ramchandra; Trudewind, Clemens A.; Zapp, Petra. Life cycle assessment of hydrogen production via electrolysis–a review. Journal of cleaner production, vol. 85, p. 151-163, (2014)
8. Arcílio B.S. Semente, Chapter 19 - Prospects and challenges for the green hydrogen market, Editor(s): Rohit Srivastava, Jayeeta Chattopadhyay, Diogo M.F. Santos Solar-Driven Green Hydrogen Generation and Storage, Elsevier, Pages 381-415 (2023)
9. Vazquez Pombo D, Thanopoulos S, Funkquist J et al. The First Utility Scale Hybrid Plant in Europe: The Case of Haringvliet. In Proceedings of 5th International Hybrid Power Systems Workshop. Energynautics GmbH. (2021)
10. M. Genovese, P. Fragiacomo, Hydrogen refueling station: Overview of the technological status and research enhancement, Journal of Energy Storage, Volume 61, 106758, ISSN 2352-152X (2023)
11. Amgad Elgowainy, Krishna Reddi, Erika Sutherland, Fred Joseck, Tubetrailer consolidation strategy for reducing hydrogen refueling station costs, International Journal of Hydrogen Energy, Volume 39, Issue 35, ISSN 0360-3199 (2014)
12. Canada government. Life cycle assessment of hydrogen production pathways in Canada (2022)
13. Anthony Velazquez Abad, Paul E. Dodds, Green hydrogen characterisation initiatives: Definitions, standards, guarantees of origin, and challenges, Energy Policy, Volume 138, 111300, ISSN 0301- 4215 (2020)
14. Zheng Tian, Hong Lv, Wei Zhou, Cunman Zhang, Pengfei He, Review on equipment configuration and operation process optimization of hydrogen refueling station,

International Journal of Hydrogen Energy, Volume 47, Issue 5, Pages 3033-3053, ISSN 0360-3199 (2022)

15. Parikhit Sinha, Brenor Brophy, Life cycle assessment of renewable Hydrogen for fuel cell passenger vehicles in California, Sustainable Energy Technologies and Assessments, Volume 45, ISSN 2213- 1388 (2021)
16. Omar Faye, Jerzy Szpunar, Ubong Eduok, A critical review on the current technologies for the generation, storage, and transportation of Hydrogen, International Journal of Hydrogen Energy, Volume 47, Issue 29, Pages 13771-13802, ISSN 0360-3199 (2022)
17. Sun, B., Wang, D., Lu, X. et al. Current Challenges and Opportunities Toward Understanding Hydrogen Embrittlement Mechanisms in Advanced High-Strength Steels: A Review. Acta Metall. Sin. (Engl. Lett.) 34, 741– 754 (2021)

Effects of Ammonia (NH_3) on Flame Characteristics of Methane (CH_4) External Premixed Combustion

Willyanto Anggono[1,2*], *Ivan* Christian Hernando[1], *Oegik* Soegihardjo[1], *Chandra* Waskito[1], and *Ferdinand* Ronaldo[2]

[1] Mechanical Engineering, Petra Christian University, Surabaya, Indonesia
[2] Centre for Sustainable Energy Studies, Petra Christian University, Surabaya, Indonesia

Abstract. Methane has long been proposed as an alternative to heavy hydrocarbon fuels due to its low carbon content. However, considering the effect of climate change and the need to transition to zero-carbon, ammonia has recently gained popularity as a sustainable fuel. Nevertheless, previous studies have highlighted the poor combustion of ammonia and the need to blend it with other fuels like methane. As such, this study looks for parameters from the results of the fire of external combustion of premixed methane-ammonia gas. The study was conducted by burning stoichiometry first and then using a volume ratio as the content of the methane-ammonia mixture (ammonia concentration used 0-50%). The gas that comes out of the gas cylinder was adjusted in pressure and flow before mixing and then flowed into the burner to be combusted. The flame propagating from the burner was recorded in a high-speed camera and was used to define the combustion characteristics such as the flame height, flame angle, and dimensionless flame height. The higher concentration of ammonia in the fuel mixture was found to shorten the flame height and dimensionless flame height, while the flame angle became wider.

1. Introduction

The large use of fossil fuels in meeting daily needs has led to the depletion of fossil fuels and causes air pollution which is damaging to nature. It encourages more combustion studies to focus on environmentally friendly alternative fuels. The purpose of this study is to use ammonia gas mixed with methane gas to become a fuel [1-5]. One such fuel is biogas, which is an alternative renewable fuel source from animal wastes and is comprised of mostly methane (35% - 65%). Biogas also contains various contaminants such as carbon dioxide and nitrogen, followed by trace amounts of hydrogen gas, oxygen, and H_2S. Methane is an organic component that is easily combustible, while carbon dioxide and nitrogen are considered inert gases, meaning they won't play a role in combustion. Furthermore, the heat value of biogases depends on the amount of methane present in the biogas. [6-8]. Ammonia (NH_3) is another alternative energy source to replace fossil fuels. They can be easily produced

* Corresponding author: willy@petra.ac.id

and it does not have any carbon content which leads to less pollution [9]. However, some study has noted the poor combustion characteristics of ammonia, such as low laminar flame speed, high ignition temperature, and NO_x by-products [10-12].

A previous study was done by Xia et al. The study is about the co-combustion of coal/ammonia, in which they investigated the underlying mechanism of the difference in flame propagation speeds between co-combustion and ammonia-only combustion under different ammonia compositions [12]. This study used a burner to study the laminar flame of ammonia/N_2/O_2 in the presence of extra air. This study showed that the laminar burning velocity of ammonia got as high as 36.1 cm/s when the oxygen concentration was at 35% with the stoichiometric ratio at 1:1 [13]. A study conducted by Vinod et al. uses an ammonia gas combined with gasoline and methane. This study uses a single-cylinder engine with a carburetor [14]. A numerical study about the combination of ammonia, hydrogen, and methane fuels was done by Bayramoğlu et al. The study shows that the results can reduce combustion emissions [15].

The purpose of this study is to analyze the effect of NH_3 blended with methane, by burning the stoichiometric mixture of CH_4 and NH_3 (0 to 50% of fuel), and oxygen on the burner with a nozzle diameter of 5 mm. The burner is then connected to a tube from the oxygen tank and another tube from the fuel tank. A regulator and flowmeter are attached to each of the gas tanks to measure the flow rate of gases to the burner. A valve is used to control the flow rate of the gases. Then, the flame propagation generated from the combustion will be recorded. The recording will use a high-speed camera. The data from a high-speed camera will be processed through a computer.

2. Experimental method

The experimental variable was the concentration of NH_3 in the fuel (CH_4 – NH_3), in which the mixture of ammonia was tested at 0% up to 50% in increments of 10%. The experimental schematic is shown in Figure 1.

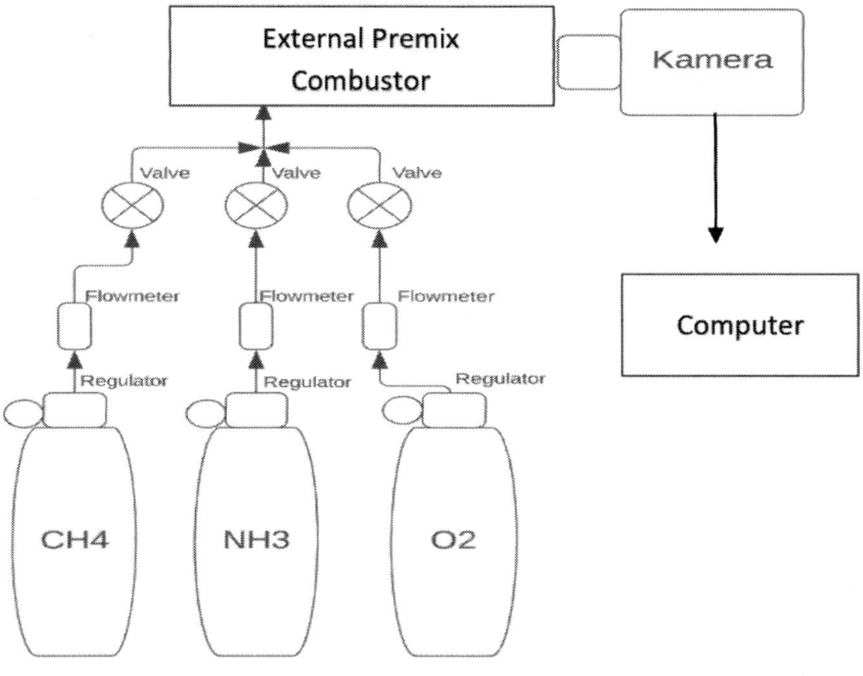

Fig. 1. Experimental schematic.

 https://doi.org/10.1051/e3sconf/202453005007

The purpose of this experiment is to assess the effects of ammonia as an additive in the external combustion of premixed biogas. The experiment is carried out by burning the stoichiometric mixture of CH_4-NH_3 (mixture ranging from 0-50%) and oxygen in which the gases were passed through the nozzle with a diameter of 5 mm. The tube from the oxygen tank, biogas tank, and ammonia gas tank will be connected to the burner. There are three regulators and three flowmeters to monitor the flow rate of the gas in each tank. Valves were used to control the flow rate of the gases.

In each of the trials, every mixture of fuel and oxygen flowed simultaneously to the burner which they will be sparked by the spark plug after they flowed for 10 seconds. Every flowmeter was checked in order to ensure that the volumetric flow rate of the gases corresponds to the stoichiometric volumetric flow rate. After that, high-speed cameras were used to take a picture of the flames from the combustion. The procedures were repeated again for different nitrogen concentrations. The experiment was carried atmospheric pressure and room temperature (20^{0}C).

After performing the experiments, the flame recording was post-processed in the computer to determine the flame angle, flame height, and dimensionless flame height. In calculating the dimensionless flame height, the measured flame height was normalized with the nozzle diameter of 5 mm.

3. Results

In studies of the effects of ammonia addition on the external combustion characteristics of methane and oxygen, the results that can be seen from experiments are flame angle, flame height, and dimensional flame height marks of 0% to 50% combustion of ammonia addition to the mixture of methane and oxygen fuel. The image of the experimental flame can be seen in Figure 2.

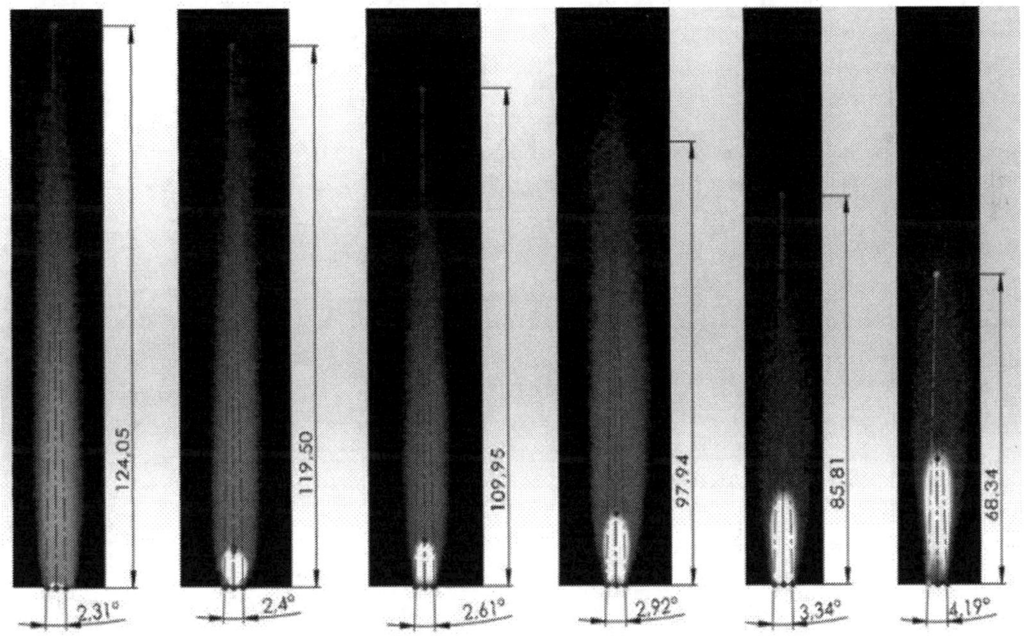

0% 10% 20% 30% 40% 50%

Fig. 2. Images of the combustion flames with different concentrations of ammonia (0-50% with 10% increments from left to right, respectively).

 https://doi.org/10.1051/e3sconf/202453005007

From the experimental results (Table 1), it can be concluded that the higher concentration of ammonia in the fuel mixture will make the flame height and dimensionless flame height shorter. Furthermore, the flame angle will also be bigger as the amount of NH_3 in the fuel mixture increases. These results can be seen in Figure 3 which shows the plot of the flame angle with different concentrations of NH_3 in the mixture from 0% to 50%.

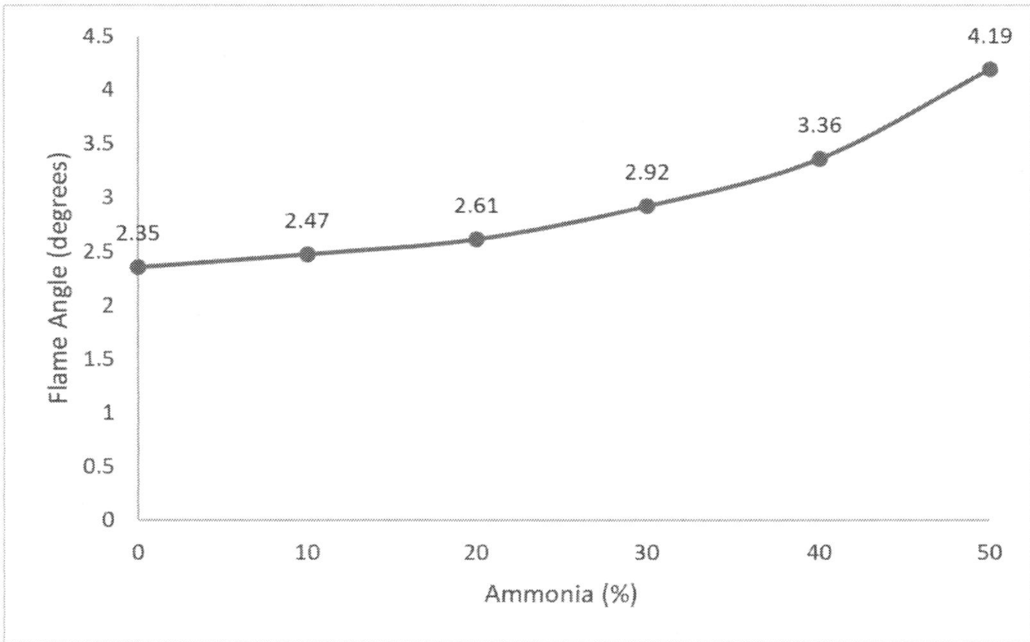

Fig. 3. The flame angle at various concentration of NH_3 (0-50%).

The flame angle graph in Figure 3 tends to increase with increasing ammonia gas concentration, the largest difference appears in ammonia concentration 40%-50% with the largest difference of 0.83 degrees, followed by concentrations 0-10%, 10-20%, 20-30%, 30-40% 40-50% with results of 0.12 degrees, 0.14 degrees, 0.31 degrees, 0.44 degrees, 0.83 degrees. The smallest difference appears (0.12 degrees) between 0-10% ammonia concentrations.

Table 1. Combustion results table

Concentration		*Flowrate*			*Flame*		*Dimensionless flame height* (H/d)
NH_3 (%)	CH_4 (%)	NH_3 (L/min)	CH_4 (L/min)	O_2 (L/min)	*angle* (°)	*Height* (mm)	
0	100	0	0.5	1	2.31	124.05	24.81
10	90	0.05	0.45	0.9375	2.4	119.50	23.9
20	80	0.1	0.4	0.875	2.61	109.95	21.99
30	70	0.15	0.35	0.8125	2.92	97.94	19.086
40	60	0.2	0.3	0.75	3.34	85.71	17.142
50	50	0.25	0.25	0.6875	4.19	68.34	13.668

The trend of the flame height graph in Figure 4 decreases with increasing ammonia concentration, the biggest difference appears in ammonia concentration of 40%-50% The largest difference is 17.37 mm, followed by concentrations 0-10%, 10-20%, 20-30%, 30-40% 40-50% with results of 4.55 mm, 9.55 mm, 14.52 mm, 9.72 mm, 17.37 mm. The smallest difference (4.55 mm) arises between the concentration of 0-10% ammonia concentrations.

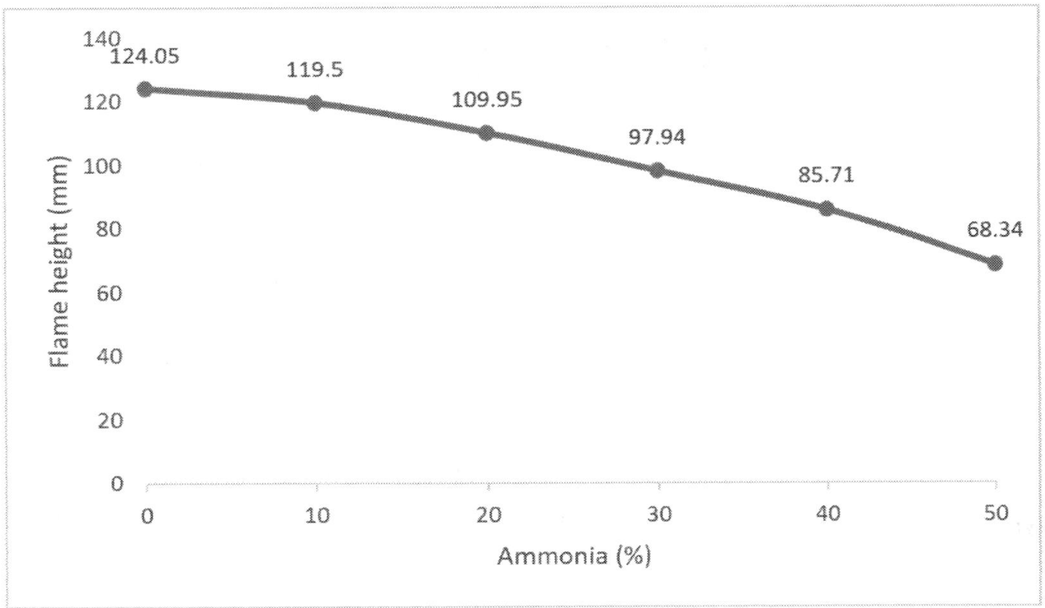

Fig. 4. Flame height plotted against NH_3 concentration (0-50%).

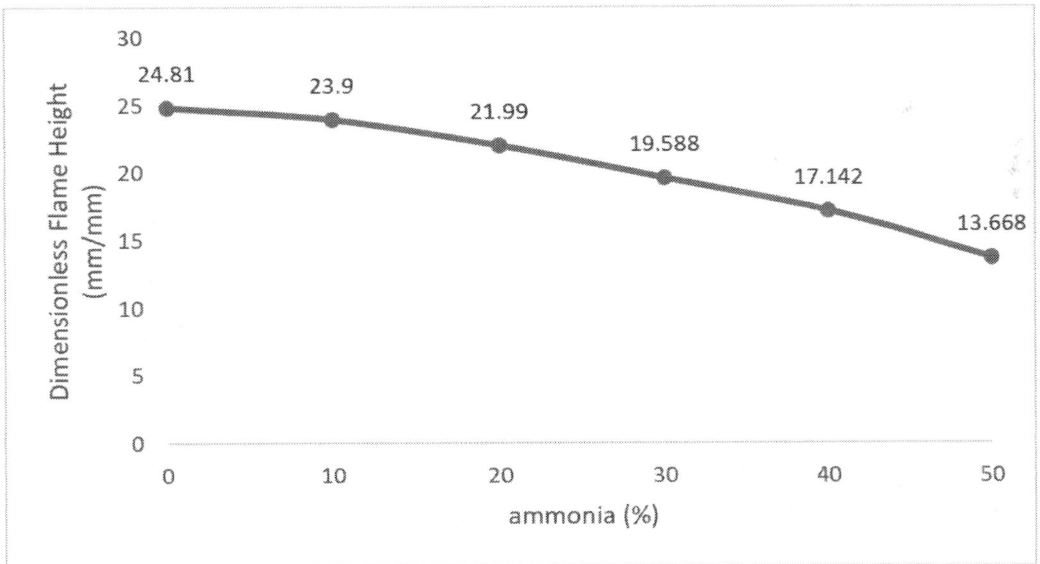

Fig. 5. Dimensionless flame height plotted against NH_3 concentration (0-50%).

The dimensionless flame height value was obtained by using the equation (H/d) that is shown in Table 1. H refers to the flame height and d refers to the nozzle diameter. The graph of dimensionless flame height in Figure 5 tends to decrease with increasing ammonia concentration, the largest difference appears in ammonia concentration 40%-50% difference of 3.474 mm, followed by concentrations 0-10%, 10-20%, 20-30%, 30-40% 40-50% with results of 0.91 mm, 1.91 mm, 2.904 mm, 1.944 mm, 3.474 mm. The smallest difference (0.91 mm) arises between the concentration of 0-10% ammonia concentrations.

The trends observed in this study can be associated with the different combustion properties of ammonia and methane. Under similar conditions, ammonia has a lower laminar flame speed than methane. In addition, ammonia also has less lower heating value [11]. The partial replacement of methane by ammonia resulted in overall less fuel energy content, and therefore less height in the flame propagation, and wider flame angle.

4. Conclusion

The addition of ammonia gas in combustion affects methane's external combustion flame characteristics were experimentally investigated. A greater concentration of ammonia results in shorter flame propagation, as shown in the flame height results when the 0% ammonia gas content produced is 124.05 mm, while the flame height when the 50% ammonia gas content reaches 68.34 mm. Accordingly, the dimensionless flame height has the same trend as the flame height. On the other hand, the greater concentration of ammonia causes the flame angle to be larger as seen from the combustion results when 0% ammonia content produces a flame angle of 2.45 degrees while ammonia content of 50% produces a flame angle of 4.19 degrees.

The author appreciates the financial support provided by Petra Christian University with grant number 10/HB-PENELITIAN/LPPM-UKP/XI/2022 for this study.

References

1. Anggono, W. et al. 2012 *Journal of Applied Sciences Research* vol 8, p 4126-4132
2. Anggono, W. et al. 2013 *Journal of Physics Conference Series* vol 423 p 1-7
3. Anggono, W. et al. 2013 *Applied Mechanics and Materials* vol 376, p 79-85
4. Kobayashi, H., Hagiwara, H., Kaneko, H. & Ogami, Y 2007 *Proceedings of the Combustion Institute* vol 31 p 1451–1458
5. Kobayashi, H., Seyama, K., Hagiwara, H. & Ogami, Y. 2005 *Proceedings of the Combustion Institute* vol 30 p 827–834
6. Anggono, W., Suprianto, F. D., Wijaya, T. P. & Tanoto, M. S. 2014 *Advanced Materials Research* 1044-1045 p 251-254
7. Anggono, W., Wardana, I. N. G., Lawes, M. & Hughes, K. J. 2014 *International Journal of Engineering and Technology* vol 5 p 4980-4987.
8. Jun. L, Shini. L, Danan. C, Rongjun. W, Noriyuki. K, Lisheng. D, Honyu. H 2021 *Forntier in Energy Research*
9. Guo, B. et al. 2021 *Automotive Experinces* vol 4 p 161-170
10. M. Koike, H. Miyagawa, T. Suzuoki, and K. Ogasawara 2016 *Journal of the Combustion Society of Japan* vol. 58 p 99-106
11. Kobayashi, H., Hayakawa, A., Somarathne, K.D. Kunkuma A., Okafor, E. C. 2019 *Proceedings of the Combustion Institute* vol 37 p 109-133
12. Y. Xia, K. Hadi, G. Hashimoto, N. Hashimoto, and O. Fujita 2020 *The Proceedings of the Thermal Engineering Conference* vol 0087

13. H. Takeishi, J. Hayashi, S. Kono, W. Arita, K. Iino, and F. Akamatsu 2015 *Transactions of the Japan* Society *of Mechanical Engineers, Series B* vol 81
14. Vinod, K. Nonavinakere, Gore, M., Liu, H., Fang, T. 2023 *Applications in Energy* and *Combustion Science* vol 16
15. Bayramoğlu, K., Bahlekeh, A., Masera, K. 2023 *International Journal of Hydrogen Energy*

E3S Web of Conferences **530**, 05008 (2024) https://doi.org/10.1051/e3sconf/202453005008
ICFEE 2024

Enhancing Building Energy and Thermal Performance in Metro Manila, Philippines through Computational Fluid Dynamics-Assisted Analysis and Optimization

Sherwin U Mina[1*] and *Reylina* Garcia Tayactac [1,2]

[1] School of Graduate Studies, Mapúa University, 658 Muralla St., Intramuros Manila, 1002 Philippines
[2] School of Mechanical, Manufacturing, and Energy Engineering

Abstract - This study presents a detailed investigation of the energy and thermal performance of buildings in the Philippines' metropolitan cities, targeting sustainability and indoor comfort against the backdrop of the country's tropical maritime climate. Utilizing a blend of occupant surveys, energy audits, climate data analysis, and building envelope assessments, the research employs Computational Fluid Dynamics (CFD) simulations to accurately evaluate energy use and thermal behaviors in residential and commercial structures. Key findings include distinct nighttime energy consumption patterns in residential buildings and the identification of climatic challenges such as high temperatures and humidity. The study proposes effective design interventions like enhanced ventilation, shading solutions, and the integration of Phase Change Materials (PCMs) for improved thermal efficiency. Significantly, the research contributes to policy development by suggesting updates in building energy codes and advocating for energy-efficient practices. Overall, the study provides actionable insights for enhancing building sustainability and occupant comfort in urban Philippine settings.

1 Introduction

The urgency of the global climate crisis, marked by escalating greenhouse gas emissions and their significant environmental impacts, calls for an in-depth examination of energy efficiency in various sectors[1]. This is particularly crucial in the built environment, where the reduction of emissions plays a critical role in mitigating climate change effects. This research focuses on the Philippines, a region undergoing rapid urbanization and facing unique challenges due to its tropical and maritime climate. The need for specific strategies to manage energy consumption and ensure occupant comfort in such climates is increasingly crucial in the context of global climate change.

* Corresponding author: sherwinmina06@gmail.com

According to the International Energy Agency (IEA), the world's total energy-related carbon emissions have risen sharply to a staggering 36.8 Gt of carbon emission to the environment by the year 2022, making the improvement of energy efficiency in buildings an imperative goal [2]. In this light, our study aims to assess and optimize the energy and thermal performance of buildings in Philippine metro cities, employing methodologies such as energy audits, building performance simulations, and the integration of renewable energy sources. The goal is to not only enhance occupant comfort and promote sustainability but also to establish guidelines for the energy-efficient renovation of existing residential buildings [3].

The backdrop of this study is the growing demand for architecture that can adapt to changing climatic conditions. The role of Computational Fluid Dynamics (CFD) emerges as a critical tool in this context, distinguished from other approaches by its efficiency in simulation across various climatic conditions. By integrating CFD with complementary methodologies, the research aims to provide a comprehensive understanding of the energy and thermal performance of buildings, contributing to sustainable, energy-efficient, and occupant-friendly building practices in the Philippines [4].

The study addresses specific issues that need to be tackled in the Philippine urban context:

- The lack of comprehensive data on building energy consumption patterns in metro cities, hindering effective energy management.
- Inadequate understanding of the impact of insulation materials and techniques on thermal performance in metropolitan areas.
- Limited use of CFD modeling to assess building airflow, temperature distribution, and heat transfer.
- Insufficient consideration of local climate conditions and the urban heat island effect in building design and evaluation.

To address these issues, the study aims to answer key questions:

- What are the specific energy consumption patterns in diverse types of buildings in metro cities of the Philippines?
- How do various insulation materials and techniques contribute to the thermal efficiency of buildings in these urban areas?
- What insights can CFD modeling and simulation studies offer to optimize airflow, temperature distribution, and heat transfer in buildings within metro cities?
- How can the effects of local climate conditions on the thermal performance of buildings in Metropolitan Manila be mitigated through design and retrofit strategies?

The significance of this study lies in its potential to contribute to energy conservation, occupant comfort, sustainability, and policy development in the Philippines. However, its limitations include the specific focus on Philippine metro cities and the potential non-transferability of findings to other regions with different climatic conditions and building practices, underscoring the need for additional research in these contexts for more globally relevant results.

2 Literature review

The Global trends in building energy consumption are alarming. The Energy Efficiency 2019 analysis report [5] indicates that 30% of global energy demand is attributable to the building sector, and this trend is expected to continue. The IEA highlighted ASEAN's space-cooling electricity use, which expanded to 80 TWh in 2020, seven times the level in 1990, and comparable to the Philippines' annual electricity use. Air conditioners and fans used 15% and

9% of ASEAN household energy, respectively. In 2040, space cooling electricity demand could reach 300 TWh, akin to the combined electricity use of Indonesia and Singapore [6]. Therefore, enhancing the thermal performance of buildings, particularly in Southeast Asian nations experiencing rapid urbanization like the Philippines, is of the utmost importance. Urbanization and Its Impact on Energy Consumption The rapid urbanization and economic development in Southeast Asia, including the Philippines, have led to increased energy consumption, raising significant environmental concerns [7]. The urban heat island (UHI) phenomenon significantly affects building energy consumption by increasing cooling demands [8]. Urban densification exacerbates already elevated temperatures, jeopardizing air quality and raising carbon emissions. Sustainable building design strategies, such as passive design techniques (natural ventilation, shading, insulation), are recognized as effective means to improve the thermal performance of buildings [9].

Achieving optimal indoor thermal comfort is crucial in building design, affecting occupant well-being and productivity. ASHRAE Standard 55 considers factors like air temperature, radiant temperature, humidity, air velocity, and clothing insulation. Environmental factors and occupant preferences play significant roles in determining thermal comfort [10, 11]. Passive and active design strategies are key in enhancing thermal comfort while optimizing energy efficiency [12].

Integrated approaches to energy and thermal performance assessment involve various tools and methods for a holistic understanding of a building's energy use and thermal behavior [13]. Building energy simulation, enhanced by Building Information Modeling (BIM), has become crucial for predicting thermal behavior and energy performance [14]. Computational Fluid Dynamics (CFD) modeling, combined with building energy simulations, is increasingly used for accurate estimation of energy requirements and enhancement of thermal comfort in buildings [15].

Despite growing awareness and studies on energy efficiency in buildings, there is a notable gap in research focusing on energy consumption patterns, thermal efficiency, and occupant comfort, particularly in Philippine urban areas. Most studies have focused on general building energy performance and thermal comfort, leaving a gap in understanding specific patterns and trends of energy consumption, the effectiveness of insulation materials and techniques, and the optimization of airflow and heat transfer within the context of the metropolitan cities of the Philippines.

3 Methodology

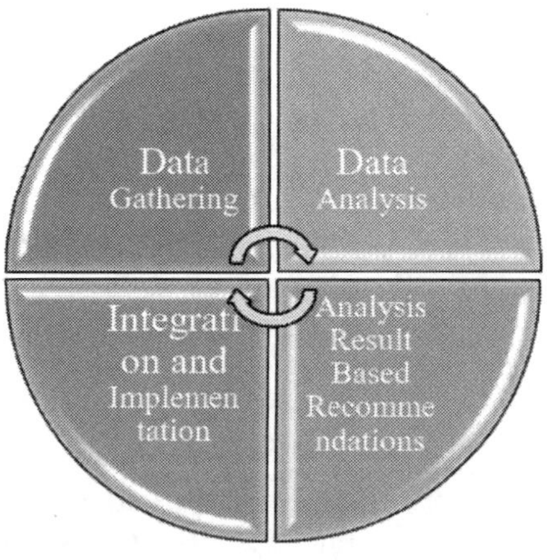

Fig. 1. Designed Research Methodology

A research strategy that is methodical and continuous was utilized to evaluate the energy and thermal performance of a residence in Quezon City and a Healthcare facility structure located situated at one of the urban cities of metropolitan Manila. Aside from embodying the qualities of mixed-occupancy structures that are prevalent in the urban regions of the Philippines, these structures were selected to symbolize the commercial and residential sectors, respectively. Data collection through occupant surveys, energy audits, walkthrough inspections, and local climate data is supplemented with comprehensive analysis through CFD simulation, recommendations are formulated based on the data analysis findings, and prior recommendations are implemented and integrated by the CFD simulation analysis results. An assessment will be conducted after the implementation and integration phases to evaluate the efficacy of the integrated processes and implemented recommendations. This evaluation will also identify potential chances for rectification, as seen in Figure 1.

3.1. Data Collection

This initial phase involves gathering comprehensive data through various methods:

- Occupant Surveys: Conducted using the ASHRAE Standard 55-2017 survey form to understand occupants' experiences and preferences regarding thermal comfort.
- Energy Audits: Assessing energy consumption patterns and identifying potential areas of energy inefficiency in the selected buildings.
- Climate Data Gathering: Collecting data on the local climate, including temperature and relative humidity, from relevant weather stations or meteorological databases.
- Building Envelope Surveys: Evaluating the components of the building envelope through walkthrough inspections and the use of data loggers to collect temperature and humidity data.

3.2. Data Analysis

Analyzing the collected data to gain insights into occupants' thermal comfort experiences and identify areas of discomfort or potential improvement. Energy audit data are analyzed to assess energy consumption patterns and identify energy-saving opportunities.

3.3. CFD Simulation

Utilizing DesignBuilder and Ansys Fluent Software for CFD simulation. This involves creating accurate models of the buildings to analyze complex physics of airflow, heat transfer, and energy exchange.

3.4. Combined Analysis

Integrating survey data, energy audit findings, local climate data, and CFD simulation results. This stage allows for the identification of correlations, patterns, and areas of improvement related to energy consumption and thermal comfort.

3.5. Recommendations and Implementation

Based on the comprehensive analysis, specific recommendations are formulated to enhance energy consumption and thermal comfort in the buildings. These recommendations are tailored to each building type, considering factors such as occupancy patterns, energy usage profiles, and local climate conditions.

The impact of the recommended energy-efficient measures on the buildings will be assessed by considering factors such as energy savings, reduction in carbon emissions, improvement in indoor air quality, and occupant satisfaction. This assessment will supply insights into the effectiveness of the measures in enhancing energy efficiency, thermal comfort, and environmental performance.

4 Results and Discussion

4.1. Urban Climate-Adaptive Building Design

In Metro Manila and other Philippine urban cities, the tropical, maritime climate demands building designs that mitigate high temperatures and humidity levels of the area with a yearly average of 31°C and 70% relative humidity as shown in Figure 2 and Figure 3.

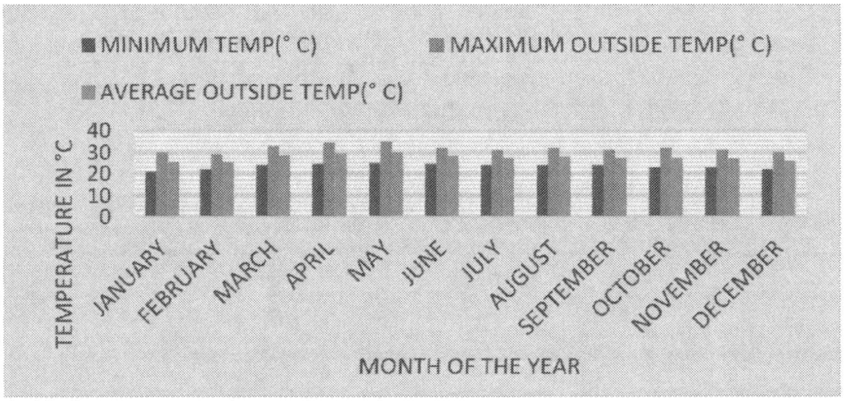

Fig. 2. Metro Manila Temperature level by month

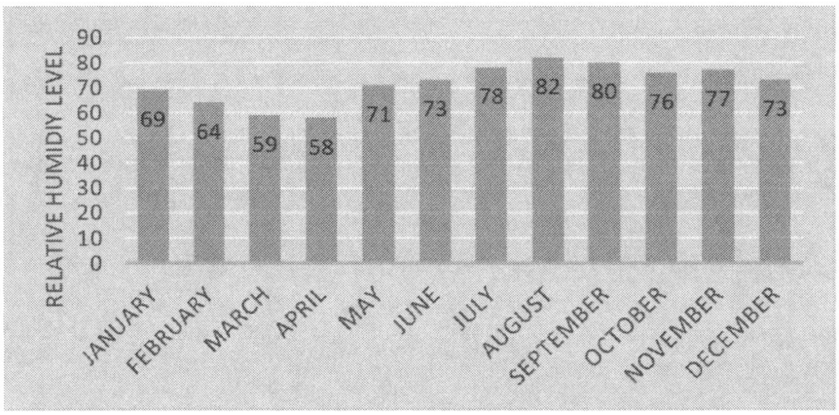

Fig. 3. Metro Manila Humidity Level (Average %) by month

4.2. Urban Energy Consumption and Comfort Preferences

As illustrated in Figure 4, residential structures in the urban environment of Metro Manila generally experience elevated electricity consumption throughout the night, but commercial buildings exhibit higher daytime electricity consumption, as depicted in Figure 5. To preserve comfort despite the urban heat island effect, Survey results show that building occupants prefer indoor temperatures between 21 and 23 degrees Celsius in commercial buildings and 22 and 24 degrees Celsius in residential areas, with ideal relative humidity between 50 and 55 %.

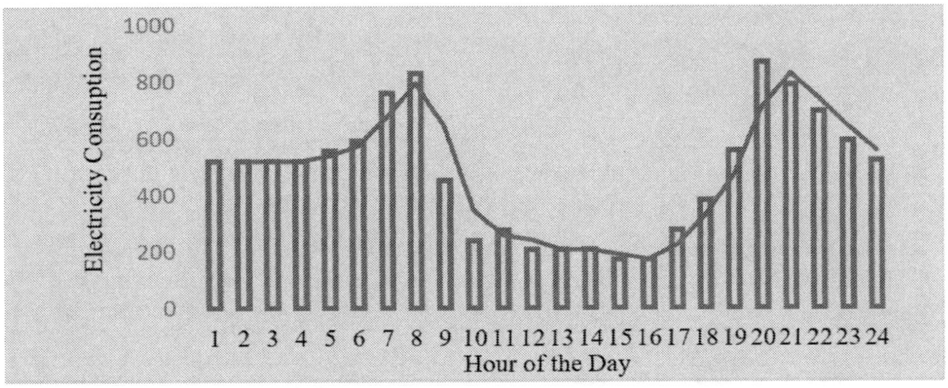

Fig. 4. Subject Residential Building Average Hourly Energy Consumption Intensity (Wh)

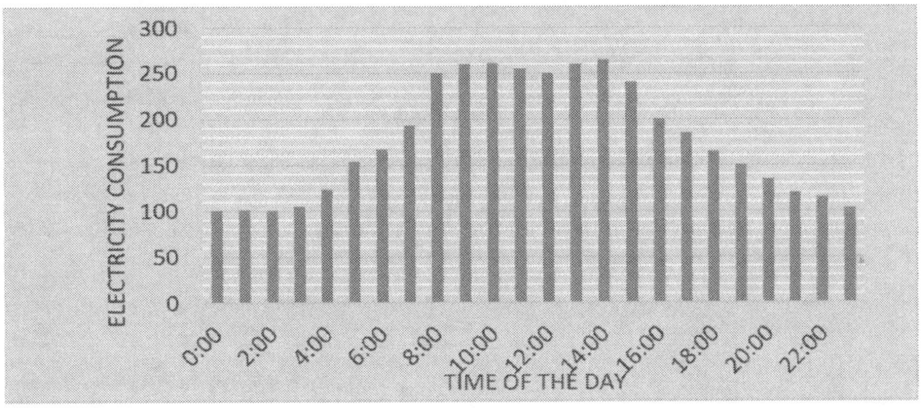

Fig. 5. Subject Commercial Building Average Hourly Energy Consumption (kWh)

4.3. CFD Simulation Insights and Suggestions for Philippine Metro Cities

The use of Computational Fluid Dynamics (CFD) simulations in this study has been particularly insightful for understanding the thermal behavior of building areas in Philippine metro cities like Metro Manila. The key findings and suggestions from these simulations include:

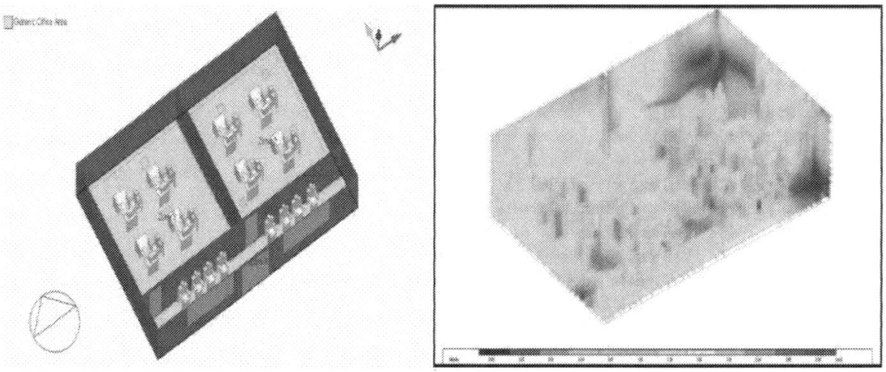

Fig. 6. A 3D model of the simulated building area (left) and area simulation results showing area position with no movement(shown in blue spots) (right).

region, allowing insight into where HVAC system design should be maximized. The CDF findings also indicate that round ducts, which are more effective than rectangular ducts, should be utilized instead of rectangular ducts, which are less efficient in terms of air distribution Figure 6 and Figure 7.

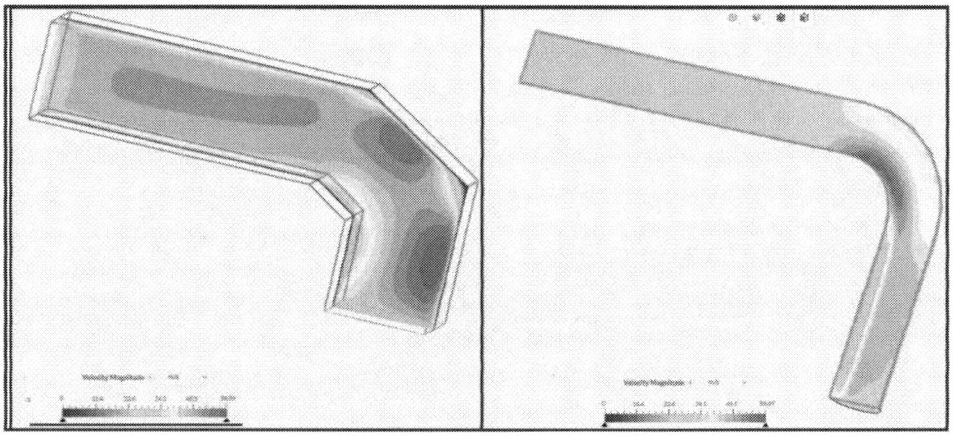

Fig.7. Simulation-based air velocity comparison between square duct(left) and round duct (right).

- The CFD modeling study results revealed locations in urban buildings with insufficient air circulation. To improve this, the diameters of air vents, diffusers, and grilles should be adjusted to their optimal dimensions based on the occupants' air requirements. Table 1 shows design parameters for an optimal diffuser throw of 45°.

Table 1. Design parameters for an optimal diffuser throw of 45°.

Velocity (m/s)	Effective Air Travel Distance (Diffuser Throw Angle, 45°)	OPTIMUM DIFFUSER DISTANCE (CtoC, m)	OPTIMUM DIFFUSER DISTANCE to wall, m
0.5	0.025	0.0875	0.03
0.7	0.05	0.175	0.05
0.9	0.083	0.2905	0.08
1.1	0.123	0.4305	0.12
1.3	0.172	0.602	0.17
1.5	0.229	0.8015	0.23
1.7	0.295	1.0325	0.30
1.9	0.368	1.288	0.37
2.1	0.45	1.575	0.45
2.3	0.539	1.8865	0.54
2.5	0.637	2.2295	0.64
2.7	0.743	2.6005	0.74
2.9	0.857	2.9995	0.86

- CFD simulations provided insights into heat transfer issues in urban buildings. Utilizing materials with superior thermal properties. The simulations also highlight the potential of PCM-incorporated walls(cement plaster, BioPCMM27/Q23, concrete block, cement plaster) in stabilizing indoor temperatures because of their lower U value compared to the traditional wall setup (cement plaster, concrete block, cement plaster) used in the Philippines as shown in Figure 8 and Figure 9. Incorporating PCMs into building materials like wallboards or ceiling tiles is proposed to augment thermal stability in varying urban climate conditions of the Philippines. Incorporating such materials in traditional wall and ceiling setups has the potential to reduce the thermal heat gain of the building from such locations up to 30%.

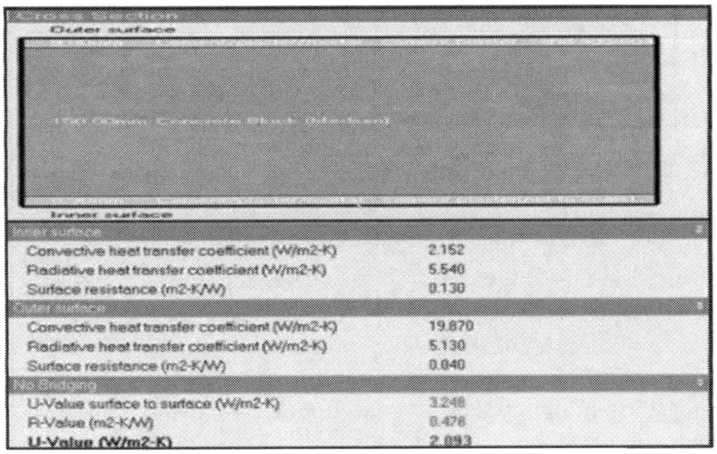

Fig. 8. U-Value of traditional wall set up in the Philippines.

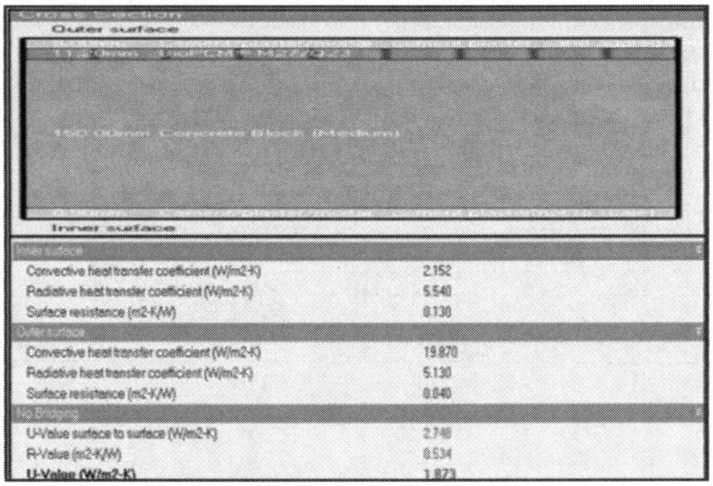

Fig. 9. U-Value of PCM Integrated wall set up.

4.4. Long-term Performance Monitoring

It's essential to monitor the long-term performance of implemented strategies in Philippine metro cities. This includes tracking energy consumption, indoor temperature, and occupant comfort to evaluate the effectiveness of the applied changes in urban building environments.

4.5. Urban Policy and Retrofitting Strategies in the Philippines

For Metro Manila and other urban areas, updating building codes to focus on energy efficiency and thermal comfort is crucial. Retrofitting initiatives for existing urban buildings, such as enhanced insulation and HVAC system upgrades, are vital for improving energy efficiency and living conditions in the Philippines' urban settings.

5 Conclusion

This research offers valuable perspectives on the design of climate-adaptive buildings in urban environments, concentrating on Metro Manila and other major urban areas in the Philippines. This emphasizes the critical nature of tackling the obstacles presented by the Philippines' tropical, marine environment, which is characterized by elevated temperatures and humidity, to promote sustainable urban growth, energy efficiency, and thermal comfort. Several significant findings of the research are outlined below:

- The research highlights the importance of adopting building designs that are responsive to the Philippines' unique climatic conditions. Implementing strategies such as the use of optimized materials and Mechanical system design improvements is imperative to mitigate the urban heat island effect and ensure effective cooling efficiency in densely populated areas.

- The distinct patterns of electricity consumption in residential and commercial buildings emphasize the need for tailored design and policy approaches. Addressing the increased nighttime energy use in residential buildings and the daytime energy demands in commercial buildings is essential for creating energy-efficient urban environments.

- The utilization of Computational Fluid Dynamics (CFD) in this study has been particularly valuable. CFD simulations have provided deeper insights into the thermal behavior and air circulation patterns within urban buildings, leading to more informed decisions in building design. The findings from CFD analyses have facilitated the optimization of airflow, ventilation, and HVAC system layouts, contributing significantly to the enhancement of indoor environmental quality in urban buildings. The study also underscores the potential of innovative materials like Phase Change Materials (PCMs) and the insights offered by CFD simulations for improving building thermal efficiency. These advancements are crucial for reducing energy consumption and enhancing occupant comfort in urban settings.

- The research advocates for updating building codes and retrofitting strategies in urban areas, emphasizing the integration of energy-efficient and thermal comfort considerations. This is crucial for transforming existing urban buildings into more sustainable and comfortable living spaces.

In summary, this research makes a significant contribution to the field of urban climate-adaptive building design in the Philippines, offering practical solutions, and policy recommendations, and highlighting the invaluable role of CFD simulations. These contributions are instrumental for the sustainable development of urban areas in a changing climate, providing a robust foundation for future advancements in urban planning and building sciences.

References

1. M. Röck *et al.*, "Embodied GHG emissions of buildings – The hidden challenge for effective climate change mitigation," *Appl Energy*, vol. **258**, Jan. 2020, doi: 10.1016/j.apenergy.2019.114107.(2020)
2. international energy agency, "CO2 Emissions in 2022," 2022. [Online]. Available: www.iea.org.(2022)
3. J. Huang, S. Wang, F. Teng, and W. Feng, "Thermal performance optimization of envelope in the energy-saving renovation of existing residential buildings," *Energy Build*, vol. **247**, Sep. 2021, doi: 10.1016/j.enbuild.2021.111103.(2021)
4. F. Lai, J. Zhou, L. Lu, M. Hasanuzzaman, and Y. Yuan, "Green building technologies in Southeast Asia: A review," *Sustainable Energy Technologies and Assessments*, vol. **55**. Elsevier Ltd, Feb. 01, 2023. doi: 10.1016/j.seta.2022.102946.(2023)
5. I. Energy Agency, "Energy Efficiency 2019." [Online]. Available: www.oecd.org/about/publishing/(2019)
6. I. Energy Agency, "Roadmap Towards Sustainable and Energy-Efficient Space Cooling in ASEAN." [Online]. Available: www.iea.org/t&c/
7. L. Belussi *et al.*, "A review of performance of zero energy buildings and energy efficiency solutions," *Journal of Building Engineering*, vol. 25. Elsevier Ltd, Sep. 01, 2019. doi: 10.1016/j.jobe.2019.100772.(2019)
8. W. L. Filho *et al.*, "sustainability Addressing the Urban Heat Islands Effect: A Cross-Country Assessment of the Role of Green Infrastructure," 2021, doi: 10.3390/su.(2021)
9. E. Cuce, F. Sher, H. Sadiq, P. M. Cuce, T. Guclu, and A. B. Besir, "Sustainable ventilation strategies in buildings: CFD research," *Sustainable Energy Technologies and Assessments*, vol. **36**, Dec. 2019, doi: 10.1016/j.seta.2019.100540.(2019)

10. T. Daikoku, Q. Fang, T. Hamada, Y. Handa, and Y. Nagai, "Importance of environmental settings for the temporal dynamics of creativity," *Think Skills Creat*, vol. **41**, Sep. 2021, doi: 10.1016/j.tsc.2021.100911.(2021)
11. S. Coccolo, D. Pearlmutter, J. Kaempf, and J. L. Scartezzini, "Thermal Comfort Maps to estimate the impact of urban greening on the outdoor human comfort," *Urban For Urban Green*, vol. **35**, pp. 91–105, Oct. 2018, doi: 10.1016/j.ufug.2018.08.007.(2018)
12. E. Mushtaha *et al.*, "The impact of passive design strategies on cooling loads of buildings in temperate climate," *Case Studies in Thermal Engineering*, vol. **28**, Dec. 2021, doi: 10.1016/j.csite.2021.101588.(2021)
13. F. Johari, G. Peronato, P. Sadeghian, X. Zhao, and J. Widén, "Urban building energy modeling: State of the art and future prospects," *Renewable and Sustainable Energy Reviews*, vol. **128**. Elsevier Ltd, Aug. 01, 2020. doi: 10.1016/j.rser.2020.109902.(2020)
14. H. Gao, C. Koch, and Y. Wu, "Building information modelling based building energy modelling: A review," *Applied Energy*, vol. **238**. Elsevier Ltd, pp. 320–343, Mar. 15, 2019. doi: 10.1016/j.apenergy.2019.01.032.(2019)
15. M. Bühler and T. Bednar, "A review on coupled building physics analyses," in *Journal of Physics: Conference Series*, IOP Publishing Ltd, Dec. 2021. doi: 10.1088/1742-6596/2069/1/012141.(2021)

Evaluating Thermal Comfort in Ward Areas of the Public Hospital in the Tropical Climates

Kittiwoot Chaloeytoy[1*], *Atima* Dubsok[1], and *Suthirat* Kittipongvises[1]

[1]Environmental Research Institute, Chulalongkorn University, Bangkok, Thailand

Abstract. Thermal comfort in hospitals is a crucial element to provide appropriate indoor air quality for health and well-being. However, the air-conditioning system usually has deficiencies to meet target criteria for the relevant thermal environmental factors. Therefore, this study aimed to evaluate thermal comfort in ward areas of the newly built public hospital in Thailand during the operation phase to examine how the building maintain comfort. The measurement equipment was installed for 24 hours in ward areas for collecting data and evaluating thermal comfort using the predicted mean vote (PMV) model. The results indicated that achieving thermal comfort could be found among inpatients and medical staff, with differential considerations on personal factors. As a few studies in hospitals in Thailand has been done, the results from this study can act as evidence to develop building performance assessments and operation strategies to optimally enhance health and well-being of the occupants in a public healthcare facility in Thailand and tropical regions.

1 Introduction

The thermal comfort in hospital environments require special attention at providing indoor air quality (IAQ) for health and well-being to patients and medical staff. The indoor environment of a mechanically ventilated hospital buildings serves as functions of prevent airborne infection risks, physiological strain on the body, while it also extends to influence patient's healing processes and overall medical outcomes [1, 2]. Appropriately designed and operated air conditioning systems should be at primarily concern to learn how hospital stakeholders perceive thermal comfort.

Thermal comfort in hospital buildings is mainly designed to accommodate patients with health and physical conditions, which can affect thermophysiology, thermal sensation, metabolism, blood flow, regulatory response, such as control of body skin [2, 3]. Moreover, hospital environments are subdivided into specifically functional areas for patient care, healthcare delivery, administration, with supply and service [4]. Specific requirements and designs are essential to improve the indoor environments for both patients and medical staff, particularly in ward areas, where the inpatient is identified as being of major concern. It is expected to adjust temperature, humidity, and air flow in preservation of individual's health [5].

* Corresponding author: kittwoot.c@chula.ac.th

To enhance the occupant's comfort and satisfaction within the indoor environment, standards for air temperature and humidity in healthcare facilities had been declared by the Guidelines for Design and Construction of Hospital and Health Care Facilities by the American Institute of Architects Academy of Architecture for Health (AIA) and the Facility Guidelines Institute (FGI) [6] at 22.2-24.4 ◦C, with relative humidity (RH) 30–60% in general waiting areas and exam rooms. In Thailand, the standard for the thermal environment was introduced by the Department of Health Service Support, Ministry of Public Health (MOPH). According to their Standard for Environment Sanitation and Safety in Hospitals [7], the recommended air temperature in the waiting area of an outpatient area was 20.0-25.0 ◦C, at 50-70%RH and 20.0-25.0 ◦C, at 60%RH for an exam room. There was an effort to adjust the criteria to be applicable for the hot and humid context such as indicating a higher humidity range. However, the adjusting the temperature may not correspond to the actual requirement for the thermal comfort of occupants [8].

In tropical regions, it is required further discussion with concerning on high demand healthcare services [8], since a few studies in hospitals in the tropics has been done. For example, Chew examined on thermal comfort and revealed that the neutral temperature for Malaysian hospitals is 26.4 ◦C, and a comfortable temperature range between 25.3-28.2 ◦C [9], whereas Azizpour suggested a neutral temperature of 23.4 ◦C and a comfort range between 21.2-25.5 ◦C [10]. Koutroumpi suggested that thermal comfort surveys of environmental and behavioural performances have not been applied in naturally ventilated inpatient facilities [11]. In Thailand, Sattayakorn found that the preferred temperature of tropical occupants was slightly colder than neutral. The acceptable temperature range for the patient, the visitor, and the medical staff were at 21.8–27.9, 22.0–27.1, and 24.1–25.6 ◦C, respectively [8].

One of the MOPH's policies is to standardize indoor air quality of health service system and healthcare facilities. Hospital building's standards and guidelines were set to develop the building performance for public and government hospitals. However, the operated air-conditioning system usually has deficiencies to meet target criteria for the relevant thermal environmental factors from the before and during the building commissioning phase. Particular attention is required to draw the nature of the interaction between buildings and their occupants. Thus, this study aims to evaluate thermal comfort in ward areas of the newly built public hospital in Thailand during the operation phase to examine how the building maintain their occupant's comfort. Systematic evaluation about buildings in use from the perspective of thermal environment and personal factors were incorporated using the predicted mean vote (PMV) model. The results found in this study can reflect the dynamic nature of the relationship between the people, the climate, and the buildings, which help develop building performance assessments and operation strategies to optimally enhance health and well-being in a healthcare facility in Thailand and tropical regions.

2 Methodology

For this study, the public hospital ward building involving an air-conditioning operation were selected for the field measurement. It was a 5,848 m2 concreted-structured building with 8 storeys, located in Chiang Rai, Thailand (19°54'30.89"N and 99°49'57"E). The building reflected generic environments in a typical ward building from the MOPH's policies, which was commonly found in the suburban context of Thailand. The targeted area was 32 multi-units ward of 320 m2, utilized on the 4th floor of the building, served as healthcare functions for inpatients under the medicine treatments such as post-stroke, with the 8:00 to 20:00 h shifted staff support, such as nurse and doctor.

The field measurement was held for 24 hours during working hours, local time (UTC + 7), based on a collaboration between stakeholders and occupants under the post-occupancy stage. The measurement involved collecting thermal variables for the PMV model based on the Fanger's study [12]. The PMV was developed for thermal comfort evaluations within environmental and personal factors and has been verified in field studies in different contexts, including referring in ASHRAE 55: Thermal Environmental Conditions for Human Occupancy [13]. In terms of environmental factors, the set of automatic data loggers (named as T&D TR-74, TR-52) with a globe ball by diameter of 0.15 m. was installed in the center of the ward and the nurse station area at the working plane or bedding level (approximately 1.10 m. from the floor) to measure air temperature, mean radiant temperature (MRT), and humidity, as reference point A and B presented in Figure 1. An anemometer (named as TL 1000) was also set on a tripod to measure the air velocity. All the devices were set up in an automatic recording mode with time intervals of 1 minute, then presented and formatted in hourly data for observing trend of thermal comfort.

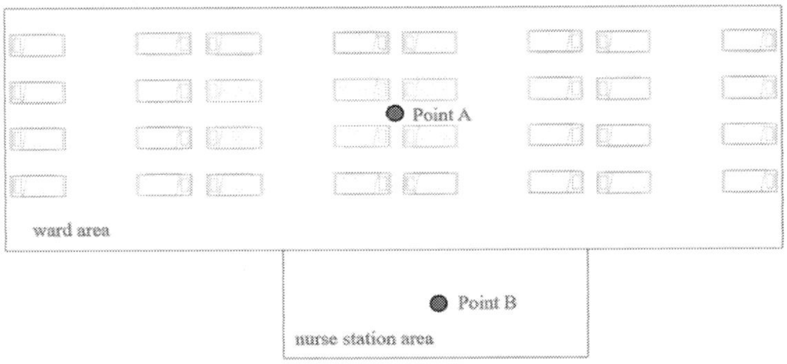

Fig. 1. Multi-unit ward layout of the public hospital in this study, with reference points.

The personal factors related to metabolic rate and clothing insulation were estimated from observation on the occupant's clothing and their activities. The metabolic rate was determined for nurse and doctor as 1.4 met by visual observing their physical examination activities in standing positions, while the inpatient was 0.8 met for reclining in lying positions, as mentioned in ASHRAE 55-2013 [13]. Clothing insulation values were derived by observing and applying the equation from ISO 9920 (2007). The nurses and the doctors normally wear a uniform, but the doctors were additionally covered with a hospital gown while in wards, their clothes were marked as 0.49 and 0.66 clo, respectively. Meanwhile, the inpatients clothes marked as 1.90 clo with considerations on each body part in reclining or supine decubitus position, alongside the quilt covering [14]. In the nurse station context, the area without patient care services mainly contributed to perform a seated task. The metabolic rate of the nurses and the doctors were identified as 1.2. The nurse's clothing remained unchanged, while the doctors always took their hospital gown off, leaving them to 0.49 clo in this area. As presented in Table 1., the measured air temperature, globe temperature, and air velocity can be combined to calculate the operative temperature with MRT, and then incorporated with the personal factors for evaluation using the PMV model [13].

Table 1. Thermal comfort variables from the field measurement in ward areas in the hospital

Thermal comfort variables	Ward area			Nurse station area	
	Inpatient	Nurse	Doctor	Nurse	Doctor
Air temperature (C)	22.4-26.0			22.3-24.8	
Relative humidity (%)	52-80			56-73	
Mean radiant temperature (C)	22.3-28.3			22.9-23.1	
Air velocity (m/s)	0.17-0.23			0.11	
Clothing insulation (clo)	1.90	0.49	0.66	0.49	0.48
Metabolic rate (met)	0.7	1.4	1.4	1.2	1.2

3 Results and Discussions

Fig. 2 illustrates the thermal comfort variables from the field measurement. In the ward area, the temperature ranged from 22.4 to 26.0 C. Meanwhile, the relative humidity ranged from 52 to 80%. Their mean and standard deviation revealed that thermal variables presented consistency. However, since the air-condition operation strategies suggest being turned off during 4.00 to 8.00 for maintaining mechanical efficiency while partially opening the windows to get some fresh air outside, this could lead to temperature fluctuation in the ward area from external thermal conditions. Any operation strategies should be carefully discussed on seeking feedback from the occupant's health as well as providing feedback throughout a building's performance from its initiative to actions. In the nurse station, the temperature ranged from 22.3 to 24.8C under fully equipped operated, but only during the working hours from 8.00 to 20.00 h. It can be observed that both the ward and the nurse station area were effectively run to operate to reach the MOPH's target at indoor air quality. The building in use should continually maintain practical strategies to control temperature and humidity at optimal levels.

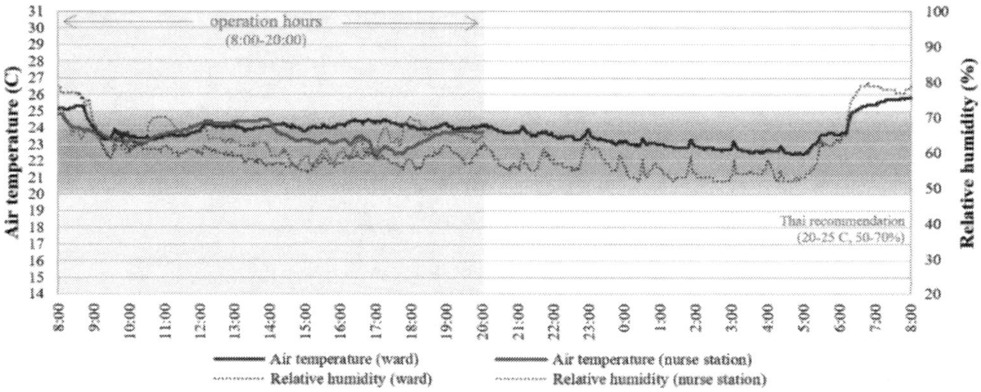

Fig. 2. Distribution of air temperature (C) and relative humidity (%) in ward areas.

Fig. 3 illustrates the PMV values in the ward and the nurse station area, based on types of occupants as the inpatients, the nurses, and the doctors. During the operation hours, it can be seen that all of them mostly achieved thermal comfort since their thermal sensation mostly fitted the comfortable range between -0.5 to 0.50 [13]. However, in ward environments, their sensation defined at different level, although they all generally perceived comfortably. Based on the PMV, comfortable range of the inpatient, at -0.01 to 0.19, was higher than the doctor and the nurse at, -0.21 to 0.20 and -0.49 and 0.02, respectively. It can be inferred that the personal factors could play a role in thermal sensation due to the greater body covering with clothing of the inpatient. Meanwhile, the nurse's sensation notably marked on their smaller clothing with high thermal exposure to environments. Moreover, focusing on the after-operation hours which only the inpatients were accommodated, the hourly downward trend of the PMV could be noticed in night-time conditions. This is in line with Sattayakorn, who found that cooling sensations of the occupants in Thai hospitals related to the longer exposure time to the hospital environment. Additionally, it is important to noted that the occupants with diverse health conditions and performing different activities should impose specific indoor environmental requirements [3]. Further study is required to meet the needs of thermal comfort of both the inpatients and medical staff towards clarification on different operation strategies in different areas.

As the thermal sensitivity for healthcare occupants in ward areas is varied, a detailed analysis for each group of occupants was examined by concerning on the PMV range between -0.5 to 0.50 with their corresponding operative temperature. Fig 4 illustrates the comfortable temperature of the inpatients, the nurses, and the doctors as average of 23.6, 24.4, and 24.3 C, respectively. In terms of the medical staff, the findings from this study are smaller than the study of Yau & Chew and Azizpour in Malaysia hospitals at 26.4 and 25.0, respectively, but greater than Sattayakorn, who marked 22.4 in Thai hospitals. However, the study on the inpatient has been a neglected area of research so far. There is still a lack of quantitative evidence and suggestions for describing the specific thermal comfort needs of hospital occupants, particularly in a tropical context. Moreover, past research suggested that the PMV prediction often mismatched the thermal sensitivity of healthcare occupants [8, 10]. The applicable models and methods from the field survey should be further investigated to observe thermal sensation, acceptability, and satisfaction in each occupant group.

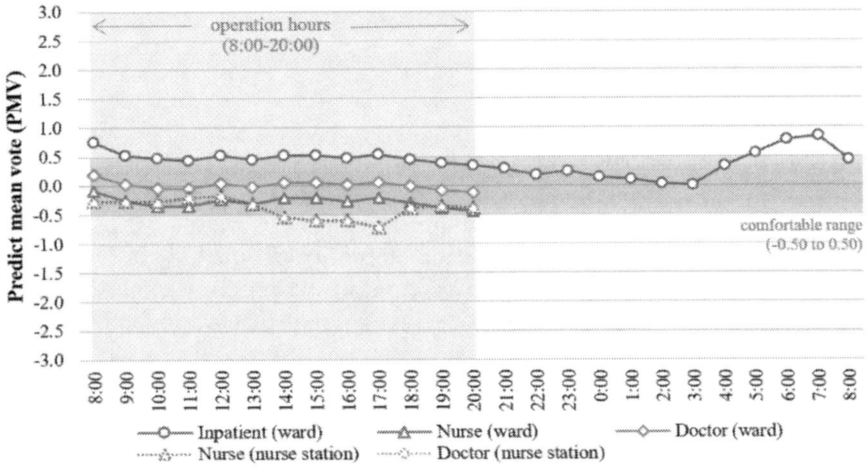

Fig. 3. Predict mean vote (PMV) of the hospital's occupants in ward areas.

In the practical terms, the results indicated that thermal comfort can be achieved under the normal operation strategies. Since the building started to fully operate by the beginning of 2022, it is envisaged that the stakeholders throughout the building phases should certainty maintain operating and practicing along this efficient and consistent manner. Different temperature set point for the inpatients, the nurses, and the doctors are worthwhile to carefully discuss for providing health and well-being, particularly the inpatients with their thermal sensitivity and adaptive responses for thermal acclimation. The insight gained from this study serve as evidence to support the standardized guideline regarding indoor air quality from the MOPH's policies, which can be adopted and applied for public or government hospitals in the suburban context of Thailand to achieve the best possible health outcomes for patients and medical staff in the long run.

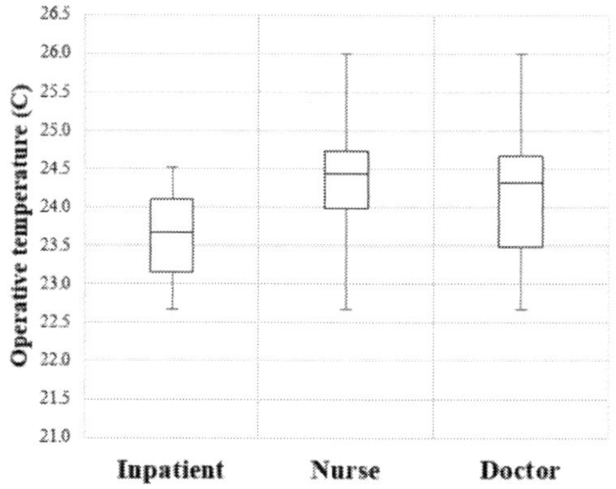

Fig. 4. Box plot of comfortable temperature range for the hospital's occupants in ward areas.

4 Conclusion

For hospital buildings, maintaining health and comfort is an issue that can never be compromised. This study aims to evaluate thermal comfort in ward areas of the newly built public hospital in Thailand during the operation phase to examine how the building with the building maintain comfort after built. Thermal comfort evaluation was performed using the PMV model. The results from the field survey suggested that achieving thermal comfort could be found among inpatients and medical staff, with differential comfortable level due to their personal factors.

The comfortable temperature of the inpatients, the nurses, and the doctors were average of 23.6, 24.4, and 24.3 C, respectively, according to the PMV model. This is in line with past research on the hospitals in the tropical context, suggesting that different healthcare occupants required different thermal conditions for their comfort. As a few studies in hospitals in the tropics has been done, the discussions on personal health conditions and activities performing should be further investigated to draw attention to specific indoor environmental requirements in different occupant's groups and different areas, particularly in ward areas with high sensitivity effects on health and well-being. Additionally, there are limitations to represent future research suggestions. The larger set of data is required for establishing a comprehensive benchmark to help identify comfort conditions. A long-term analysis with the applicable models like post-occupancy survey should be initiated to observe a wider outlook. Future research should also investigate indoor airborne infection, ventilation, and operation strategies to gain more understanding on balancing comfort and health impact to achieve the vision of quality healthcare in public or government hospitals in the tropical regions.

Acknowledgement

The authors would like to express their deepest gratitude to the building support division of the case study hospital for help and support during this period of study.

References

1. A. Shajahan, C.H. Culp, B. Williamson, Ind. A., **29**, 2 (2019)
2. J. Verheyen, N. Theys, L. Allonsius, F. Descamps, Bld. & Env., **46**, 1195-1204 (2011)
3. F. Yuan, R. Yao, S. Sadrizadeh, B. Li, G. Cao, S. Zhang, J. of Bld. Eng., **45**, 103463 (2022)

https://doi.org/10.1051/e3sconf/202453005009

4. American Society of Heating, Refrigerating and Air-Conditioning Engineers (ASHRAE). *ASHRAE Handbook-HVAC Applications (SI),* Atlanta, US (2019)
5. R.M. Smith, A. Rae, J. of hyg., **78**, 1, 17-26. (1977)
6. AIA/FGI. *Guidelines for Design and Construction of Hospital and Health Care Facilities. The American Institute of Architects Academy of Architecture for Health and the Facility Guidelines Institute. U.S. Department of Health & Human Services*, The American Institute of Architects Press, Washington, D.C, USA (2006)
7. Department of Health Service Support, Ministry of Public Health, *Standard for Environment Sanitation and Safety in Hospital*, Ministry of Public Health, Bangkok, Thailand.
8. S. Sattayakorn, M. Ichinose, R. Sasaki, En. & Bld., **149**, 45-57 (2017)
9. Y. Yau, B. Chew, Ind. A., **19**, 500–510 (2009)
10. F. Azizpour, S. Moghimi, E. Salleh, S. Mat, C.H. Lim, K. Sopian, En. Bld., **64**, 317–322 (2013)
11. S K. Koutroumpi, G. J.of Eng. Sc., **6**, 2, 000632 (2020)
12. P.O. Fanger, *Thermal Comfort. Analysis and Application in Environment Engineering.* McGraw Hill Book Company, New York, USA (1982)
13. American Society of Heating, Refrigerating and Air-Conditioning Engineers (ASHRAE). *ANSI/ASHRAE Standard 55-2013 Thermal Environmental Conditions for Human Occupancy*, Atlanta, USA (2013)
14. H. Zhang, X. Xie, S. Hong, L. Haitao, En. & Blt. Env., **2**, 223-232 (2020)

 https://doi.org/10.1051/e3sconf/202453005010

Estimation of electrical energy consumed during the construction of residential houses in San Pedro Sula, Cortés, Honduras

J Cruz Davila[1*] and *H* Villatoro Flores [2]

[1]Undergraduate student, Faculty of Engineering, Universidad Tecnológica Centroamericana, San Pedro Sula, Honduras
[2]Lecturer, Faculty of Engineering, Universidad Tecnológica Centroamericana, San Pedro Sula, Honduras

Abstract: There are several technical and non-technical losses in the transmission and consumption of electrical energy in Honduras. Currently, the electrical energy consumed in the construction of residential houses is not charged. It is necessary to estimate the amount of electricity consumed at the time of construction of residential houses in order to be able to charge for it. In this research, permits to install meters and the collection of national data on construction areas were carried out. The installation of CL100 FM2 meters in each house. And from the data collected, a formula was generated using a scatter plot that related electrical energy consumption to square meters built. This formula was applied locally and nationally to calculate the electrical energy consumed and not billed. The research also revealed the millions that ENEE did not receive because the meters were not installed. This research will serve to estimate the calculation of the electric energy to be used for the construction of a residential house and thus be able to charge at the time of obtaining a construction permit.

1 Introduction

According to ENEE reports, it is known that 38% of the energy generated is not billed, either due to technical or non-technical losses [1]. In the year 2022, ENEE's annual losses amount to 20,788.7 million lempiras, which is estimated at 8,472.6 million dollars, of which technical and non-technical losses represent 38% in the year 2022. According to the Secretary of Energy, Erick Tejada, "We find this company with a historical debt of 75 billion lempiras, which represents 10.5% of the gross domestic product, is 50% of the government's fiscal deficit and represents the main hole of the State [2].

G. Gonzalez and L. Figueroa (2006), analyzed the non-technical losses of C.A. and showed the techniques and tools used by the company ELEVAL, among the tools that stand out are control devices and detection of alterations to the meters, among the most outstanding are those known as totalizing meters [6]. U. H. Bezerra, T. M. Soares, J. P. A. Vieira, M. E. L. Tostes, A. R. R. Manito J. C. H. Paye (2018) present what is an Equivalent

* Corresponding author: icruz@unitec.edu

Operating Impedance which is a new concept for the analysis of technical and non-technical distribution losses [12], this analysis serves to effectively calculate what are the technical and non-technical losses in medium and low voltage distributions. Adding the research of

N.H.V.T.N. Nanayakkara; C.D. Udawattha and R.U. Halwatura where they mention the costs of the construction of residential houses, where it is also mentioned that the electrical energy used for the construction of houses is not so high if we compare it to the other expenses that exist such as labor, materials and construction permits [9] [10] [11].

Similar to the past researches that sought to reduce non-technical and technical losses, or in the case of the research that looked at the costs of home construction, this research will focus on determining the electrical energy required for the construction of residential homes [15]. The methodology to be employed will involve obtaining permits and data, installing the CL100 FM2 single-phase meter, recording the electrical energy consumption of both houses on a weekly basis, to determine the electrical energy consumption according to their construction phases [7].

After registering all the required electric energy, a formula will be created to estimate the electric energy required at the moment of building a residential house, according to the amount of square meters of the house to be built, in the same way the formula will be used to determine the amount of electric energy that was not billed in the year 2022 in San Pedro Sula and in Honduras [8].

2 Context

Due to the lack of investment in electric energy distribution, the National Electric Energy Company (ENEE) allowed electric energy losses to increase to a percentage that in my opinion is high, since having losses of 38% means that ENEE continues to get into debt and from there they have a problem for their sustainability. According to ECLAC, in 2013 Honduras presented the highest level of losses in Central America, reaching 31.2%; Costa Rica, on the other hand, presented the lowest percentage of losses with 11.6%, and the average for the region was 17.1%. [4].

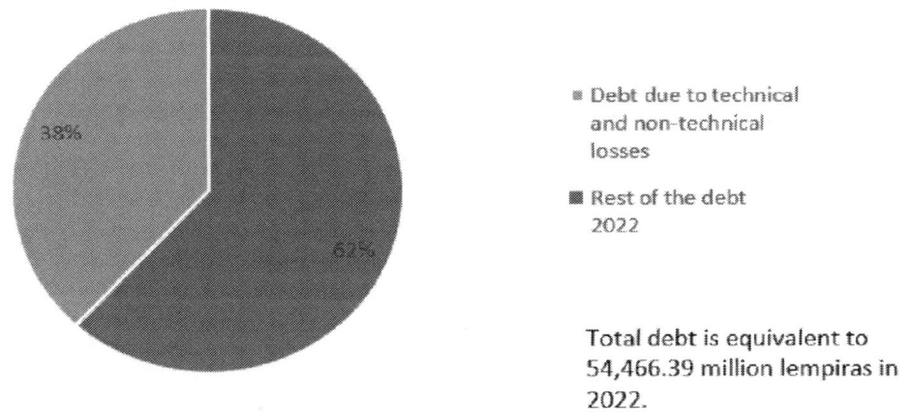

Fig. 1. ENEE total debt 2022

Source: Own elaboration based on [3].

In the year 2022, ENEE's annual losses amount to 20,788.7 million lempiras, which is estimated at 8,472.6 million dollars, of which technical and non-technical losses represent 38% in the year 2022. According to the Secretary of Energy, Erick Tejada, "We found this company with a historical debt of 75 billion lempiras, which represents 10.5% of the gross domestic product, 50% of the government's fiscal deficit and represents the main hole of the State [5]. Within this percentage of non-technical and technical losses are the electricity not billed for the construction of houses.

3 Methodology

The research approach is quantitative, aimed at determining the amount of electrical energy consumed during the construction of houses in San Pedro Sula, Honduras. The variables of study are the square meters of construction and the consumption of electrical energy in the construction of houses. To carry out the research, two main instruments were used: a single-phase electric energy meter called CL100 FM2 and an Excel spreadsheet.

The meter was installed prior to the start of house construction and was used to measure electricity consumption throughout the process. The methodology involved obtaining permits and data, which was challenging due to the reluctance of some builders to cooperate. Finally, the necessary permits were obtained to measure two houses, house number one has a construction area of 142.8234 square meters and house number two has a construction area of 110 square meters. In addition, information on square meters of construction was requested nationwide through the Institute for Access to Public Information (Instituto de Acceso a la Información Pública) [13].

Electrical energy measurements began in March-April and were taken weekly in one of the houses, allowing detailed monitoring of consumption at each stage of construction. Information on house plans was limited for privacy reasons. Data obtained from the meters were continuously recorded, and a civil engineer with construction experience was involved to better understand the construction process.

The data analysis was executed using Excel, generating a scatter plot and a formula to estimate the electrical energy consumption per square meter built in San Pedro Sula and Honduras. This methodology would allow for an automatic update in the graph as more house data is obtained in future research.

Similarly, the formula can be used to estimate the amount of money that was not billed in the year 2022 by estimating the amount of electricity used and then multiplying it by the low voltage tariff.

4 Results and Analysis

In this section we will analyze the results of the amount of kWh needed for the construction of the houses. We will create a formula and estimate the amount of electricity that was used in the year 2022 in Honduras and thus calculate the amount of money that was not billed in that year.

First, we will look at the amount of electrical energy used in house number one, in which we were able to calculate the electrical energy consumed for each phase of construction [14].

4.1 Electrical Energy consumed by construction phases in house number one

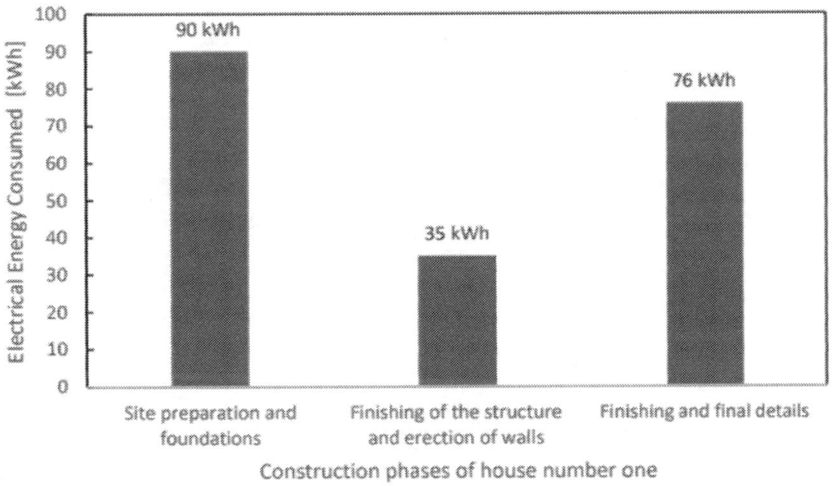

Fig. 2. Electrical Energy consumed by construction phases

Source: Own elaboration

Phase one was the phase that consumed the most electrical energy, as can be seen in Figure 2. In this stage, more energy was consumed because the land where this house was built was uneven, so it was necessary to raise the foundation with steel beams, which is why the welding machine was used more. This tool consumes a high amount of electrical energy compared to the other tools used in the construction of houses; this stage lasted six weeks. In stage two, only 35 kWh were consumed, since this stage only involved the erection of walls and the construction of the second-floor slab; this stage lasted seven weeks. And to finish the construction of the house in phase three, 76 kWh were consumed, in this phase the details and finishes of both floor and walls of division inside the house were finalized, this phase lasted eight weeks. This phase was the most time consuming since they are smaller details that require more time. In total, the construction of house number one used 201 kWh in a period of 5 to 6 months.

In Figure 3, as can be seen, house number one consumed more electrical energy for its construction because it has more square meters of construction. In the same way we can analyze that more electrical energy was consumed in this house due to its structure and foundation condition, since in house number one the welder was used more for the lifting of the structure due to the slope of the land. We can also calculate with the data obtained the amount of kWh needed for each square meter of construction.

4.2 Electricity consumption per household

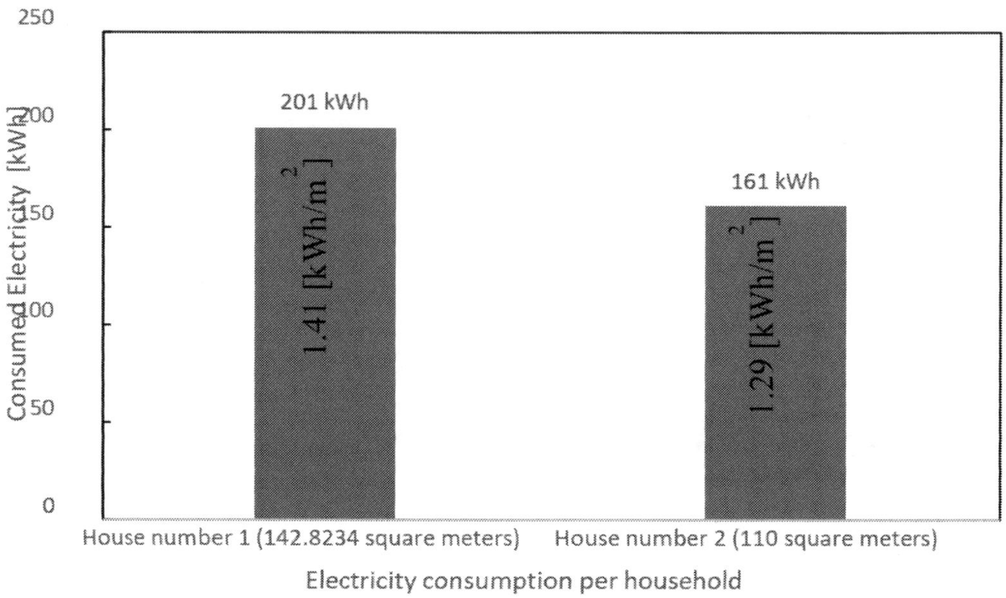

Fig. 3. Electricity consumption per household

Source: Own elaboration

4.3 Scatter plot and formula obtained

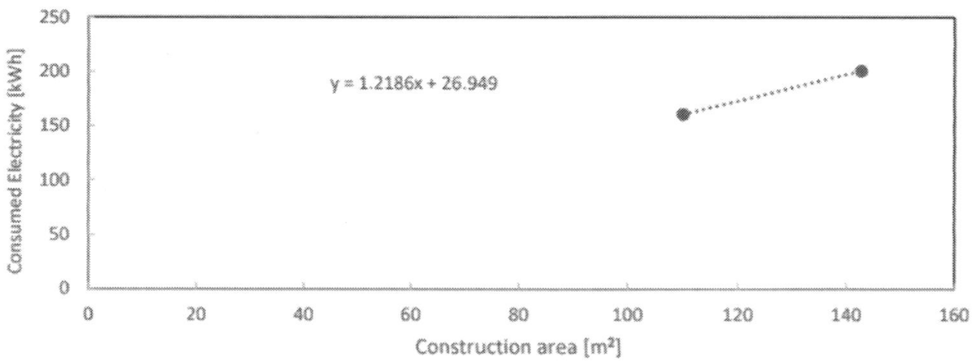

Fig. 4. Scatter plot

Source: Own elaboration

In the previous graph you can see what a formula is, which has as a result y = 1.2186 (x) + 26.949. This base formula will help us to calculate how much electrical energy was consumed in San Pedro Sula and Honduras. The next step will be to estimate the electrical energy consumed in San Pedro Sula and Honduras based on the square meters of construction in San Pedro Sula and Honduras. We know that in San Pedro Sula there are

319,243.80 square meters of construction and nationwide there are 1,520,600 square meters. Knowing the square meters, we would only have to insert the data in the formula to know the electric energy consumed and not billed. Then, based on the results, we will calculate how much money was not billed because the meters were not installed before construction began. Based on that result we will calculate the percentage of electric energy that was consumed and not billed in the year 2022 and that turned out to be attributed to technical and non-technical losses.

4.4 Electricity consumed in the year 2022

By applying the obtained formula of y= 1.2186 (x) + 26.949, with the data from San Pedro Sula, the result of unbilled kWh consumed is 389,057.44 kWh in San Pedro Sula alone. This was achieved by replacing the x in the formula with the number of square meters of construction in San Pedro Sula. In terms of money, in the city alone, L. 2,267,232.25 or its equivalent in dollars of $92,502.34 was not billed. At the national level, approximately 1,853,030.11 kWh were not billed, which is equivalent to

L. 10,798,532.96 or its equivalent in dollars of $ 440,576.62.

The electricity consumption in Honduras for the whole year 2022 in the residential sector was 2,802,718.1 MWh, of which 38% was registered as technical and non-technical losses. To make a more exact comparison we will use the amount of money that was billed in the residential sector in the whole year 2022, since the exact amount in the San Pedro Sula sector is not mentioned. In the year 2022, 15,719.75 million lempiras were billed in the residential sector alone, according to information from ENEE's statistical bulletin for the year 2022. Based on this value and knowing that at the national level 10.80 million lempiras were not billed approximately or $ 440,576.62 which would represent 0.70% of what was billed at the national level. We may think that this is not so much but considering the current state of ENEE's debt, it is not possible that all this electric energy is not being billed.

The amount of unbilled electric energy that was used for the construction of residential houses, compared to the electric energy that was lost and accounted for as technical and non-technical losses at the national level. The result was that 0.2% of this unbilled electric energy is due to the fact that there are no devices to measure the electric energy at the time of starting the construction of the residential

houses. As a final point we can mention that, for each square meter built in the residential sector, based on the results of the two houses, 1.35 kWh of electrical energy is required for the construction of each square meter.

5 Conclusions

A formula that estimates the electrical energy required for the construction of a house. Measurements of two residential houses in Honduras allowed us to estimate and create a formula that, according to the square meters of construction of the houses, to estimate how much energy was or will be required at the time of construction. We also used this formula to determine how much electricity was not billed in San Pedro Sula and in Honduras in the year 2022. The most important results of this thesis are the following:

1. The amount of square meters of construction in San Pedro Sula in 2022 was 319,243.80 m2.
2. The formula generated with the scatter plot was y= 1.2186 (x) + 26.949.
3. Applying the formula we obtained as a result that the total electrical energy used in San Pedro Sula was of 389,057.44 kWh and in Honduras in the year 2022 was 1,853,030.11 kWh.

4. In San Pedro Sula alone, a total of L. 2,267,232.25 or its equivalent in dollars of $92,502.34 kWh was not billed. At the national level, approximately L. 10,798,532.96 or its equivalent in dollars of $ 440,576.62 was not billed.

The main limitation of this work was the number of houses that could be obtained to install the meters and record the amount of electricity required for the construction of the houses. Therefore, the results are the estimation of these two particular houses, so in order to obtain a better formula, a greater number of houses would be needed to create a more accurate formula.

Despite this limitation, the consumption of these houses was successfully recorded and in one of them it was possible to determine the amount of electricity required for each phase of construction. This research can be used to be able to charge the electric energy that will be consumed in the construction of the houses at the time of obtaining their respective construction permits.

References

1. SIELHO | IAIP. (n.d.). https://sielho.iaip.gob.hn/ (Agosto 15,2023)
2. Caribe, C. E. P. a. L. Y. E. (2023, January 23). Https://www.cepal.org/es/publicaciones/47924-la- energia-America-Latina. https://www.cepal.org/es/publicaciones/47924-la-energia-america-latina- caribe-acceso-renovabilidad-eficiencia
3. E. Empresa Nacional de Energía Eléctrica, “Boletín Estadístico Diciembre 2022,” 2023.
4. E. Diaz Madrid, “Honduras en cifras,” 2022. https://presencia.unah.edu.hn/noticias/honduras-en- cifras/ (accessed Agosto 15, 2023).
5. Empresa Energía Honduras, “Consumo Eléctrico por Municipio,” 2023.
6. Alves, R. R., Casanova, P., Quirogas, E., Ravelo, O., & Gimenez, W. (2006). Reduction of Non- Technical Losses by Modernization and Updating of Measurement Systems. Estimation of Electrical Energy Used for Houses Construction. https://doi.org/10.1109/tdcla.2006.311590
7. G. Gonzalez and L. Figueroa, "Strategic Plan for the Control and Reduction of Non-Technical Losses applied in C.A. Energia Eléctrica de Valencia," 2006 IEEE/PES Transmission & Distribution Conference and Exposition: Latin America, Caracas, Venezuela, 2006, pp. 1-6, doi: 10.1109/TDCLA.2006.311491.
8. Rigoberto, O. C. M. (1997, July 1). Pérdidas técnicas en un sistema eléctrico de distribución. http://bibdigital.epn.edu.ec/handle/15000/9778
9. M. Madrigal, J. J. Rico and L. Uzcategui, "Estimation of Non-Technical Energy Losses in Electrical Distribution Systems," in IEEE Latin America Transactions, vol. 15, no. 8, pp. 1447-1452, 2017, doi: 10.1109/TLA.2017.7994791.
10. A. C. Rodrigues, A. S. Costa and D. Issicaba, "Identification of non-technical losses in distribution systems via state estimation and geometric tests," 2018 Simposio Brasileiro de Sistemas Eletricos (SBSE), Niteroi, Brazil, 2018, pp. 1-6, doi: 10.1109/SBSE.2018.8395675.
11. Zaghwan, A.; Gunawan, I. Resolving Energy Losses Caused by End-Users in Electrical Grid Systems. Designs 2021, 5, 23. https://doi.org/10.3390/designs5010023
12. U. H. Bezerra, T. M. Soares, J. P. A. Vieira, M. E. L. Tostes, A. R. R. Manito and J. C. H. Paye, "Equivalent operational impedance: A new approach to calculate technical and non-technical losses in electric distribution systems," 2018 Simposio Brasileiro de

Sistemas Eletricos (SBSE), Niteroi, Brazil, 2018, pp. 1-6, doi: 10.1109/SBSE.2018.8395709.

13. Leitão, N.C.; Lorente, D.B. The Linkage between Economic Growth, Renewable Energy, Tourism, CO2 Emissions, and International Trade: The Evidence for the European Union. Energies 2020, 13, 4838. https://doi.org/10.3390/en13184838
14. H. Lasso, C. Ascanio and M. Guglia, "A Model for Calculating Technical Losses in the Secondary Energy Distribution Network," 2006 IEEE/PES Transmission & Distribution Conference and Exposition: Latin America, Caracas, Venezuela, 2006, pp. 1-6, doi: 10.1109/TDCLA.2006.311651.
15. A. Shah, W. Mesbah and A. T. Al-Awami, "An Algorithm for Detaching Technical Losses from Non-Technical Losses in Distribution Systems," 2021 IEEE Power & Energy Society Innovative Smart Grid Technologies Conference (ISGT), Washington, DC, USA, 2021, pp. 1-5, doi: 10.1109/ISGT49243.2021.9372255.

AUTHOR INDEX

EDP Sciences – Web of Conferences
17, Avenue du Hoggar
Parc d'Activité de Courtabouf
BP 112
F-91944 Les Ulis Cedex A
France

ISBN 978-1-7138-9805-4

CURRAN ASSOCIATES INC.
proceedings.com